Environmental Alpha

Founded in 1807, John Wiley & Sons is the oldest independent publishing company in the United States. With offices in North America, Europe, Australia, and Asia, Wiley is globally committed to developing and marketing print and electronic products and services for our customers' professional and personal knowledge and understanding.

The Wiley Finance series contains books written specifically for finance and investment professionals as well as sophisticated individual investors and their financial advisors. Book topics range from portfolio management to e-commerce, risk management, financial engineering, valuation, and financial instrument analysis, as well as much more.

For a list of available titles, visit our Web site at www.WileyFinance.com.

Environmental Alpha

Institutional Investors and
Climate Change

ANGELO A. CALVELLO

Phil,
Your friendship has
underpinned my thinking
and writing for over 30 years,
Thanks for your counsel,
criticism, and companion.

Angelo

WILEY

John Wiley & Sons, Inc.

For general information on our other products and services or for technical support, please contact our Customer Care Department within the United States at (800) 762-2974, outside the United States at (317) 572-3993 or fax (317) 572-4002.

Wiley also publishes its books in a variety of electronic formats. Some content that appears in print may not be available in electronic books. For more information about Wiley products, visit our web site at www.wiley.com.

Library of Congress Cataloging-in-Publication Data:

Environmental alpha : institutional investors and climate change / [edited by] Angelo A. Calvello.
    p.   cm. – (Wiley finance series)
  Includes bibliographical references and index.
  ISBN 978-0-470-29062-0 (cloth)
  1. Environmental economics. 2. Climatic changes. I. Calvello, Angelo A., 1956-
  HC79.E5E5749 2010
  363.738'746–dc22

                                                    2009017150

Printed in the United States of America

10 9 8 7 6 5 4 3 2 1

To Lisa, Giana, Joseph, and Michael.
Tolle Lege, Tolle Lege

Climate change, as a fundamental alteration of global reality, requires another paradigm shift in the calculation of risk, and a consequent response—only this time the stakes are much higher. The flow of capital has world-changing force, and now that carbon has caught up with fiduciaries, they have the opportunity—and the undeniable duty—to help lead us away from a remorseless tragedy of universal ruin.

—Michael Northrop and David Sassoon, "Climate Change Now a Fiduciary Duty—and Opportunity," 2006

Contents

Preface

The genesis of this book could be traced to a meeting I had with Kerry Brick, manager of pension investments for Cargill, in 2007. We were discussing a new fund that my then-employer was offering, a fund that was to build and operate integrated coal mines/power plants in China's Shanxi province. The fund's return would be tied to the sale of the electricity generated by the power plant and—here's where it gets interesting—the sale of the carbon credits the fund would earn because of the environmentally efficient manner in which it mined the coal. The operation was structured so that it would be certified by and registered with the United Nations Framework Convention of Climate Change as a Clean Development Mechanism project, allowing it to receive emissions credits (certified emission reduction credits, or CERs). Each CER signifies an emission reduction of one tonne of CO_2 equivalent and could be traded in various market-based systems.

The fund offered a play on emerging markets, China, commodity prices, and energy, but at the end of the day it was the carbon credits that clearly got Kerry's attention. Carbon was a new and real source of return. He—and other institutional investors with whom I later spoke—quickly understood that this strategy potentially offered attractive risk-adjusted returns because skillful managers could gain and exploit material advantages. In a word, they saw that this strategy offered an environmental alpha—a skill-based return resulting from an investment in assets whose value is primarily driven by climate change-related issues. But they also saw that the source of these returns—climate change—presented them with new and profound risks.

As I left Kerry's office, I felt like the prisoner in Plato's "Allegory of the Cave," who has been freed of his chains. That conversation allowed me to see that I was previously living in a shadow world that failed to fully reveal the investment risks and opportunities presented by changes in climate. Since that time, I have continued my ascent out of the shadow world, toward a higher form of investing—environmental investing. And like Plato's prisoner, it took my eyes quite a while to adjust to the light. But I soon understood that climate change would require institutional investors to rethink how they manage and invest the assets under their control. More specifically, I came to understand that environmental investment strategies could offer excess returns and diversification and that climate change was the driver of these benefits.

I knew I had to return to the cave to share this intellectual (and commercial) insight with those remaining captive in the shadow world. I wanted to create a resource that would help fiduciaries like Kerry move beyond their intuitive appreciation of environmental investments and thoroughly understand key topics that institutional investors need to understand in order to prudently direct and commit capital in the time of climate change.

This book marks the beginning of my return to the cave. I knew I could not return alone. The topics of climate change and environmental investing were simply too overwhelming for me to manage on my own. So I searched for portfolio managers, academics, investment consultants, institutional investors, and other industry participants who had made this same ascent and solicited their help in writing this book.

THE NATURE OF THE CLIMATE CHANGE CHALLENGE

Understanding the nature of the challenge begins with a definition of environmental investing and environmental alpha. The definition of environmental investing is developing before our eyes, partly because of the newness of the inquiry and partly because of the nature of the subject matter itself. After discussions with people in different disciplines and with different perspectives, I have chosen a definition that focuses on the drivers of return instead of content of the opportunity set. Environmental investing is simply too dynamic to be defined by investment approach or function. This book defines environmental investing as *investing directly or indirectly in assets whose values are affected primarily by climate change.* Environmental investing can be further described as a type of thematic investing, and the theme that drives/underlies environmental investing is climate change.

The climate change theme, like all other investment themes, is impacted by macroeconomic and broad sociopolitical issues, but the specific thematic factors affecting environmental investments—the drivers of returns—can be reduced to the following four climate change–related factors:

1. Science
2. Economics
3. Policy and regulation
4. Technology

The climate change theme can be narrowly defined as "The scientific community has reached a strong consensus regarding the science of global climate change. The world is undoubtedly warming. This warming is largely the result of emissions of carbon dioxide and other greenhouse gases from human activities including industrial processes, fossil fuel combustion, and changes in land use, such as deforestation" (Pew Center on Climate Change 2009). In order to solve the climate change problem, we need to mitigate greenhouse gas (GHG) emissions to a specific level within a defined time period and adapt to the possible consequences of climate changes already in the pipeline. Mitigating and adapting to climate change will require changes in behavior and the development and dispersion of an evolving portfolio of market-based technologies.

However, it is critical that readers understand that the climate change theme is actually much bigger than simply "global warming." In this book, climate change is a "cluster concept" and encompasses both the above-mentioned narrow definition and such direct environmental topics as (Kiernan 2009):

- Water quality
- Air pollution
- Waste management
- Deforestation and land degradation
- Chemical and toxic emissions
- Biodiversity loss
- Depletion of the ozone layer
- Quality of fisheries and oceans

And such broader social issues as:

- Energy supply
- National security
- Human development
- Population growth
- Global changes in demographics
- Poverty and income disparity
- Public health
- Human rights
- Labor rights
- Human resource management

In order to fully understand the risks and opportunities arising from climate change and to properly assess and access the environmental investment universe, investors should understand the scientific definition while recognizing the cross-disciplinary meaning of climate change.

More importantly, investors should recognize that "climate change" is the concept chosen to express this cluster of topics because while each of the related topics represents a critical issue, only climate change comes with the exigency of time. Action unquestionably needs to be taken with regard to the other issues, but only climate change *demands* timely action be taken if we are to avoid an ecological "tipping level . . . a measure of the long-term climate forcing that humanity must aim to stay beneath to avoid large climate impacts" (Hansen et al. 2008). Said another way, where the other topics might come with a moral imperative, climate change comes with a temporal imperative.

For investors, this means the market-based solutions used to solve the climate change problem—environmental investments—could not only mitigate GHGs and help us adapt to changes in climate, they could produce broader results that might help remediate these other related issues. For example, developing clean power sources (like wind or solar) that could eventually be used at scale will not only result in reduced GHGs emissions but could also result in reduced air pollution, reduced chemical and toxic emissions, improved water quality, improved public health, and potentially improved working conditions. Such clean technologies also improve a country's native energy supply and could make it less reliant on foreign sources of energy, thereby improving national security.

Climate change, broadly considered, reveals other attributes of the climate change theme: it is a long-term, secular theme that cuts across asset classes, investment approaches, styles, and geographies, making it the mother of all themes.

It's important to recognize that climate change is not some distant, abstract risk; it is upon us now. "Companies are already being impacted in financial terms due to the effects of climate change on their costs, revenues, assets or liabilities" (Mercer Investment Consulting 2005). Also, "believing" in climate change is not a prerequisite condition for environmental investing. Whether or not an investor believes in climate change, events are occurring that will give rise to significant risks that will impact existing portfolios and to new investment opportunities that could benefit these same portfolios. Environmental alpha opportunities, like Plato's Forms, exist independent of investors' beliefs. In some ways, many of these opportunities exist even independent of climate change itself. For example, we still face limited supplies of fossil fuels and need to find alternative sources of energy. Developing and implementing new clean replacement technologies makes sense from the standpoint of energy supply, national security, economic efficiencies, and clean air. Additionally, the opportunity to earn significant risk-adjusted returns will continue to arise. Science will continue to produce new ideas that will be transformed into market-based solutions. Policy will continue to be developed and implemented that will shape market-based solutions and impact the companies' performance and behavior.

So investors do not need to believe in climate change; they simply need to understand that this theme will materially affect their current portfolio and investment decisions for years to come.

Correlatively, while climate change might be a value-charged topic, environmental investing as discussed in this book has no necessary connection to investors' values or worldview and it does not necessarily involve extra-financial considerations such as social benefits. There is no need to include such value judgments in the investment process. Environmental investing is about economics, not ethics; alpha, not absolution. (See Chapter 6 for Matthew Kiernan's delineation of this issue.) For this reason, environmental investing is not "similar to SRI in that both are strategies driven by investors desire to achieve financial return while maximizing social good" (Chhabra 2008). Maximizing social good is neither a sufficient nor necessary condition of environmental investing. For this reason, environmental investing, in spite of its name, cannot be lumped into another subset of socially responsible investment (SRI), ESG—environmental, social, and governance.

Let me be clear: there is nothing inherently problematic about SRI or ESG, but classifying environmental investing as a type of SRI raises all kinds of specious issues:

- Definitional confusion.
- Investment returns are constrained.
- Fiduciary responsibility is compromised.
- Incorporating environmental issues into existing investment approaches is a challenge (Taylor and Donald 2007).

These issues could unnecessarily distract investors from examining the risks and considering the genuine benefits of environmental investing. However, climate change does require investors to take a more active role in the management of their assets. Because a company's exposure to climate risks is not always readily available, institutional investors cannot easily determine the risks inherent in their portfolio. Investors have a duty to understand as much about the risks—climate-related or otherwise—as possible so they can make informed decisions to "meet the real needs of our members and beneficiaries" (Mercer Investment Consulting 2005).

This could require institutional investors to collaborate with their peers and industry organizations, asset managers, and governmental agencies to create uniform reporting standards on climate risks. (These ideas will be discussed in Chapters 5 and 14.)

A TAXONOMY OF ENVIRONMENTAL INVESTMENTS

Environmental investments break down into five major classifications:

1. Carbon
2. Land use, land-use change, and forestry (LULUCF)
3. Clean technology
4. Sustainable property
5. Water

This taxonomy—and the details of each category—will be discussed in detail in Chapters 7 through 13. Briefly, though, the rationale for this taxonomic scheme is the following:

- Carbon is the currency of climate change.
- LULUCF, clean tech, and sustainable property represent market-based initiatives to mitigate emissions from sources such as agriculture, forestry, power, transportation, industry, and built property, which collectively account for almost all the sources of global anthropogenic GHGs.
- Water is a primordial resource that will be significantly impacted by climate change.

Regardless of the classification, it is critical to understand that the climate change theme provides the opportunity for alpha in each of these investment classifications—hence the name of the book, *Environmental Alpha,* or alpha derived from investment opportunities arising from climate change.

Environmental alpha refers to the return earned as a result of manager skill, as opposed to return earned from market exposure (beta). It exists because the drivers of return—science, economics, policy, and technology—are dynamic, opaque, and complex, making it possible for certain investors to gain advantages that they could skillfully exploit to generate a return that is not attributable to market exposure. These advantages could be access to and understanding of relevant information (e.g., better understanding of Chinese climate policy, new "clean" coal technologies, or the working of the UN Clean Development Mechanisms) or access to scarce resources (like the optimal site for a wind farm).

In this sense, environmental alpha is like other alphas. It also shares two other attributes: environmental alpha should have a low correlation with the market, and it is subject to three universal constraints: alpha is scarce, transitory, and capacity constrained. However, because the source of the alpha arises out of the phenomenon of climate change, the potential for continually evolving alpha opportunities is great.

> *All Alpha factors will fade into the background eventually....*
> *Therefore, it becomes a question of how long the trend can last.*
> *Given the 40–50 year investment horizon and the size of the*
> *problem—$45 trillion of investment needed in energy markets*
> *alone—we believe that climate change will remain the source of*
> *identifiable Alpha for many years ahead.*
>
> **"Investing in Climate Change 2009," DB Advisors, 2008**

INTENDED AUDIENCE

This book has value for all investors but it is specifically written for institutional investors such as defined pension funds, endowments, foundations, insurance companies, and superannuation funds, organizations with "delegated investment responsibilities" (Taylor and Donald 2007). Institutional investors differ from individual investors in that they oversee and invest assets on behalf of an enterprise/organization or group of individuals. Many operate under trust law, which imposes a fiduciary obligation on the trustees and the advisors to these funds. While the nature and scope of this obligation varies from jurisdiction to jurisdiction, in general "... any investment strategy chosen by a trustee must be founded on objective evidence, which has been rigorously analyzed and carefully considered by the trustee" (Taylor and Donald 2007).

As fiduciaries, institutional investors are required to make informed and reasonable decisions. This requires that they understand the material risks facing their portfolio and suggests that they likewise understand new investments available to them. This was precisely what was proving to be problematic for institutional investors when they considered the China carbon fund and other environmental investments. These investment opportunities presented them with a variety of material but unfamiliar issues, all related to climate change: the current and potential Chinese and international climate policy issues; how the UN CRM worked and whether the fund would be able to comply with its standards; the workings of the international carbon markets; and transformative clean technologies. More generally, investors all seemed to need help understanding four main topics covered in the book:

- The nature of the challenge and the opportunity presented by climate change
- Climate change and fiduciary duty

- The opportunity set and associated challenges
- Practical matters

This book is structured to provide institutional investors with information on all of these topics.

Environmental Alpha is intended for institutional investors because climate change will impact them more profoundly and comprehensively than other investors. This is because institutional investors are universal owners, which is defined as:

> . . . *a large financial institution, such as a pension or mutual fund, which owns securities in a broad cross-section of the economy. Because of the diversified portfolio of stocks, bonds and other asset classes, investment returns (especially long-term ones) will be affected by the positive and negative externalities generated by the entities in which the universal owner invests. Being external means they are not controlled by the entity and therefore can be viewed in terms of potential risk (for negative externalities) or opportunity (for positive ones).*
>
> **"Universal Ownership: Exploring Opportunities and Challenges,"**
> **Mercer Consulting Conference Report, 2006**

Climate change is a negative externality or risk that will broadly impact every asset class, sector, industry, and investment in different ways and at different times. "The fact of climate change is unlike any other 'risk factor' that our modern financial system has ever confronted. It contains no reciprocal or alternative opportunity. It is a universal threat that will spare no nation, no market, and no industry" (Northrop and Sassoon 2006).

Professor Sir Graeme Davies, Chairman, Universities Superannuation Scheme Ltd. phrased it as follows:

> *The question that you may be asking is why should a pension fund be interested in a long-term issue like climate change, when many of us live or die by quarterly or yearly performance data? Given that this is the case, why does USS as a pension fund believe that we should be addressing climate change as an issue for our fund? There are two reasons: Firstly, we are universal owners. Secondly, we need to meet the real needs of our members and beneficiaries.*
>
> **"A Climate for Change: A Trustee's Guide to Understanding and Addressing Climate Risk," Mercer Consulting, 2005**

Institutional investors are the intended audience also because their size and goals allow them to positively affect the climate change problem. Environmental investments are ultimately market-based solutions that attempt to mitigate GHG emissions or adapt to the impact of climate change. Because of the scope of the problem, the market-based solutions require an enormous amount of investment capital in order to be effectively developed and deployed at a meaningful scale. Some estimate that the capital cost alone of transforming the U.S. power sector would be on the order of $10 trillion, or one year of U.S. gross domestic product (GDP) (see Chapter 9). Governments will not be able to single-handedly provide the necessary capital, especially during financially trying times like we are experiencing today. Institutional investors—with over $40 trillion in assets under management—are an obvious source of this capital, especially as environmental investment opportunities cut across their portfolio allocations and offer the possibility of robust uncorrelated returns.

Environmental Alpha seeks to create an alignment of interests: institutional investors need to manage a significant externality and to find skill-based returns that will help them achieve their investment objectives; society needs large, long-term sources of investment capital to fund the development and dispersion of the transformation technology that will help solve the climate problem.

STRUCTURE OF THE BOOK

This book is a compilation of essays tied together by a common goal of providing institutional investors with the information they need to prudently assess and respond to the risks and opportunities associated with climate change. Its division into four parts corresponds with the four major areas of need mentioned earlier:

1. The nature of the challenge and the opportunity
2. Climate change and fiduciary duty
3. The opportunity set and challenges
4. Practical issues

Part One: The Climate Change Challenge

Part One provides the critical background information investors need to understand the phenomenon of climate change and, more specifically, the four basic drivers of returns of environmental investing: science, economics, policy and regulation, and technology.

In Chapter 1, Dr. Richard Betts, head of Climate Impacts at the Met Office Hadley Centre, presents the science of climate change. Dimitri Zenghelis, chief economist of the Cisco Systems' Climate Change Practice, extends this discussion in Chapter 2, where he examines the economic consequences of climate science. In Chapter 3, David Gardiner, principal of David Gardiner & Associates, then explains the myriad of current and potential climate change policies intended to shape our individual and collective responses to climate changes and how they could potentially impact environmental investments.

Part Two: Climate Change and Fiduciary Duty

Climate change presents institutional investors with a new set of risks, which, because institutional investors tend to be universal owners, could potentially impact their entire portfolio.

Mindy Lubber, president of Ceres, begins Part Two with a discussion of the four major climate change risks facing institutional investors: physical, regulatory, reputational, and litigation. Paul Watchman uses Chapter 5 to extend his groundbreaking work in the United Nations Environment Programme Financial Initiative Report, "A Legal Framework for the Integration of Environmental, Social and Governance Issues into Institutional Investment" (or what has come to be called the Freshfields report), and makes clear that climate change is the overarching risk facing fiduciaries and that fiduciaries must incorporate environmental issues into their investment decisions. In Chapter 6, Innovest CEO Matthew Kiernan's analysis of Carbon Beta™ dispels the commonly held view that environmental investing is simply another flavor of SRI, and in so doing frees investors from the constraining yoke of SRI.

Part Three: Environmental Investing

With this foundational information, the book moves to environmental investing proper.

In Chapter 7, I provide the taxonomy of environmental investing, as well as delineate the idiosyncratic challenges of environmental investing that give rise to the opportunities to generate environmental alpha. In Chapter 8, Mark Fulton and Bruce Kahn of Deutsche Asset Management's Climate Group examine the opportunities set in more detail and situate environmental investments in a portfolio context.

Experts in each of the major classifications then provide a detailed explanation of the specific climate change–based investment thesis and opportunity set associated with each category. In Chapter 9, Jurgen Weiss of

Watermark Economics and Veronique Bugnion of Point Carbon begin by examining the carbon markets and the related alpha opportunities. In Chapter 10, Charles Palmer and Stefanie Engel of the Institute for Environmental Decisions link this discussion of the carbon markets to specific financing solutions that seek to reduce emissions from deforestation and degradation (REDD), and Martin Berg of Merrill Lynch explains the investment opportunities associated with REDD activities. Russell Read and John Preston of C Change Investments next delineate clean tech investing and discuss clean tech's coming of age. In Chapter 12, Tim Dixon, professor of real estate and director of the Oxford Institute for Sustainable Development (OISD) in the School of the Built Environment at Oxford Brookes University, United Kingdom, takes on the challenge of defining and explaining the relationship between climate change and sustainable commercial property as a source of environmental alpha.

Part Three concludes with Rod Parsley, portfolio manager of the Perella Weinberg Oasis Fund, and his colleague Hua Liu, deconstructing the water market from an investor's perspective.

Part Four: Practical Considerations

This understanding of the categories of environmental investing leads to a discussion of three practical issues. In Chapter 14, Danyelle Guyatt, principal in Mercer's Responsible Investment team, explores how institutional investors could most effectively/efficiently confront/respond to the global, transgenerational challenge of climate change and convincingly suggests that collaboration is the model of choice. In Chapter 15, Nick Hoskins and Martin Batt of the Virtuous Circle tackle the larger issue facing institutional investors, the convergence of climate change, corporate agendas, and environmental investing.

The book concludes with my suggestions of how institutional investors could practically approach and exercise their fiduciary duties in this time of climate change.

FINAL THOUGHTS

Climate change is upon us. "The stakes, for all life on the planet, surpass those of any previous crisis. The greatest danger is continued ignorance and denial, which could make tragic consequences unavoidable" (Hansen et al. 2008). It requires a response, especially from fiduciaries. This book hopes to act as a cairn on the path out of the cave.

ANGELO A. CALVELLO

REFERENCES

Chhabra, R. 2008. Environmental investing: Is the grass always greener? JP Morgan, *Investment Analytics and Consulting Newsletter.* April.

DB Advisors. 2008. *Investing in climate change 2009: Necessity and opportunity in turbulent times.* Deutsche Bank Group. October. www.dbadvisors.com/deam/stat/globalResearch/climatechange_full_paper.pdf.

Hansen, J.M. Sato, P. Kharecha, D. Beerling, R. Berner, V. Masson-Delmotte, et al. 2008. Target atmospheric CO_2: Where should humanity aim? *Open Atmospheric Science Journal* 2 (November): 217–231. www.bentham.org/open/toascj/openaccess2.htm. Accessed March 1, 2009.

Kiernan, M. 2008. *Investing in a sustainable world: Why green is the new color of money on Wall Street.* New York: AMACOM.

Mercer Investment Consulting. 2006. *Universal Ownership: exploring opportunities and challenges.* Conference Report. Saint Mary's College of California, April 10–11.

Mercer Investment Consulting. 2005. *A climate for change: A trustee's guide to understanding and addressing climate risk.* The Carbon Trust. www.thecarbontrust.co.uk/trustees.

Northrop, M., and D. Sassoon. 2006. "Climate Change Now a Fiduciary Duty—and Opportunity." *The Environmental Forum.* September/October.

Pew Center on Climate Change. *Global warming basic introduction.* www.pewclimate.org/global-warming-basics/about. Accessed March 8, 2009.

Taylor, N., and S. Donald. 2007. *Sustainable investing: Marrying sustainability concerns with the quest for financial returns for superannuation trustees.* Russell Investment Management, August.

Acknowledgments

This book is truly a collaborative effort. I'd like to thank each of my contributors for their substantial efforts, patience, shared vision, and help in the editorial process. Editing is not a skill that comes easily, especially to a guy who started in the business as a floor trader. I tried to bite but not leave teeth marks. No study is immune to errors; for those, I alone am responsible.

In addition to the named contributors, others have supported me in this endeavor: my mother confessor, Lia Abady; my facilitator, Rob Challis; my advisor, Tony Ryan; and my booster, Bill Weldon. Others also have contributed their special skills. Emma Baldock was an invaluable conduit to ideas and sources. Stuart Mason, Chris Suedbeck, Jerome Malmquist, and their colleagues at the University of Minnesota generously supported the creation of the case study. Matthew Kiernan acted as a matchmaker in more than one case. Bob Jaeger provided regular doses of realism that challenged me to rethink some basic assumptions. Matt Bassista witnessed the genesis of this project and his suggestions helped shape the scope and content, especially with regard to the water chapter. Steven Howard provided valuable direction on how to present the science of climate change. Jamey Sharpe and Phil Ruden offered their insights into practical fiduciary issues. Philip Payne, Bob Ratliff, and Christian Gunter helped create the sustainable property case study. And the friends who have helped me with my thoughts, design, writing, and proofreading: Sara DiVillo, Mandy Kristufek, Valerie Oseguera, and especially Phil Schneden, who has been with me on this journey since Proviso East.

The people at John Wiley & Sons contributed greatly to this project. I'd like to thank my editor, Pamela van Giessen, for believing in me—again— and Emilie Herman, Kate Wood, and Rosanne Lugtu for their patient and gentle support.

Finally, I want to thank my wife, Lisa, for her unwavering confidence and support, and my children, Giana, Joe, and Michael, for believing in me and giving me the courage to take on this project. To them and all of those who have helped, namaste.

A. A. C.

Introduction to Climate Change Issues and Consequences

CHAPTER 1

The Science of Climate Change

Richard A. Betts, PhD

Climate change is a complex scientific problem, but its implications could have major consequences for the human species and indeed the rest of the world. Moreover, human actions to reduce climate change and adapt to its effects could also have major consequences. In order to make informed decisions about our responses to the issue, we require robust scientific understanding of the issue and the likely consequences of our actions, or at least some grasp of the range of potential consequences if we are unable to be certain.

This chapter reviews the latest scientific conclusions about recent climate change and its causes, and discusses the implications of different levels and timing of emissions reductions. Climate science issues relating to adaptation to climate change are also discussed. Most of this chapter is grounded in the science described in detail in the volume "The Physical Science Basis" in the *Fourth Assessment Report of the Intergovernmental Panel on Climate Change* (IPCC 2007), which is widely known as "AR4."[1] More recent work from the Met Office Hadley Centre is also discussed.

HUMAN-CAUSED CLIMATE CHANGE: THE EVIDENCE

A vast body of evidence demonstrates that the world is becoming warmer and that this is not a natural phenomenon—beyond reasonable doubt, humans are to blame. By gathering data from a wide range of sources, from weather stations to tree rings, from ice cores to computer models, we can

Richard A. Betts is head of climate impacts at the Met Office Hadley Centre.

clearly see that the climate has already moved out of its previous natural state. This forms the bedrock of evidence that the human species is already influencing its own environment and can now choose whether to continue to increase this influence or reduce it.

The World Is Warming

One of the iconic measures of climate change is the global average temperature near the surface. This can be established for roughly the last 150 years from a worldwide network of weather stations on land and observations made aboard ships. In some places the observed temperature record extends farther back, but before 1860 the worldwide coverage was not sufficiently dense to provide a credible global average of thermometer-based measurements. The records show that global average temperature has risen by more than 0.7 degrees Celsius since the start of the twentieth century (Figure 1.1). The rise has not been steady; before 1940 there was a warming of around 0.3 degrees Celsius, then there was a cooling of approximately 0.2 degrees Celsius until 1950, followed by a renewed warming of 0.13 degrees Celsius per decade since then.

The world has been warmer over the last decade than at any time since measurements began. This warming is observed over the oceans as well as over land, suggesting that it is a truly global phenomenon and not a conglomeration of "local" warmings caused by some small-scale process such as the urban heat island effect. In AR4, the IPCC concluded that "warming of the climate system is unequivocal."

This Warming Is Unusual

But what about before our global network of thermometers was established? In the history of the Earth, 150 years is not long, and to establish whether the current warming is unusual we need to know more about temperatures farther into the past. Temperatures can be estimated from a variety of "proxy" evidence such as the patterns of growth in the rings of ancient trees, the distribution of particular species as indicated by their pollen found in the soil, and the chemical composition of air bubbles trapped in ancient ice. A number of independent studies have used these lines of evidence to reconstruct northern-hemisphere temperatures over the last 1,000 years or more (Figure 1.1), and while they do not agree with each other perfectly, they all indicate that the temperatures of the last 50 years are likely to be the highest in any 50-year period in the last 1,300 years. (The IPCC indicates an outcome or result as "very likely" if expert judgment assesses the probability

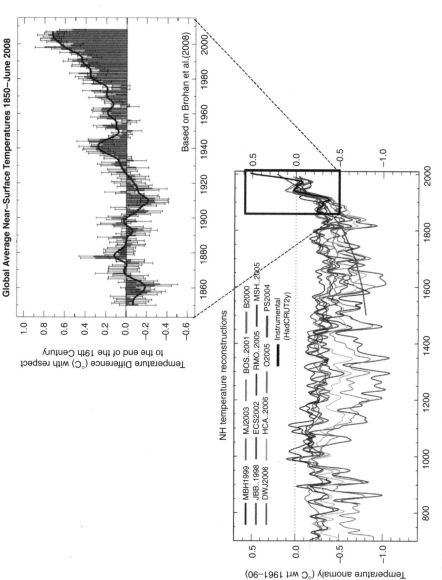

FIGURE 1.1 Historical records and reconstructions of past climate change and variability. *Sources:* IPCC, 2007 (main figure), Met Office, British Crown Copyright (inset).

of occurrence to be greater than 90 percent. "Likely" is defined as a greater than 66 percent probability.)

The millennial reconstructions in Figure 1.1 include the famous "Hockey Stick" graph of Mann, Bradley, and Hughes (1999) (here labeled MBH99), which was targeted by climate skeptics following its publication in the IPCC Third Assessment Report. Although subsequent studies and reanalysis led to slightly different reconstructions of past temperatures, all studies show the common feature of the twentieth-century warming being highly unusual in comparison with the previous millennium.

Other Changes Are Consistent with Warming

Changes have also been seen in other aspects of the climate system. Snow cover and mountain glaciers have shrunk, and some melting of the Greenland and Antarctic ice sheets has been measured. Global average sea level rose by approximately 17 cm over the twentieth century, partly because of the additional water in the ocean basins arising from the melting of ice on land, and partly because water expands when it warms. Patterns of precipitation (rainfall and snowfall) have also changed, with parts of North and South America, Europe, and northern and central Asia becoming wetter while the Sahel, southern Africa, the Mediterranean, and southern Asia have become drier. Intense rainfall events have become more frequent. In Europe, Asia, and North America, growing seasons have extended, with flowers emerging and trees coming into leaf several days earlier in the year than in the mid-twentieth century.

Carbon Dioxide and Other Greenhouse Gases

Some gases such as water vapor, carbon dioxide (CO_2), methane, and nitrous oxide are known as greenhouse gases (GHGs). They absorb and re-emit some of the heat radiation given off by the Earth's surface, hence warming the lower atmosphere. These gases occur naturally in the atmosphere, and without their warming presence in the atmosphere the Earth's average surface temperature would be around −20 degrees Celsius. In terms of its contribution to the natural greenhouse effect, the most important is water vapor, followed by CO_2. However, the concentrations of CO_2, methane, and nitrous oxide have shown a rapid increase over the past 200 years, and this increase is accelerating (Figure 1.2). Records extending back 10,000 years have been obtained by analyzing the chemical composition of air bubbles trapped in ancient ice (Figure 1.2, main panels), and in the 1950s, actual instrumental measurements of CO_2 and other GHGs began to be taken. These are now routinely taken at a number of locations around the world,

Changes in Greenhouse Gases
from Ice-Core and Modern Data

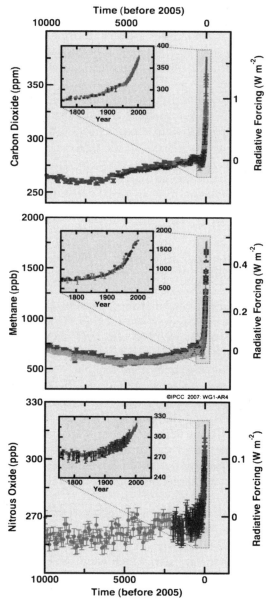

FIGURE 1.2 Historical records of the atmospheric concentrations of three major greenhouse gases: CO_2, methane, and nitrous oxide.
Source: IPCC, 2007.

including Mauna Loa observatory in Hawaii, the South Pole, and Cape Grim in Tasmania (Figure 1.2, insets). Together, these records show that the concentrations of these greenhouse gases are all rising rapidly, and that these rises are unique in the past 10,000 years. It is this increasing concentration of these gases, rather than the background levels, that is of relevance to the issue of climate change.

Changes in the atmospheric concentration of CO_2 are a consequence of shifts in the balance of very large flows of carbon into and out of the atmosphere. In the natural system, CO_2 is produced through the chemical combination of carbon with oxygen in the process of respiration, most of which is a natural part of biological processes in soils, vegetation, plankton, and animals, such as breathing and decay. Some CO_2 is also produced by combustion of vegetation—in other words, forest fires. These are natural sources of CO_2. Opposing these are the processes that take up CO_2 from the atmosphere, such as photosynthesis by plants, which combines CO_2 with water to produce sugars that store the energy absorbed from the sun during the process. Further uptake of CO_2 involves the dissolving of CO_2 in water (mostly in the oceans) from which it can then be extracted by plankton. These latter processes provide sinks of CO_2. All life on earth contains carbon, and indeed the vegetation, soil microbes, plankton, and higher life forms collectively store more carbon than is present in the atmosphere. Very large quantities of carbon are also held as dissolved CO_2 in seawater. So the "carbon cycle" as described here consists of flows of carbon between large stores in the atmosphere, ocean, and life.

The rates of release and removal of CO_2 by these processes are affected by environmental conditions, such as the concentration of CO_2 in the atmosphere, the temperature, and the availability of moisture for plant growth. Hypothetically, if environmental conditions were unchanging, the natural flows of CO_2 into and out of the atmosphere would balance out and the atmospheric CO_2 concentration would not be changed. Even though the flows in each direction are large, in a balanced system they could in theory cancel each other out.

However, the balance is currently being upset by the addition of further sources of CO_2, particularly the combustion of fossil fuels—coal, oil, and natural gas. These substances are the remains of vegetation and plankton from millions of years in the past, which were buried underground or beneath the ocean floor. The carbon stored within them, as described above, becomes coal and oil. The energy that was stored as sugars as a result of photosynthesis is held within these substances and their concentrated form means that a small quantity of coal or oil can hold considerable quantities of energy that can then be released when the coal or oil is burned, hence their value as fuels. However, burning of these fuels rapidly releases CO_2

that was previously removed over millions of years and held away from the atmosphere underground. This is therefore an extra release of CO_2 that tips the balance between sinks and sources.

Greenhouse Gas Concentrations Are Increasing Due to Our Emissions

Further records of air bubbles in ancient ice show us that CO_2 is now at its highest concentration for more than 650,000 years. We know for certain that combustion produces CO_2, and we know for certain that we have been burning fossil fuels at ever-increasing rates over the last 200 to 300 years. From international records of energy consumption, it is easy to calculate the quantity of CO_2 that has been produced by burning fossil fuels; by 2000 this had reached around billion tonnes of CO_2 per year, having risen by over 1 percent per year over the previous 20 years. In total, we have emitted over a trillion tonnes of CO_2 since fossil fuel burning began, which is more than enough to account for the rise in CO_2 in the atmosphere. Indeed, the rate of rise of CO_2 in the atmosphere is only about 60 percent of the rate of emission from fossil fuel burning (Figure 1.3a); some of the additional CO_2 is absorbed through an increase in photosynthesis by vegetation and an increased dissolving of CO_2 in ocean waters (Figure 1.3b). It is worth noting

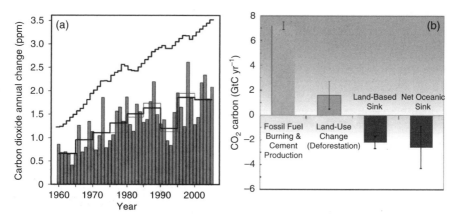

FIGURE 1.3 The role of carbon sinks in buffering us from the full effect of our CO_2 emissions. (a) Annual emissions of CO_2 from fossil fuel burning (upper stepped line), and annual change in atmospheric CO_2 concentration (lower bars). (b) Emissions from fossil fuel and deforestation, partly offset by uptake of carbon by land ecosystems and the oceans.
Source: IPCC, 2007.

that this uptake is affected by the climate itself, so if this changes as a result of climate change, this would be a feedback on the atmospheric CO_2 rise.

Moreover, we have also been clearing the world's forests to make way for farmland or degrading them through logging, and since forests lock up carbon within their biomass, their removal (which again often involves combustion) inevitably leads to a further release of CO_2 to the atmosphere. Therefore, we are certain that the observed increase in CO_2 in the atmosphere has resulted from a combination of burning fossil fuels and deforestation. It is less easy to determine the exact contribution of each of these sources, but the current best estimate is that fossil fuel burning has contributed approximately three quarters of the current excess of CO_2 above preindustrial levels, and deforestation has provided approximately the remaining quarter. Cement production also makes a small contribution.

Methane and nitrous oxide, two other GHGs, are also at record high levels. Most methane emissions and one third of nitrous oxide emissions are from human activities, largely agriculture.

Humans have also introduced new GHGs of their own, the "halocarbons" such as chlorofluorocarbons or CFCs (which, incidentally, have also damaged the ozone layer in the stratosphere). While CO_2, methane, nitrous oxide, and the halocarbons by themselves do not contribute as much to the overall greenhouse effect as water vapor, they are increasing because of human activity. However, water vapor is not directly affected by humans to any appreciable degree. The fact that CO_2 and the other human-affected GHGs are increasing leads to an enhancement of the greenhouse effect and hence a warming influence on climate.

The Observed Warming Is Very Likely to Be Human Caused

Because an unusual rise in temperatures across the globe has coincided with a unique and man-made rise in the concentration of gases known to exert a warming influence on the Earth, this suggests a role of human-induced GHG concentrations in this warming. However, while this provides strong circumstantial evidence for a human influence on climate, it does not provide a rigorous scientific test of the theory. There are many other processes that could cause climate change and have done so at various times in the history of the Earth—for example, changes in the output of energy from the sun, or changes in the Earth's orbit or the tilt of its axis, which affect how much of the sun's energy is received by the Earth and its distribution across the Earth's surface. Large volcanic eruptions can inject very large quantities of aerosol high into the atmosphere, where they can spread around the globe and cool the Earth by blocking solar radiation. Also, as well as these "externally forced" variations in the energy received from the sun, natural "internal" variability in the climate such as shifts in ocean currents and wind patterns can lead to warmer and cooler periods over years, decades, and longer. To be confident in the causes of the current warming, and hence make predictions about the future, it is necessary to go beyond mere correlations and do more rigorous scientific studies.

The established scientific method for explaining a phenomenon is to carry out a controlled experiment in which two (or more) samples are examined, with one sample being subject to a deliberate change while another is held unchanged. Clearly, this method cannot be applied to the Earth, since we have only one! However, it is possible to construct a "virtual Earth" using well-established laws of physics and measured chemical and biological processes, and conduct controlled experiments on this instead. Such a virtual Earth takes the form of a computer model, which brings together a vast array of understanding of the Earth's atmosphere, oceans, and life, and represents the physical, chemical, and biological processes in the form of mathematical equations solved by a computer program. For example, the models simulate the global patterns of wind; the flows and changes of water, energy, and chemical compounds between atmosphere, land, and ocean; and the biological processes that affect these. They simulate the cycling of water through precipitation on the land and ocean, the flow of rivers to the ocean, evaporation back to the atmosphere from land and ocean, the condensation of water vapor back to liquid water ready for precipitation again, and the freezing and melting of ice at various points in the water cycle. The models simulate the energy that the planet receives from the sun, the proportions that are reflected back to space by clouds or that reach the surface, and the proportions of the latter that are absorbed by the surface or reflected. They also simulate the emission of energy from the Earth's surface and the proportion that is absorbed in the atmosphere through the "greenhouse effect"; again, clouds are also important here, as they also reduce the loss of energy to space. These processes are affected by the chemical composition of the atmosphere, particularly the concentrations of GHGs such as water vapor, CO_2, and methane, and again this chemical composition is simulated by the models. Large-scale ecosystem changes such as deforestation are also included in many models, and these affect the absorption of energy at the surface, the evaporation of water, and the release and uptake of CO_2 and some other GHGs.

The equations themselves have been established through careful observations, measurements, and experiments both in laboratories and in the outside world, by countless scientists from Isaac Newton onward. They are central to other aspects of physics, chemistry, and biology, as well as being the building blocks of computer models of climate. The climate models bring these equations together to provide an integrated view of the workings of the planet, and once again the models are tested and refined by comparison against observations. The very same computer models are used to provide weather forecasts on a daily basis, and the fact that such models are now able to generally provide accurate weather forecasts lends confidence to the idea that they are reasonably good representations of how the world works.

With our virtual Earth, we can now "play God" and subject our mathematical planet to changes such as increases in the concentrations of GHGs and aerosols, changes in forest cover, and changes in the energy received from the sun. We can examine the effects of these acting together and in isolation from each other (Figure 1.4). Climate models can be used to estimate how the climate of the twentieth century should have evolved in the absence of human influence, driven only by natural forcings (changes in the sun and volcanoes) (pink bands in Figure 1.4). These simulations do not account for the warming seen in the last decades of the twentieth century (black lines in Figure 1.4). The climate models agree with the observed past climate change only when they additionally include the human-induced forcing of increasing concentrations of GHGs and aerosols (blue bands in Figure 1.4), hence suggesting that the observed warming is human induced.

We can also examine a climate state with no external influences, to assess the magnitude and rate of the year-to-year and decade-to-decade natural cycles of warming and cooling associated with shifts in wind patterns and ocean circulation. By comparing a variety of such simulations with the observed record of past climate, and seeing which model setup agrees best with reality, we can establish further evidence for the causes of climate change. This technique is more sophisticated than simply comparing year-by-year global average temperatures; the geographical patterns of change are also compared, allowing the "climate fingerprint" of different causes of change to be examined against the "fingerprint" of the real change, and climate change is only attributed to a particular cause (or set of causes) if the fingerprints agree within established bounds of statistical significance. While this technique obviously relies on the model's being realistic, it should be remembered that the models are grounded in well-established science and are tested against other data.

Such "detection and attribution" studies have been carried out by a number of independent groups of climate scientists around the world and all agree that the rise in temperature observed over the last 30 to 40 years cannot be explained without the rise in GHG concentrations (Figure 1.4). If only natural factors are taken into account, the computer models do not produce a warming over this period. Although the energy received from the sun (solar irradiance) has increased slightly relative to preindustrial times, the warming influence of this is less than one thirteenth of that due to the total effect of man-made changes. Moreover, solar irradiance has not increased since the 1970s and so cannot account for the warming seen more recently.

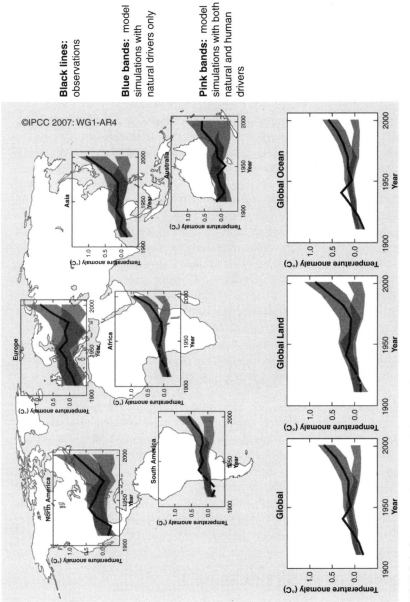

Global and Continental Temperature Change

Black lines: observations

Blue bands: model simulations with natural drivers only

Pink bands: model simulations with both natural and human drivers

©IPCC 2007: WG1-AR4

FIGURE 1.4 Using climate models to explain the observed rise in temperature (black lines). Simulations without human influence on climate (pink bands) do not reproduce the observed changes; this requires simulations with human influence through greenhouse gases and aerosol emissions (blue bands).
Source: IPCC, 2007.

Internal variability in the climate system does not appear to produce such rates of warming. The conclusion of the IPCC AR4 is therefore that "most of the observed increase in globally averaged temperatures since the mid-twentieth century is very likely due to the observed increase in anthropogenic GHG concentrations" (IPCC 2007).

We Are Shielded from the Full Warming by Other Pollution

At the same time as increasing GHG concentrations, we are also increasing the already-vast number of particles in the atmosphere. These particles are technically known as "aerosols," but are not to be confused with the spray cans of the same name. As with GHGs, some kinds of aerosols occur naturally, examples being dust, volcanic ash, sea salt, spora from plants and dimethylsulphide (DMS) from plankton. However, many are produced by human activity, mostly from burning fossil fuels. Burning wood also produces aerosols, so forest clearance and its use as fuel both increase aerosol concentrations. Additionally, desertification can increase the release of dust.

Aerosols have complex effects on climate. One effect is a cooling, since many of the particles cause some of the sun's energy to be reflected back into space, either directly through their own brightness or indirectly by increasing the brightness or lifetime of clouds. But aerosols can also absorb some of the sun's energy or some of that given off by the Earth's surface, with both processes giving rise to a warming effect. The overall effect of all aerosols is to exert a cooling influence, although the precise strength of this cooling is less certain than the warming effect of GHGs since it is more difficult to measure.

Somewhat ironically, the full rate of warming due to GHGs has not been realized because of the corresponding increase in aerosol concentrations, which is exerting a cooling effect to partly offset the greenhouse warming. Moreover, the rate of rise of CO_2 in the atmosphere is only about half of the rate of emissions from fossil fuel burning because some of the CO_2 is being absorbed by the world's vegetation and ocean waters (see Figure 1.3b). Therefore, we have been buffered from the full effect of our GHG emissions, partly by a service provided by the biosphere, and partly by a further consequence of our own pollution.

OTHER HUMAN INFLUENCES ON CLIMATE

While human-caused increases in GHGs are the main influence on current climate change, with aerosol pollution having a secondary influence, humans are also exerting other effects on climate through our transformation of the land surface and our release of energy into the environment. While land cover change and urbanization do not contribute significantly to global warming, they do influence climate at a local level, especially in the most densely inhabited parts of the world. These effects therefore need to be accounted for when assessing the impacts of future climate change on human health and our society's infrastructure.

Land Cover Change

The world's vegetation cover plays other vital roles in the climate system. As well as providing carbon sinks and stores, land ecosystems also affect climate through their influence on the character of the Earth's surface. For example, forested landscape absorbs more of the sun's energy than unforested land, so it can exert a warming influence in comparison with the unforested land, which reflects more of the sun's energy back to space. However, forests also evaporate water more than unforested land, so they can also exert a cooling influence. The relative importance of these effects depends on the local background climate. In regions where snow lies for much of the year, the difference in albedo between forests and open land are more accentuated; if you have ever flown over Canada or Siberia in winter, you may have seen how the forests stand out as black against the bright, white snow fields. In these cold regions, relatively little evaporation is occurring and the albedo effect dominates, so forests exert an overall warming influence. In contrast, tropical forests are in regions of very high evaporation and the difference between their albedo and that of snow-free tropical grasslands is not as significant, so the overall effect of tropical forests is to cool their local environment. Tropical forests also recycle significant quantities of rainwater back to the atmosphere, which maintains high rainfall in those regions.

Changes in land cover through deforestation and afforestation therefore affect climate through the changes in these physical properties of the land surface. In particular, tropical deforestation exerts two warming influences on climate, the first through the release of CO_2 to the atmosphere, adding to the enhanced greenhouse effect, and the second by reducing evaporation (and the associated cloud cover).

Urban Effects

Urban areas affect their own local climate both through the physical properties of the landscape and partly from the release of heat into the environment by the use of energy for human activities such as heating buildings and powering appliances and vehicles. As a result, urban areas tend to be warmer than their rural surroundings, a phenomenon known as the "urban heat island" effect. The contribution of urban heat islands to the global average temperature rise is negligible; cities cover an estimated 0.046 percent of the Earth's surface (Loveland et al. 2000), so the aggregate effect of all the local urban heat islands to the global average temperature is small. Averaged over the entire globe, the heat flux from urban areas is estimated as 0.03 watts per square meter of the Earth's surface (Nakićenović, Grübler, and McDonald 1998), which is less than 1 percent of the total perturbation to the Earth's energy balance ("radiative forcing") through human-induced increases in GHGs. Weather stations in or near urban areas are excluded from the records used to monitor global average temperature, in order to avoid contaminating the record with urban effects. Moreover, warming is also observed in sea surface temperatures thousands of kilometers from any city, so it is clear that urban effects are not a significant contributor to global warming.

However, urban effects do need to be taken into account if projections of local climate change are required, for example, for planning for adaptation to climate change. The local effects of human energy production can be very large indeed; in central London, the heat release is approximately 60 W m^{-2}, and daytime values in central Tokyo typically exceed 400 W m^{-2}, with a maximum of 1,590 W m^{-2} in winter (Ichinose, Shimodozono, and Hanaki 1999). Therefore, plans to adapt urban infrastructure or protect human health in a changing climate need to consider urban effects as well as greenhouse-forced climate change.

CHALLENGES IN PROJECTING FUTURE CLIMATE CHANGE

The implication of the evidence is that continued deforestation and burning of fossil fuels will inevitably lead to further changes in climate. The complexity of the climate system is such that the extent of such warming is difficult to predict, but the same computer models that are used to attribute climate change to its causes can be used to provide an estimate of future warming if provided with scenarios of GHG and aerosol emissions.

There are large uncertainties in the rate and extent of future warming, as is evident from the fact that the IPCC gives a range of figures for the warming rather than a single number. While there is sufficient confidence in our understanding of the climate system to make generalized projections for the future, there are still major gaps in our scientific understanding. Therefore, the approach is more one of assessing risks of particular changes and their rates, rather than making firm predictions.

The uncertainties in predictions of climate change arise from four main reasons:

1. The unknown future of GHG emissions. To a large extent this depends on whether action is taken to halt and reverse the current trend for increasing emissions. However, even if "business-as-usual" continued, it is not clear what this would mean for emissions—even in the absence of specific climate policy, emissions will depend on factors such as population growth, development of technology, and the nature and condition of the global economy.
2. Uncertainties in translating emissions into GHG concentrations, especially since this depends on the ecosystem's service of carbon reabsorption described earlier, which itself is affected by climate change.
3. Uncertainties in the response of the global climate to a given change in GHG concentrations.
4. Uncertainties in climate change and variability at local scales.

Emissions Scenarios

A key factor for future climate change will be the quantity of GHG emissions. These will depend on the population, their lifestyle, and the way this is supported by the production of energy and the use of the land. A large population whose lifestyle demands high energy consumption and the farming of large areas of land, in a world with its main energy source being fossil fuel consumption, will inevitably produce more GHG emissions than a smaller population requiring less land and energy and deriving the latter from nonfossil sources. These factors could vary in a multitude of ways; the international community is already examining how energy demand and production can be modified to cause lower emissions, but the implementation of this will depend on both the international political process and the actions of individuals. Even if no specific action is taken to reduce emissions, the future rates of emissions are uncertain since the future changes in population, technology, and economic state are difficult if not impossible to forecast. Therefore, rather than make predictions of future emissions, climate science examines a range of plausible scenarios in order to examine the implications of each scenario and inform decisions on reducing emissions and/or dealing with their consequences.

The IPCC's climate models have generally used a set of scenarios from the Special Report on Emissions Scenarios (SRES; Nakićenović et al. 2000). These scenarios were grounded in plausible story lines of the human socioeconomic future, with differences in economy, technology, and population but no explicit inclusion of emissions reductions policies. A large number of

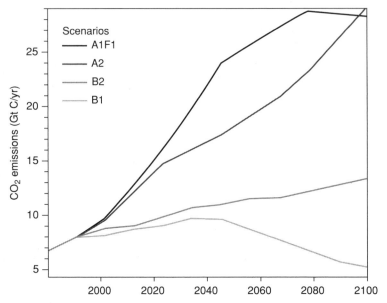

FIGURE 1.5 Four of the IPCC SRES CO_2 emissions scenarios used to drive climate models out to 2100, expressing emissions as gigatonnes (billion tonnes) of carbon per year (Nakićenović et al., 2000).
Source: Met Office, British Crown Copyright.

these scenarios were developed in the mid-1990s, and four particular scenarios shown in Figure 1.5 illustrate the range of futures assessed and have been used to drive climate models to assess their climatic consequences. These scenarios extend out to 2100 and vary widely in their projected emissions by that time (Figure 1.5), although none of them include a reduction in emissions through climate policy. The A1FI scenario describes a future world of very rapid economic growth, global population that peaks in midcentury and declines thereafter, with convergence among regions and decreasing global differences in per capita income. New technologies are introduced rapidly, but with a continued intensive use of fossil fuels. The B1 scenario describes the same pattern of population change as A1FI but with much greater emphasis on clean and resource-efficient technologies, with global solutions to economic, social, and environmental sustainability and improved equity. The A2 scenario describes a heterogeneous world with a continuously increasing population, regionally oriented economic development, and fragmented per capita economic growth and technological change. The B2 scenario also features ongoing population growth but at a lower rate than A2, and with less rapid and more diverse technological change than A1FI and B1. As with B1, B2 is oriented toward environmental protection and social equity, but focuses on local and regional levels.

So far, actual emissions have been near the upper end of the range suggested by the scenarios,[2] but emissions reductions are now high on the international political agenda. There is a need to examine a wider range of scenarios, both above and below the main SRES range, in order to help determine the effects of different levels of emissions cuts or of potential further acceleration of emissions. New scenarios will be used in the IPCC Fifth Assessment Report in order to assess the consequences of different levels of ongoing emissions, from a high rate of "business as usual" to scenarios of deep emissions cuts.

Translating Emissions to Concentrations: Importance of a Weakening Natural Carbon Sink

It is important to recognize the distinction between GHG emissions and the concentration of GHGs in the atmosphere. These are not the same thing and they do relate to each other in a simple way. CO_2 does not undergo chemical reactions in the atmosphere, so emissions to the atmosphere are only countered by removal by vegetation growth, dissolving in ocean waters or rock weathering. These are slow processes, so a large proportion of a given quantity of emissions remains in the atmosphere for decades to centuries. The concentration of CO_2 in the atmosphere at any time depends not only on the recent emissions but also the history of emissions and removals over many previous decades. A key consequence of this is that changes in emissions, such as reductions that may result from international agreements, are unlikely to significantly affect the ongoing rise in CO_2 concentration for many years. Since global temperatures take decades to respond in full to a given change in GHG concentrations, we are therefore largely committed to ongoing CO_2 rise and warming for some time. However, beyond a few decades away, the rate of CO_2 rise and warming would be impacted by any emissions cuts that began now. Therefore, while some further change is already unavoidable, action to reduce emissions in the near future could still reduce climate change in the longer term.

A key uncertainty in translating emissions into concentrations arises from uncertainty in the strength of the land and ocean carbon sinks. Since these sinks currently slow the rate of CO_2 rise, the rise in concentrations would be more rapid if the natural carbon reabsorption service is weakened. There is evidence that this may occur through three processes: photosynthesis ceasing to increase as CO_2 rises, some areas of forest dying as a result of a drying climate, and plant respiration and rates of decay in the soil increasing more rapidly with warming. At current levels, an increase in CO_2

concentration leads to an increase in photosynthesis, resulting in an uptake of CO_2 from the atmosphere by plants across the world. This process partly offsets our emissions of CO_2 and helps to slow the rise of CO_2 concentrations in the atmosphere. However, experimental studies demonstrate that this process has a limit, and the increase in photosynthesis becomes smaller and smaller as CO_2 concentrations become higher. The ecosystem service of an uptake of CO_2 therefore becomes weaker. Moreover, climate models project changes in local climates, which themselves could affect the ability of global vegetation to buffer us against the CO_2 rise. One key example is the possible reduction in rainfall over Amazonia. About 1,500 mm of rainfall per year are required to support a rainforest, and some models project the rainfall to fall well below this level during the second half of the twenty-first century. Such a drying would increase the frequency of forest fires and prevent the subsequent regrowth of dense rainforest vegetation. The carbon currently locked up within the forest biomass would therefore be released to the atmosphere, accelerating the CO_2 rise.

Finally, experimental studies suggest that the process of decay in the soil becomes more rapid under warming temperatures. Decay involves a release of CO_2 to the atmosphere, and since the world's soils currently contain more than twice as much carbon as the atmosphere, such an increase in decay worldwide could potentially be a very large feedback on the CO_2 rise and climate warming.

Just How Responsive *Is* the Climate System? Uncertainties in Climate Sensitivity

Another source of uncertainty in projecting future climate change arises from difficulties in establishing the response of global temperatures to a given rise in GHG concentrations. This may seem like an easy problem to solve because we can measure GHG concentrations and temperatures well enough to be very confident that both are increasing. In theory, the problem can be expressed as a simple equation with three terms: radiative forcing, climate sensitivity, and climate response. In this conceptual model, radiative forcing is the driving force of change, in the form of a change imposed on the Earth's energy balance. The climate sensitivity is the change in global average temperature to a given radiative forcing, representing how responsive the climate is to an imposed change. The climate response is the final result. However, although we can estimate the radiative forcing due to GHGs quite precisely, the problem is made more difficult by the other contributions to the overall radiative forcing such as aerosols. As described above, the extent of the cooling influence of aerosols has not yet been established with much precision, which means that the net radiative forcing of GHGs and aerosols together is also not very well known. Therefore, although we know the final climate response from our temperature measurements, we do not precisely know the forcing causing this response and therefore cannot estimate climate sensitivity precisely, either. Different climate models give different estimates of climate sensitivity, often related to the strength of cloud feedbacks.

A key issue arising from this is that we do not know the extent to which aerosols are masking the true effect of GHGs. If the current aerosol cooling is strong, this implies that the greenhouse warming that it is partly counteracting must also be strong. This is an issue because aerosol emissions can be reduced more easily than GHGs, and also aerosol particles are washed out of the atmosphere within days, whereas CO_2 is gradually taken up by the ocean and land ecosystems over decades and centuries. Therefore, it would be relatively easy to reduce the aerosol cooling effect, and therefore unveil the full warming effect of GHGs.

Butterflies and Pooh Sticks: Challenges in Predicting Regional Climate Change

A further difficulty for climate scientists arises from the inherently chaotic nature of the atmosphere and oceans. While the processes in the atmosphere and oceans are governed by the laws of physics, these laws include those of "chaos theory" in which very small differences at the start of a process can lead to enormous differences in the final outcome. This was illustrated by Ed Lorenz with the classic "butterfly effect," asking "Does the flap of a butterfly's wings in Brazil set off a tornado in Texas?" Tiny changes in wind, such as would be caused by the butterfly's wings, can influence other, larger wind flows, which in turn influence still larger winds until some major event is affected. While a tornado could be "set off" from such a small start, equally another tornado could be prevented from happening.

The reason this is relevant to weather and climate modeling is that it is impossible to measure and model the atmosphere to infinite precision, and even the small approximations necessary in the measurements and models can lead to differences in the outcome. Minuscule changes in the numbers put into a model lead to different results, so this poses a limit on what is predictable, at least in terms of day-to-day weather. This is why weather forecasts for individual days are currently given only up to about a week ahead—any farther ahead and the effects of today's butterflies are magnified so much as make precise forecasting impossible. Since we do not include butterflies in our measurements of today's weather, we cannot hope to capture their effects when they become large enough to matter. And even if we could know where all the butterflies are and how hard they are flapping, there would still be some further level of detail we cannot capture.

But while day-to-day local weather cannot be forecast very far ahead, average or likely conditions over longer times and larger scales can be more predictable when large, slow oceanic changes or large drivers of change from outside are involved. A good example is the game of "Pooh Sticks"—we can throw several sticks into a river and know that they will all be carried downstream with the current, and run to the other side of the bridge to watch them come out, but it is much harder (if not impossible) to predict which one will win the race because that depends on the individual swirlings in the water. Similarly, we can be confident that trapping more heat radiation from the Earth by increased GHG concentrations will cause a general warming on average over the coming decades, and can make projections of the general nature of changes in the atmospheric circulation. However, the local details of weather depend on smaller-scale processes that again are much harder to predict even on average, let alone for particular years or months in the future. Understanding these processes and using them to improve predictability is a major focus for climate research.

FUTURE CLIMATE CHANGE WITH AND WITHOUT EMISSIONS REDUCTIONS

So far, the main focus of IPCC climate projections has been to assess the long-term impacts of climate change over the coming century if no action is taken to reduce emissions, in order to assess the consequences of "business as usual" and establish whether there is indeed a need to reduce emissions. More recently, with the need for emissions reductions now widely accepted, there are new requirements to assess the consequences of different levels of emissions cuts, in order to inform climate policy negotiations in more detail. Again, these studies focus on the long-term timescales of the middle to the end of the century. However, with adaptation to climate change also becoming recognized as necessary even if emissions are reduced, there is also a further need to assess the local-scale climate changes to which we need to adapt. In many cases, decisions for adaptation planning require projections of changes and variability in the nearer term, over the next few years or decades, as these are the timescales for which planning needs to begin now. Consequently, climate models are now being used for an increasing variety of projections, to inform an expanding portfolio of decisions.

Projections of Change if No Action Is Taken

The IPCC SRES emissions scenarios have been widely used as input to climate models to project likely further global warming over the coming century if no action is taken to reduce emissions; these models project warming

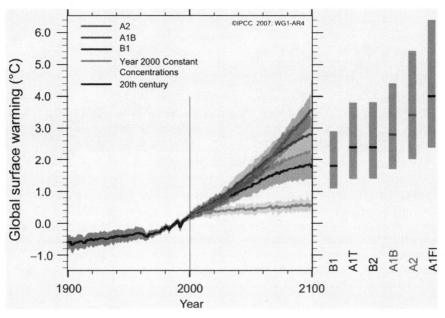

FIGURE 1.6　Projected changes in global mean temperature resulting from the main IPCC SRES emissions scenarios.
Source: IPCC, 2007.

of between 1.0 and 6.4 degrees Celsius by the end of the twenty-first century (Figure 1.6). The range of results reflects the uncertainties in "business as usual" emissions, uncertainties in translating emissions to concentrations, and uncertainties in the climate response to a given level of GHG concentration increases.

Sea level is projected to rise by between 28 cm and 58 cm, and snow and sea ice cover are projected to shrink. Some models suggest that the Arctic could be ice free in late summer by the latter part of the twenty-first century. Heat waves and extreme rainfall events are projected to increase. Although the number of tropical cyclones is tentatively projected to decrease, their intensity is projected to increase.

These changes would not be spread uniformly around the world. Some regions will warm faster than the global average, while others will warm less. Faster warming is expected near the poles, as the melting snow and sea ice exposes the darker underlying land and ocean surfaces, which absorb more of the sun's radiation instead of reflecting it back to space as bright ice and snow do. Indeed, such "polar amplification" of global warming is already being seen. Changes in precipitation are also expected to vary from place to place (Figure 1.7).

Projected Patterns of Precipitation Changes

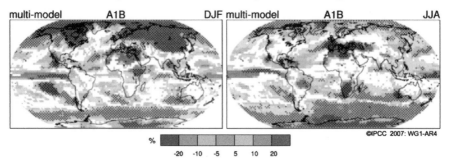

FIGURE 1.7   Changes in precipitation projected for the 2080s by a number of climate models, including an indication of the level of agreement. Shadings show where more than 66 percent of models agree on the sign of the change, white areas show where less than 66 percent agree. Black dots show where 90 percent of models agree on the sign of the change. Contours show the average change projected by all models. Results are shown for the seasons of December-January-February (DJF, left) and June-July-August (JJA, right).
Source: IPCC, 2007.

In the high-latitude regions (central and northern regions of Europe, Asia, and North America) the year-round average precipitation is projected to increase, while in most subtropical land regions the precipitation is projected to decrease by as much as 20 percent. Some climate models (but not all) project a particularly strong decrease in rainfall in Amazonia, due to changes in atmospheric circulation caused by particular patterns of warming in the north Atlantic and equatorial east Pacific oceans.

The projections used for long-term assessments represent the overall trend of global mean temperatures, but not the precise year-to-year variations. While the models do include year-to-year variations that are realistic in a statistical sense—they simulate relatively warmer and cooler years with about the right frequency—they are not expected to be realistic for individual years. Progress in forecasting shorter-term changes for informing adaptation, including year-to-year variations, is discussed below.

Informing Mitigation: Assessing the Impact of Different Emissions Cuts

With AR4 having established that ongoing greenhouse emissions would cause major climate changes, the policy focus is now turning to the question of how deeply and how quickly emissions should be reduced.

Jason Lowe at the Met Office Hadley Centre used a simple climate model calibrated against more complex models to address issues relevant to this question (Met Office 2007). Many climate policy makers and stakeholders focus on a global mean temperature rise of 2 degrees Celsius relative to preindustrial as a threshold to avoid surpassing; Lowe estimated the likelihood of exceeding this 2 degrees Celsius rise with different illustrative scenarios of emissions reductions, shown in Figure 1.8.

The simple model estimated the uncertainty in the climate response to given GHG concentrations, and also included feedbacks between climate change and the carbon cycle as described earlier. In one scenario, emissions reductions begin in 2012 and are reduced to zero in the 2060s. The model suggested that the most likely result was that a global warming of 2 degrees Celsius would be avoided with this hypothetical scenario; however, the model also suggested a significant probability of exceeding 2 degrees Celsius. A key result was that the peak warming was significantly delayed after the peak emissions; in the "best estimate," the peak in global mean temperature occurred in the 2060s, approximately half a century after the peak in emissions.

In the maximum sensitivity case, ongoing warming continued until the end of the simulation at 2200, as a result of ongoing release of carbon from natural stores as a consequence of climate change itself. As described earlier, climate change itself is altering the balance of uptake and release of carbon by ecosystems and the ocean waters. The net sink of carbon is weakening and is expected to continue to do so in the future, particularly as warming continues to increase the release of carbon from soils. This means that stabilizing and recovering atmospheric CO_2 concentrations is likely to be more difficult than would otherwise be expected, because once the CO_2 is in the atmosphere the resulting climate change could weaken the natural carbon sink and hence reduce the removal of CO_2 from the atmosphere (Jones, Cox, and Huntingford 2006). So, despite a reduction in emissions, if climate change leads to large carbon cycle feedbacks, CO_2 concentrations could continue to rise even while emissions are being reduced.

In a second scenario, emissions cuts begin in 2036 and are reduced to zero by the 2080s. Again, peak warming lags behind peak emissions by several decades, with the world continuing to warm until the 2080s and the "best estimate" showing 2 degrees Celsius being exceeded. Uncertainties in the warming are significant, again largely due to uncertainties in carbon cycle feedbacks, and while the "best estimate" scenario shows temperatures declining over the twenty-second century, the upper estimate shows ongoing warming, which reaches 7 degrees Celsius by 2200. As the strength of the carbon sink decreases as the climate warms, stabilization of concentrations requires increasingly large emissions cuts to compensate for the weaker sink.

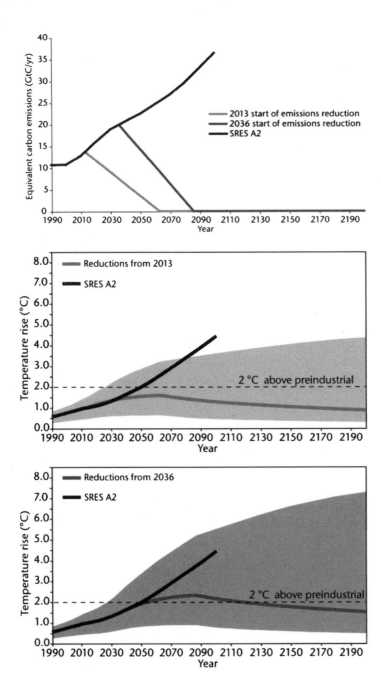

FIGURE 1.8 Simple climate model study of the effects of emissions reductions begun earlier and later. Lines show best estimates of global temperature changes, bands show estimated uncertainties due to uncertainties in the resulting CO_2 concentrations and climate change. *Source:* Met Office, British Crown Copyright.

The implication of this illustrative study is that delays in emissions cuts increase the commitment to large warming and also the risk of feedbacks becoming large. There is a lag of several decades between instigating emissions cuts and seeing the benefit of these cuts, so it is not possible to wait to see the outcome of past emissions before deciding on the level of emissions reductions. Some further changes are already inevitable, but there is still a possibility that global warming could remain below 2 degrees Celsius—but this will require early action in reducing emissions significantly. The longer the delay in reducing emissions, the greater the likelihood of exceeding 2 degrees Celsius warming.

Having seen the importance of carbon sinks for stabilizing CO_2 concentrations, it is also worth noting that forests are an important carbon sink, so deforestation also makes stabilization and recovery of CO_2 concentrations more difficult. Not only does deforestation add to emissions, it reduces the removal of CO_2 from the atmosphere by decreasing the size of the forest carbon sink (Betts et al. 2008). The farther deforestation is allowed to advance, the weaker the carbon sink and the more deeply fossil fuel emissions will need to be reduced.

Informing Adaptation: Forecasting Unavoidable Climate Change

As is evident from the preceding discussion, some further climate change is already unavoidable simply due to the laws of physics. Note, for example, the bottom line in Figure 1.6, which shows a simulation of the ongoing rise in temperature that would occur even if CO_2 concentrations had ceased to rise in the year 2000. The Earth takes time to respond fully to a change in GHG concentrations, so it is still "catching up" with the increased concentrations over the twentieth century. Even more climate change seems inevitable for social, political, and economic reasons: An immediate total shutting down of emissions would require an abandonment of the modern lifestyle overnight. Note again in Figure 1.6 how the projected climate changes are similar for all the main SRES emissions scenarios up until approximately 2040, despite the large differences in actual emissions by then (Figure 1.5). This is because, as described above, concentrations are affected more by previous emissions than current emissions, and because the Earth's temperature also takes time to fully respond. We already appear to be seeing impacts of climate change; many plant species are flowering or leafing out earlier in the year due to warmer springs (Fischlin et al. 2007), and it has been estimated that European summer temperatures as high as those experienced in the heat wave of 2003 are probably now twice as likely (Stott, Stone, and Allen 2004). Therefore, with further changes already in the pipeline, there is a need to adapt. Those who plan ahead are more likely to be able to minimize damages and even exploit opportunities, and in competitive fields, those who act first would be expected to gain the edge.

In an ideal world, adaptation plans would be based on robust forecasts of the potential impacts of climate change on a specific activity in a specific place at a specific time. For example, a decision on siting a new piece of infrastructure such as a dam or a power station would take account of the climate changes projected for that site over the lifetime of the investment, which could be several decades or longer. In the nearer term, a commodity trader may wish to know which crops will do well and which will fail in the coming season or next few years. In practice, these kinds of questions push the current scientific capability to its absolute limits, but it seems that useful advice can be given even with this aspect of the science in its infancy.

Climate prediction in the near term (years to a few decades) is made more difficult by the fact that natural variations are still relatively important on this timescale, whereas on multidecadal timescales we expect the increased greenhouse warming to lead to long-term changes that are greater than the natural year-to-year variability. This poses a huge challenge when attempting to forecast on these timescales in order to inform adaptation. This area of climate science is only just beginning to demonstrate predictive skill, but nevertheless some skill is there. The key lies in good measurements of the current state of the climate (especially ocean temperatures), including the direction of any trends in the system, in order to set off the forecast in the right direction. Returning to the Pooh Sticks analogy, imagine that you are about to drop your sticks into the river and you see a large, slow-moving swirl of water making its way down the river. You can be reasonably confident that a stick dropped on the arm of the swirl moving in the same direction of the main stream flow will probably move faster than a stick dropped in the arm of the swirl moving against the main flow. Similarly, measurements of ocean temperatures can help capture the natural variations within the climate system and allow them to be factored into the models. Forecasts of global temperatures on decadal timescales have been shown to be credible, with techniques developed by Smith et al. (2007) in the Met Office Hadley Centre having been used to reproduce decadal temperature changes over the 1980s–1990s and 1990s–2000s using only the information that would have been available to forecasters at the start of those periods. Moreover, a forecast issued by Smith et al. in 2005 predicted that global temperatures would stall over the following few years, and only then begin to rise again. Since 2005, the rise in temperature has indeed stalled, with a La Niña event in 2007–2008 being a major feature temporarily counteracting the greenhouse-forced warming. However, Smith et al. forecast the warming to begin again before the end of the decade, and they forecast the warming

DePreSys : from March 2007

IPCC AR4

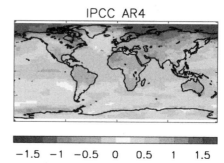

−1.5 −1 −0.5 0 0.5 1 1.5

FIGURE 1.9 Forecast of temperature changes by 2020 using the Met Office Hadley Centre decadal forecasting system starting from the actual observed state in 2007 (left) and standard climate models initialized from the long-term average climate (right). Initialization from the current observed state rather than the long-term average improves the ability to forecast year-by-year variations. *Source:* Met Office, British Crown Copyright.

between 2005 and 2014 to be 0.3 degrees Celsius. This is half of the warming observed over the twentieth century. Figure 1.9 shows that this forecast suggests different rates of warming to those projected by the standard climate models, which do not set their forecasts in motion using observations of recent trends in ocean temperature.

Decadal forecasting is still in its early stages, and significant challenges remain, especially for forecasting of local rather than global changes. However, rapid progress is being made, and such forecasts are likely to be invaluable to adaptation planners in the coming years.

SUMMARY

It is clear that the world is now warming at a highly unusual rate and, consequently, snow and ice are melting, sea levels are rising, and patterns of rainfall are changing. This is very likely to be the result of the very high atmospheric concentrations of CO_2 and other GHGs, which are undeniably a result of deforestation and the burning of fossil fuels by humans.

The effects of our CO_2 emissions have not been fully realized, partly because of a feedback mechanism of carbon uptake by the biosphere and partly because fossil fuel burning also produces aerosol particles, which exert a cooling effect by reflecting sunlight. Continued emissions of GHGs are confidently expected to lead to further warming, resulting in further sea level rise and rainfall pattern changes with consequent impacts on society. The precise magnitude and nature of future changes is difficult to predict, partly because feedback mechanisms (which have buffered us from the full effect of our emissions) may weaken by an uncertain extent in the future. While some level of change is now inevitable, and hence must be faced up to and adapted to, there is still the possibility of reducing the magnitude of future change later in this century by reducing emissions of GHGs.

NOTES

1. IPCC Assessment Reports are prepared by a large international team of leading scientists, with several rounds of very extensive peer review, and provide a thorough assessment of the recent peer-reviewed literature. Reports and further information are available online at www.ipcc.ch.
2. Suggestions that actual emissions have been greater than the IPCC SRES range are based on a comparison with the averages of different versions of the scenarios from different sources (Raupach et al., 2007) rather than with the individual scenarios that were actually used in climate models. The emissions scenarios were generated from plausible "story lines" of future population, gross domestic products, energy sources, technology, and other socioeconomic factors, and several different versions of each scenario were produced by a number of Integrated Assessment Models (IAMs). From the group of versions of each scenario, one particular version was chosen as the "marker scenario" and was used to drive climate models for the IPCC Third and Fourth Assessment Reports (IPCC, 2001, 2007). Raupach et al. compared actual emissions with the averages of the versions of each scenario, rather than with the marker scenarios. Actual emissions have been within the range of the marker scenarios (van Vuuren and Riahi 2008).

REFERENCES

Betts, R., A. Gornall, J. Hughes, N. Kaye, D. McNeall, and A. Wiltshire. 2008. Forests and emissions: A contribution to the Eliasch Review. Exeter, UK: Met Office. www.metoffice.gov.uk/climatechange/policymakers/policy/eliaschreview.html.
Fischlin, A., G. F. Midgley, J. Price, R. Leemans, B. Gopal, C. Turley, et al. 2007. Ecosystems, their properties, goods and services. In *Climate change 2007: Impacts, adaptation and vulnerability. Contribution of Working Group II to the Fourth Assessment Report of the Intergovernmental Panel on Climate Change.* Cambridge, UK, and New York: Cambridge University Press. www.ipcc.ch.
Ichinose T., K. Shimodozono and K. Hanaki. 1999. Impact of anthropogenic heat on urban climate in Tokyo. *Atmospheric Environment* 33:3897–3909.
IPCC. 2001. *Climate change 2001: The physical science basis. Contribution of Working Group 1 to the Third Assessment Report of the Intergovernmental Panel on Climate Change.* Cambridge, UK, and New York: Cambridge University Press. www.ipcc.ch.

IPCC. 2007. *Climate change 2007: The physical science basis. Contribution of Working Group 1 to the Fourth Assessment Report of the Intergovernmental Panel on Climate Change.* Solomon, S., D. Qin, M. Manning, Z. Chen, M. Marquis, K. B. Averyt, et al., eds. Cambridge, UK, and New York: Cambridge University Press. www.ipcc.ch.

Jones, C. D., P. M. Cox, and C. Huntingford. 2006. Impact of climate–carbon cycle feedbacks on emission scenarios to achieve stabilisation. In Schellnhuber, H. J., W. Cramer, N. Nakićenović, T. Wigley, and G. C. Yohe, eds., *Avoiding dangerous climate change.* Cambridge, UK: Cambridge University Press.

Loveland T. R., B. C. Reed, J. F. Brown, B. Ohlen, Z. Zhu, Y. Yang, et al. 2000. Development of a global land cover characteristics database and IGBP Discover from 1 km AVHRR data. *International Journal of Remote Sensing* 21:1303–1330.

Mann, M. E., R. S. Bradley, and M. K. Hughes. 1999. Northern hemisphere temperatures during the past millennium: Inferences, uncertainties, and limitations. *Geophysical Research Letters* 26:759–762.

Met Office. 2007. *Climate research at the Met Office Hadley Centre: Informing government policy in the future.* Exeter, UK: Met Office. www.metoffice.gov.uk/publications/brochures/clim_res_had_fut_pol.pdf.

Nakićenović, N., A. Grübler, and A. McDonald, eds. 1998. *Global energy perspectives.* New York: Cambridge University Press.

Nakićenović N., J. Alcamo, G. Davis, B. de Vries, J. Fenhann, S. Gaffin, et al. 2000. *IPCC special report on emissions scenarios.* Cambridge, UK, and New York: Cambridge University Press. www.ipcc.ch.

Smith, D. M., S. Cusack, A. W. Colman, C. K. Follans, G. R. Harris, and J. M. Murphy. 2007. Improved surface temperature prediction for the coming decade from a global climate model. *Science* 317:796–799.

Raupach, M. R., G. Marland, P. Ciais, C. Le Quéré, J. C. Canadell, G. Klepper, et al. 2007. Global and regional drivers of accelerating CO_2 emissions. *Proceedings of the National Academy of Sciences* 104:10288–10293.

Stott, P. A., D. A. Stone, and M. R. Allen. 2004. Human contribution to the European heat wave of 2003. *Nature* 432:610–614.

van Vuuren, D., and K. Riahi. 2008. Do recent emission trends imply higher emissions forever? *Climatic Change* 9:1, 237–248.

CHAPTER 2

The Economics of the Climate Change Challenge

Dimitri Zenghelis

Climate change is a global problem unparalleled in scale. It requires international collaboration that is equally unprecedented, stretching across sectors, countries, and disciplines. It is a pressingly urgent problem, too. The world has only a few years to establish a strong and credible global deal for action to reduce emissions if it is to avoid large risks of severe damage to the planet and to the prospects for sustained growth and development. In 2006, I jointly authored the *Stern Review* with Lord Stern and a small team of experts. Since then, the latest scientific evidence suggests we probably underestimated the overall magnitude of risks associated with climate change.

On the one hand, even after accounting for the impacts of the recent global economic recession, underlying emissions look likely to be higher than many previous studies argued. These include the Intergovernmental Panel on Climate Change and the *Stern Review*. On the other hand, scientists warn of greater risks, and more severe and rapid effects of global warming, than were previously assumed. Declining carbon sinks, faster melting of summer arctic ice, and the effects of particulate emissions masking underlying warming all point to growing risks of dangerous outcomes.

The challenge cannot be underestimated, and it is now clear that piecemeal, incremental change will not suffice to stabilize temperatures at levels

Dimitri Zenghelis is chief economist of the Climate Change practice in the Global Public Sector organization, Senior Visiting Fellow at the Grantham Institute on climate change, and Associate Fellow at the Royal Institute of International Affairs (Chatham House).

that safely reduce the risks from climate change. Instead, meeting the challenge requires radically transformed development paths on the back of a technological revolution.

The United Nations Framework Convention on Climate Change (UNFCCC) 15th Conference of the Parties in Copenhagen at the end of 2009 will be crucial in designing the architecture for a post-2012 successor policy framework to Kyoto. Different technologies and different policy instruments will need to be applied to different sectors and countries. However, it is important that all the different initiatives add up to delivering the overall objective. This requires an agreement on the global emissions targets, the role of developing countries in mitigation and trading, international emissions trading, financing emissions reductions from deforestation, technology, and adaptation. There is also a need to look ahead to the challenges of implementation and the nature of global institutions needed to manage the task. The first half of this chapter looks at the case for action and, updating the *Stern Review*, examines how economics can help inform the appropriate pathway for emissions. One by one, I will try and explain why none of the reasons given for holding back on action to reduce emissions stacks up to the evidence. The second half then assesses the policy context required to bring about an effective, efficient, and equitable reduction in greenhouse gases (GHGs).

CONSIDERING THE CASE FOR ACTION

The central conclusion of the *Stern Review* was that the costs of action on climate change will be less than the costs of inaction. Two years on, this conclusion has attained broad acceptance among most policy makers, but is still not without critics. Although they are in a declining minority, there are still some who deny the importance of strong and urgent action on climate change, offering one of, or a combination of, the following arguments.

First, there are those who deny the scientific link between human activities and global warming, yet the weight of evidence makes this hard to countenance. The basic science of the greenhouse effect goes back to the nineteenth century. The February 2007 *Fourth Assessment Report of the Intergovernmental Panel on Climate Change* (IPCC 2007) sets out the evidence in a very convincing and clear way. There is now very little justification for believing that the remaining areas of uncertainty imply that current knowledge is inadequate as a basis for drawing conclusions for policy. The value of the risks associated with inaction is enormous, with the possibility of catastrophes of unprecedented magnitude.

The most recent report of the IPCC shows significant risks of global temperature increases of more than 5 degrees Celsius, relative to preindustrial times, by the next century if we do not act to curb emissions. This does not sound like much—5 degrees is far less than the kind of outdoor temperature variation most of us experience every day. However, this measures global average temperatures and masks substantial differences between temperature variations across oceans and continents. Continental areas, where most of us live, can expect to see much larger temperature increases.

To put this global average into context, the last time the world was five degrees warmer was 35 to 55 million years ago when the world was characterized by swampy forest and there were alligators close to the North Pole. The last time the world was five degrees cooler was during the last ice age around 10,000 to 12,000 years ago when giant ice sheets stretched as far south as New York and central England. The argument here is not about crocodiles or ice sheets. It is about redrawing the geography of how and where people live at unprecedented speed. If temperatures were to rise by 5 degrees or more in the next century, the level of gross domestic product (GDP) probably would be set back decades, with billions suffering from hunger, water stress, mass migration, and conflict. Figure 2.1 outlines some

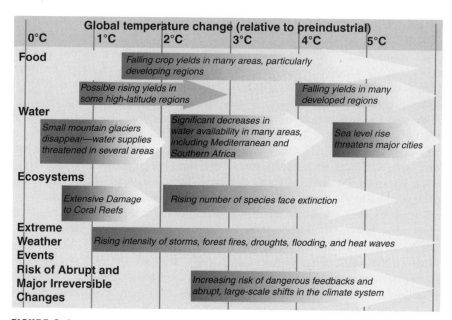

FIGURE 2.1   Projected impacts of climate change.
Source: The economics of climate change: The Stern Review. Cambridge, UK: Cambridge University Press, 2007.

key impacts corresponding with higher average global temperatures across five dimensions.

The Costs of Inaction

The consequences of delayed action could be catastrophic, with major socioeconomic and geopolitical implications. Many, perhaps most, regions of the world are likely to suffer from further water stress. Areas that are already relatively dry, such as the Mediterranean basin and parts of southern Africa and South America, are likely to experience additional decreases in water availability. In contrast, southern Asia and parts of northern Europe and Russia are likely to experience increases. Declining crop yields, especially in Africa, could leave hundreds of millions without sufficient food. There are serious risks to coastal areas, with increasing pressures for protection in Southeast Asia as well as in coastal cities such as New York, Cairo, Mumbai, and London. As a result, thousands of millions of people could become permanently displaced, prompting massive emigration waves. This scenario is likely to exacerbate conflicts in already unstable areas of the world and may have serious implications for energy supply from major oil exporting countries in increasingly unstable regions.

When economists first began to model the likely impacts of climate change, many assumed a world characterized by relative certainty with impacts on things they could measure: impacts on agricultural output and the costs of some adaptive measures. Few accounted for things that were harder to measure, such as human health; the environment; and the costs of conflict, refugees, and migration. By assuming certain impacts, they also did not allow for risks of catastrophic events in their calculus. These are listed in the fifth row of Figure 2.1 and cover risks of thresholds and discontinuities such as accelerated methane emissions from melting permafrost, Amazonian die-back, acidification of the oceans, and rapid melting of the Greenland and west Antarctic ice sheets and faster-than-expected erosion of the earth's capacity to reabsorb carbon. It is no surprise, therefore, that early estimates of the impacts showed rather small effects, often expressed in terms of a few percentage points of GDP. In fact, many of these economists acknowledged that their results should not be taken to cover the totality of climate change impacts but only a small subset.

Others, however, persisted in using numbers that seemed at odds with what the scientists were saying, and brandished them as a gauge of the overall impacts of climate change. For example, Bill Nordhaus somehow gets from IPCC temperatures of 3 degrees by 2100 and 5.3 degrees by 2200, to impacts of only a couple of percent and 8 percent of GDP, respectively. He notes in his recent book (Nordhaus 2008) that "the sub-models used in the DICE model cannot produce the regional, industrial, and temporal details that are generated by the large specialized models," but it is at the regional level where the real human damage is estimated. By underplaying the regional and local risks, Nordhaus predicts that by 2200 world GDP will be set back a mere four years if we do nothing, relative to a baseline without climate change damages, thus effectively trivializing the dangers from climate change.

A second group opposing climate action are those who, while accepting the science of anthropogenic climate change, argue that the human species is very adaptable and can make itself comfortable whatever the climatic consequences. Our ability to adapt to changes in the climate is indeed likely to increase as countries become richer and more technologically advanced and climate adaptation is unavoidable. However, the scale and speed of the change, if we carry on as we are, is predicted by scientists to likely rewrite the physical and human geography of the planet within a century or two, requiring human mass migration on an unprecedented scale. Moreover, the costs of adaptation will rise at a faster rate than the temperature, especially as populations are forced to inhabit increasingly hazardous areas. Some changes in climate are not amenable to technological improvement in the same way that, say, energy generation or fighting diseases is. Finally, adaptation, although necessary, cannot substitute mitigation, as only the latter reduces uncertainties. Without effective mitigation, it will be difficult and costly to plan exactly what it is we have to adapt to.

The Costs of Action

A third group of climate action skeptics consists of those who believe the costs involved in reducing emissions are just too great and are despondent about the challenge of mitigating climate change. This seems hard to reconcile with the importance of GHG-related costs in the economy as a whole. Even energy and transport account for less than a tenth of whole economy costs, and the low emissions replacement technologies are often highly competitive, and in a number of cases (such as energy efficiency), they actually lower costs. A range of studies from data-driven input-output models, technology models, and full behavioral macroeconomic models suggest the likely costs of urgent and coordinated action is around 1 percent or 2 percent of GDP by midcentury, give or take a few percentage points, depending on the evolution of technologies and the price of fossil fuels in particular (see Stern 2006, Chapter 10). This is not a trivial amount; it represents a very significant change in the patterns of energy investment toward low-carbon energy technology, but after a one-off cost increase, growth should not be affected. In the absence of these factors, or were action to be delayed, the costs could be significantly higher. The opportunities for business cannot be underestimated. The International Energy Agency (IEA) reckons that roughly $1,000 billion per annum will be invested in energy infrastructure over the next few decades (IEA 2008). As government policies shift toward ensuring that these investments are carbon constrained, firms, institutions, and states that move earliest to develop new technologies are likely to capture new competitiveness opportunities and make substantial market gains.

However, as we argued in the *Stern Review,* the decision on whether to act now hinges on the question of irreversible outcomes and risks. Decisions made today will have potentially large and irreversible consequences in terms of climate change impacts; this is not true to the same extent of GHG abatement costs. Moreover, policy makers can keep cost estimates under review and revise policy in the light of new information. By contrast, the impact from global warming will become increasingly costly to reverse.

How to Value Impacts Far Away in Time: The Infamous "Discount Rate"

Finally, there are those who accept the science of climate change and the likelihood that it will inflict heavy costs, but simply do not care much for what happens in the future beyond the next few decades. This brings on the infamous discussion about discounting. Rightly, there has been much focus on ethical judgment made in assessing the impact of climate. I say "rightly" because these cannot be avoided. Like it or not (mostly not), climate change affects people of different ages and incomes at different points in time. How we value these people will be central to the assessment of the impacts (see Sen and Williams 1982).

The question of "discounting" lies at the heart of any evaluation of the expected future events resulting from rising global temperatures. Discounting, when applied to consumption, is the process of defining the present value of a unit of consumption at some future date. The reduction in the value of a unit's consumption from one year to the next is determined by the appropriate annual discount rate. From an individual's point of view, we all prefer jam today to jam tomorrow, and at some point we expect to die. Consequently, we discount heavily in our own lives. But from society's point of view, the question of how the policy maker should value the consequences of today's actions on future generations needs to move beyond the innate impatience of individuals. This has long been accepted by economists involved in public policy, but is often forgotten by economists working in narrower fields (Stern 2007).

There have traditionally been two reasons why economists and philosophers apply a discount rate at the social rather than individual level to society's future income. First, future costs are discounted because the future world will be richer and better able to afford them. The loss of "happiness" from a dollar of forgone income (e.g., from reduced access to water or coastal flooding) will, in general, be smaller for rich people than for those struggling to make a living. The intuition is simple to grasp: the "happiness" loss from a dollar's worth of lost consumption to Bill Gates will be less than the same loss to a hungry street child, where it could mean the difference between life and death. So, if future generations will be richer, then taking action now to save them from unpleasant impacts will hold a lower claim on current resources than taking action now to avoid the very same affects on poorer people today. Logically, this means we need to discount away future impacts in our assessment of the present value of climate change damages, but only because we expect the future to be better off than the present.

So far so good. But the problem of uncertainty surrounding the precise nature and extent of the impacts makes the process of assessing the costs of inaction rather complex. We just don't know how much richer future generations will be in a world with climate change, so we cannot know the single correct discount rate. Some economists, such as Nordhaus, sidestep this problem by working off deterministic or fixed projections of the damages. Sadly, this ignores a key part of the story, which is that scientists tell us there is a possibility of some pretty catastrophic outcomes if we carry on as we are. Particularly destructive impacts, such as large-scale flooding, widespread droughts, and intense storms and irreversible runaway climate change, *could* render some future generations much poorer than the average expectation and *poorer* even than current generations, wiping out the benefits of economic growth. To evaluate these impacts would require lower discount rates; indeed, in the case where generations are poorer than today, it would require *negative* discounting under the principles outlined earlier.

The choice here is between uncertain paths with radically different implications for the planet, so it makes no sense to apply a single high discount rate, such as Nordhaus's 4.5 percent to 5.5 percent, regardless of whether the future losses being assessed are trivial or devastating. Such an approach systematically undervalues the worst-case events. A discount rate of 5 percent is so high that a dollar's worth of lost consumption in 2150 is valued as akin to only 0.02 *cents* lost today, regardless of the scale of the catastrophe. By trivializing future losses in this way, the case for action is seriously diminished by assumption.

Failure to systematically apply the concept of diminishing extra happiness from every extra dollar therefore illuminates the deficiencies of many economists' assessment of risk. In the *Stern Review*, applying this basic economic concept meant that we attached greater weight to the worst outcomes, reflecting precisely the fact that we worry about them more. Our aversion to catastrophic events that can render us poorer is why most of us insure our houses, even though we know the insurance companies make money from the odds they offer us. By applying a greater weight (i.e., less discounting) to nasty events, the Stern approach will reflect this automatically.

The approach rightly adopted in the *Stern Review* applied discounting precisely on the principle that future generations may be richer or poorer to extents we cannot be certain about. We used a probabilistic array of projections, each valued using a unique discount factor reflecting its impact on future incomes. In fact, the *Stern Review* went further by applying additional discounting to cover extreme risks such as the world succumbing to asteroids, plague, or nuclear Armageddon. If future generations cannot be guaranteed to exist, it seems inappropriate to value them on a par with the current generation, which clearly does.

However, that is as far as we went on discounting. The *Stern Review* rejected *further* discounting to discriminate against future generations purely on the basis of birth dates—a process known as *pure time* discounting. This is the second reason for discounting: the passage of time itself. This is distinct from discounting because of income differences or discounting because of the risk of future extinction, both of which can be expressed quantitatively as in the *Stern Review*. Pure time discounting is rooted in the economist's desire to reflect people's preferences, and people seem to be impatient in many of the things that they do. But climate change is such a long-term social problem that it is inappropriate to use personal telescopic preferences as the basis to determine policy. Why should we treat the welfare of current generations on an equal basis, but apply a different treatment to the welfare of generations born next year or the year after?

Many economists propose pure time discrimination because they assume that market interest rates of return reveal social preference for future versus present reward so that one can use them to back out our ethical values from the markets. But this is misguided. When the financial system is working normally, market rates are determined by the cumulative actions of individuals in their saving, consumption, and investment decisions. They do not reveal societal ethical preferences. A century of mainstream economic literature from Marshall and Pigou to Arrow and Mirrlees has recognized such logic to be wrong, except under implausible circumstances, including that all markets work perfectly and all consumers are represented.[1] Other economists, from Ramsey to Solow, Keynes, and Sen, reject pure time discounting as arbitrary, holding no ethical basis for long-term public policy choices (see Anand and Sen 2000; Mirrlees and Stern 1972; Pigou 1932, 24–25; Ramsey 1928, 543; and Solow 1974, 9, to name a few).[2]

Pure time discounting makes policy intertemporally inconsistent. We take action that affects future generations in ways that fit with our ethical judgments today, but as the clock rolls forward, our actions are consistently revealed to be wrong. In the climate change context, it would be like having to explain to a society in 2200 facing desertification, drought, coastal inundation, famine, war, and runaway climate change why we did nothing in the early part of the twenty-first century to prevent this? We would have to answer along the lines of "Yes, we knew the risks, but we applied pure time discounting to reflect personal preferences and I'm afraid this gave your suffering a weight of 0.001 percent relative to ours and so we felt the right thing to do was to leave you to it. Good luck!"

Even if we tried to invest conventionally at market rates and later tried to buy off future environmental damages, the costs of action will have risen sharply because stocks of GHGs—based on current IPCC estimates—would be so large they could tip the global climate into dangerous and irreversible changes (see Hoel and Sterner 2007; and Sterner and Persson 2007). Similarly, consumers would place a higher value on the environment so that compensating them for a given environmental deterioration would be more costly than it would today.

This is a complex area normally articulated by means of algebraic equations. But I have tried to explain the intuition in words, as discounting is of vital importance when assessing the claims future generations have on resources today. To sum up, presupposing the conclusions of a study by applying an arbitrarily high discount rate is unhelpful. It is a technical application of the view that "because we don't care about the future, we don't care about climate change." A 5 percent discount rate values the well-being of a life in the middle of the next century at a fraction of a percentage point of the value of a life today. Even a catastrophe afflicting those generations would have little bearing on our conscience. If you impose that sort of ethical judgment to start with, then you can see why it is easy to reject action to limit climate change.

Summing up the Case for Action

Having examined and responded to the arguments denying the importance of strong and urgent action on climate change, it is not necessary to understand the science or economics of climate change in great detail to have a view on the case for robust and urgent action. The basic question is essentially simple: is it worth paying 1 percent or 2 percent of GDP to avoid the additional risks of higher emissions as described in the story told by the scientists? The answer hinges on whether this payment is appropriate to insure against potentially catastrophic global risks, bearing in mind that it is less than what most societies currently commit to health care or defense?

Most of us who have examined the issue argue that this is a price worth paying. After all, if the science is wrong and we invest 1 percent or 2 percent of GDP in reducing emissions for a few decades, then the main outcome is that we will have more technologies with real value for energy security promoting the efficiency of production and reducing other types of pollution. However, if we do not invest the 1 percent or 2 percent and the science is right, then it is likely to be impossible to undo the severe damages that will follow. The case for strong and timely action, supported by well-designed economic policies, is overwhelming. We turn now to the policy framework required to address the problem.

THE POLICY FRAMEWORK

A global policy must satisfy three principles if it is to find international support: it must be effective, on the scale required; efficient, in controlling costs; and equitable, to take account of the double inequity where poor countries are hit earliest and hardest and rich countries have greater responsibility for past emissions. At the same time, with the welcome rapid growth of some parts of the developing world, it is crucial that they participate if the "deep cuts" agreed at the UN Climate Change Conference in Bali in 2007 are to be achieved. Sound policy and international collaboration can deliver strong and clean growth for all at reasonable cost. Weak or delayed action will eventually choke off growth and is a far more costly option.

Emissions Targets

The starting point for a global deal needs to be an agreement on the appropriate stabilization target for the stock of atmospheric GHG concentrations. The world currently emits about 45 gigatonnes (GT) of CO_2-equivalent (CO_2e) GHGs. By most accounts, if nothing is done to change producer and consumer behavior, emissions are likely to broadly double by midcentury to over 80 GT per annum. GHGs endure in the atmosphere for years, decades, and even centuries. This means that, as with any stock-flow system, such as, say, a water tank, inflows must equal outflows for the level of the stock to stay constant. So in order to for concentrations to stop growing and stabilize, GHG emissions must fall to levels consistent with the earth's natural ability to reabsorb them. This number is itself a function of concentrations, but, in general, it means emissions will need to fall from current levels to something between 0 and 10 GT a year. Figure 2.2 shows a number of emissions pathways consistent with ultimate stabilization at 450, 500, and 550 ppm CO_2 equivalent.

Annual Emissions

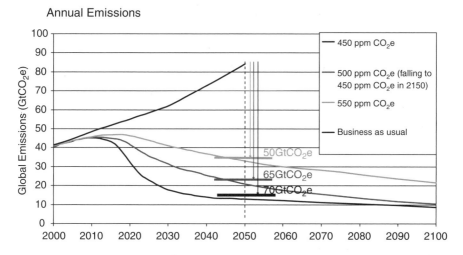

Stabilizing below 450 ppm CO$_2$e would require emissions to peak by 2010 with 6 to 10% p.a. decline thereafter.

Stabilizing below 550 ppm CO$_2$e means emissions peak in 2020 decline annually by 1 to 2.5%.

A 10-year delay almost doubles the annual rate of decline required.

FIGURE 2.2 Delaying mitigation is dangerous and costly.
Source: The economics of climate change: The Stern Review. Cambridge, UK: Cambridge University Press, 2007.

The latest science suggests that a target to stabilise GHG concentrations between 450 to 500 ppmv CO$_2$e (parts per million by volume including equivalent concentrations of non-CO$_2$ gases) would balance the projected global risks and costs. Stabilizing below 450 ppm CO$_2$e would require emissions to peak within the next few years, with annual declines of 6 percent to 10 percent thereafter. This would be very expensive, requiring new technologies to be introduced before they have matured. Stabilization at 550 ppmv CO$_2$e would seem, given current scientific understanding, to be unduly risky. A target of 500 ppmv CO$_2$e is achievable at reasonable cost, if policy frameworks are efficient in delivering early and coordinated action. Had we started acting strongly 10 years ago, meeting a 450 ppm CO$_2$e target—which gives the world a 50/50 chance of keeping temperatures below 2 degrees—might have been affordable. That opportunity has now been missed. If the world delays and vacillates another 10 years, then meeting 550 ppmv CO$_2$e target may be prohibitively expensive and 550 ppmv CO$_2$e is a very risky place to be.

The challenge of achieving a target of 500 ppmv CO_2e should not be understated: it requires a halving of GHG emissions by 2050 relative to 1990 levels (as discussed at the G8+5 summit in Heiligendamm) from around 40 GT CO_2e to 20 GT CO_2e. By comparison, business as usual would see global emissions rise above 80 GT CO_2e by 2050. To achieve the goal of a 50 percent emissions cut relative to 1990 levels by 2050, most electricity production would need to be decarbonized, while emissions from transport, land use, buildings, and industry would need to be cut sharply. Globally, per-capita emissions would need to fall, as a matter of simple arithmetic, to around 2 tons (T) per capita CO_2e as a world average from around 7 tons per capita now. Moreover, there would be very little scope for deviation. In terms of actual emissions, if one or two large countries were to manage even slightly higher targets, it would be difficult to see how other countries could achieve emissions targets close to or even below zero.

The Role of Developed Countries

Achieving a per-capita target of 2 T CO_2e by 2050 at affordable cost would require early and coordinated action. Most developed countries (Annex I) currently emit around 10 T to 12 T CO_2e per capita, with a cluster (including the United States) at nearly double that. These economies would therefore need to cut their physical emissions by *at least* 80 percent by 2050, implying that stringent emissions targets need to be taken on immediately, including interim emissions targets of 20 percent to 40 percent by 2020. The implications for GHG intensity of output are even stronger. Developed economies are expected to grow two- or threefold over this time period. To reduce their total emissions by a factor of 5, they would need to reduce emissions per unit of output by a factor of 10 to 15 from 1990 levels.

The Role of Developing Countries

Developing countries will have a central role to play in a global deal (see Stern 2008). By 2050, 8 billion of the world's expected 9 billion people will live in what is currently the developing world. Hence, the role of developing countries will need to be at the heart of the process of designing a global deal.

Indeed, a reduction of global emissions of 50 percent relative to 1990 levels by 2050 simply cannot be achieved without per-capita emissions in developing and developed countries averaging near 2 T CO_2e. This does not mean that developing and developed country allowances need to converge on 2 T CO_2e; indeed, there is a strong ethical case for developing countries taking on much lower per-capita allowances, perhaps at or below zero, taking account of historic emissions into the stock of GHGs.

Nevertheless, achieving the necessary reduction in physical emissions requires developing countries formulating credible action plans on this basis, committing to binding national targets to be adopted by 2020, and meanwhile participating actively in carbon markets and in their further development. Before developing countries can commit, developed countries must be able to demonstrate to the developing world that low-carbon, high economic growth is possible; that financial flows to countries with cheap opportunities to abate GHGs can be substantial; and that low-carbon technologies will be affordable, available, and shared.

Countries with strong emissions growth, such as China and India, will need to plan to limit and reduce emissions within the next 10 to 20 years. For this, they will require global cooperation, and they are unlikely to be able or willing to achieve these ambitious reductions without substantial technological and financial support and opportunities to innovate and ultimately export low-GHG technologies.

Until developing countries are ready to take on binding national targets, I propose a one-sided trading regime that rewards developing countries for reducing emissions, but does not punish them for failing to do so. In the absence of binding caps, this is likely to happen through an expansion of existing baseline-and-credit schemes such as the Clean Development Mechanism (CDM). However, the CDM in its current form is not able to generate the financial and technological flows needed. Internal U.K. government estimates using the global carbon finance model (GLOCAF 2008) suggest that climate stabilization undertaken at lowest cost would imply annual carbon flows of US$20 to US$75 billion by 2020 and up to US$100 billion by 2030. By comparison, the capacity of the current CDM is about 400 project registrations per year, resulting in new financial flows of perhaps US$6 billion at current carbon prices.

Moreover, the project-by-project nature of the CDM, and the measurement of emission reductions against an unobservable, project-specific baseline impose substantial transaction costs in terms of validation, verification, and independent scrutiny. The CDM regulatory process currently takes about 300 days, on average, from validation to registration. Transaction costs can easily reach US$500,000 (€ 325,000) per project (Ellis and Kamel 2007).

The parties to the Kyoto Protocol and CDM Executive Board have recognized these problems, and various proposals to remedy the situation are under consideration. However, these proposals might not be sufficient. A move from a project-based system to a wholesale approach can reduce transaction costs by alleviating regulatory complexity and scrutiny inherent in a project-based system such as the CDM. This could take the form of sector targets or programmatic emission reduction objectives. These would probably be based on efficiency targets rather than sector caps, although sector caps may be possible for globalized industries such as steel. For example, for each tonne of cement produced using less than an agreed amount of carbon, producers would be eligible to sell the difference as credits. Similarly, a large power utility might sell the sector-wide emission reductions from a reform program that includes energy efficiency and the rollout of renewable generation technology. To maintain the incentive to innovate and avoid excessive rents, benchmarks would probably need to be strengthened over time as new abatement opportunities become available.

Developing countries that commit to early action on tackling climate change stand to gain substantial benefits. In the short term, they can benefit from one-sided trading mechanisms, whereby they can profit from selling credits into carbon markets when they make measurable emissions reductions, but are not penalized for reductions they do not make. They can also benefit from codeveloping new technologies in collaboration with governments and businesses in developed countries, and gain the ability to exploit growing markets in low-carbon products and services.

International Emissions Trading: "Cap-and-Trade"

Putting a price on GHG emissions and enabling international emissions trading would be a critical factor underpinning the principles of effectiveness, efficiency, and equity in a post-2012 policy framework. It provides a powerful market signal to guide producers and consumers in order to change behavior in an efficient manner. A cap-and-trade system is appropriate to manage the risks of climate change by imposing an absolute limit on emissions, consistent with the scientific conclusions of the risk of catastrophic climate change (effectiveness). International emissions trading would also reduce the cost of mitigation by allowing emissions reductions to occur in whatever sector or country would be least costly (efficiency). Cost effectiveness is vital because it enables the world to take tougher action for given expenditure, allowing scarce resources to be spent on other goals (such as health and education).

The benefits of carbon trading could be high, generating private sector financial flows to developing countries, which could be used for low-carbon development (equity). Flows of $20 billion to $75 billion a year would be plausible if developed countries cut emissions by 20 percent to 40 percent on 1990 levels by 2020, even if only 30 percent of this were purchased from an international emissions trading scheme. Key beneficiaries would be large developing countries, such as China and India, although significant flows could also go to Africa, Latin America, and Southeast Asia, especially if forestry is included in the market.

Financing Emissions Reductions from Deforestation

Addressing forestry in a global climate change deal, and in particular deforestation and forest degradation in tropical rainforests, is essential if overall targets for stabilization of carbon emissions are to be met. According to the IPCC, "forestry" currently contributes 17.4 percent of global annual GHG emissions, the main proportion of which comes from burning or decomposition of tropical rainforests. Tropical deforestation is therefore an international problem, needing urgent international action. A global deal would need to put in place a framework to mobilize international support building on national and local governments' existing efforts and experience.

National forest protection policy should be consistent with broader national development strategies. An estimated 1.6 billion people depend on forests for their livelihoods. If trees were allocated value to remain standing, the value of leaving the trees standing would generally be much higher than that of cutting them down. Finance to reduce deforestation could then have a significant impact on poor rainforest nations and forest communities. However, until now, developing countries have not received funding from the international community on a scale sufficient to address the drivers of deforestation. Moreover, funding proposals of sufficient scale have not been seen to deal with concerns raised about leakage (where protection of one area of forest merely displaces deforestation activities to other areas that are unprotected). In the near term, support through existing multilateral funding channels such as the World Bank's Forest Carbon Partnership Facility can be scaled up. The private sector also has a large role to play in reducing deforestation, through commercializing benefits that standing forests provide. In the longer term, forests should be integrated into global carbon trading. A global compliance carbon market could be worth $100 billion in 2030, which would constitute a new source of finance. However, such a system will require investment in comprehensive technologies to monitor emissions from satellites, the air, and the ground in what is termed a "planetary skin" real-time monitoring (Eliasch 2008).

The Importance of Technology and Innovation

The challenge of significantly reducing emissions while maintaining economic growth requires a dramatic shift in technologies that determine carbon intensity of the economy. But in order for existing technologies to be fully diffused and adopted, and for new innovations to occur, critical market failures must be overcome. These include those market failures that restrict the deployment of many existing energy efficiency technologies despite rising energy prices, those blocking internalization of the costs of GHG emissions, and those that lock in high-carbon technologies due to infrastructure. Better provision of information on energy efficiency and cost minimization opportunities requires connected information technology and smart monitoring to allow consumers and producers to take advantage of the widely touted "free lunch" associated with energy efficiency gains that require up-front investment but ultimately save money and reduce emissions.

By motivating market forces and overcoming market imperfections, successful technology policy would dramatically expand the global market for low-carbon technologies, enabling prices to fall as deployment enables learning and experience to promote innovation and networks to be developed. We judge that policies are required across three time horizons. The single most important underlying policy across all horizons is creating a market for clean technologies to promote investment through carbon pricing, product and process standards, regulations, and public procurement.

1. In the near term, policies are required that diffuse existing low carbon.
2. Technologies that are currently only partly dispersed throughout the global economy. Bringing global industries up to today's best practice can save up to 10 GT of CO_2e by 2030, almost a third of the required reduction. In the current economic climate, additional investment criteria include the ability of a project quickly to boost demand, boost confidence, create jobs, and save resources. Early investment in smart grids, smart buildings, and connected public transport, as well as the development of a pervasive broadband Internet utility to promote connected and efficient work, production, and consumption should be made priorities. These investments not only promote short-run activity, they also ensure that the world locks into a sustainable infrastructure and technology path, transforming the economy and enhancing its resilience during the recovery. Projects that promote technology-led sustainable growth also help boost consumer and corporate confidence by capturing the imagination at times of deep economic recession.
3. In the medium term, policies are required to develop and scale up near commercial technologies. This set of technologies may be some 5 to 15 years away from economic viability, including carbon capture and storage (CCS), second-generation biofuels, and various forms of solar power, which together have the potential to reduce emissions by a further 10 GT of CO_2e by 2030.
4. Beyond 2030, the required cuts in carbon emissions will be achieved only through more radical shifts in technology (e.g., zero emissions power supply). These have huge potential, but will require substantial public investment in research and development.

Adapting to a Changing Climate

As global temperatures increase, all countries will need to adapt to limit the human, economic, and social impacts of climate change. Basic development is critical for building adaptive capacity, but climate change will make it more costly to deliver and sustain the Millennium Development Goals beyond 2015. Any global deal will need to commit developed countries to help developing countries adapt to the consequences of the already inflated stock of GHGs in the atmosphere. This support must be additional to existing development commitments and also to support for mitigation. Estimates of *additional* costs of adapting to climate change vary and are highly uncertain, but UNFCCC and the United Nations Development Programme estimates suggest between $25 and $100 billion a year may be required over the next 20 years. In any case, *additional* resources required to tackle climate change need to be integrated into planning and budgeting for development goals. Allocation of funding between countries will need to reflect the impact of climate change, the vulnerability to those impacts, and the capacity of governments to deliver appropriate outcomes. In addition to funding, there is a need for better access for poorer countries to markets, technology, and information to ensure that development is climate resilient.

Implementation and Institutions

Implementation is critical to the successful application of a global deal. Action needs to be well designed and sequenced. Badly designed policies can greatly increase costs. Implementation will take place in three key phases of agreement. Copenhagen in 2009, or shortly thereafter, will determine international targets, establish developed country caps, and set developing country responsibilities. The years 2010 to 2020 should be a phase of building effective and cooperative institutions for finance and technology as a basis for eventually establishing developing country caps. It will also be the time to prove that low-carbon growth is possible, with substantial financial flows linked to carbon markets and collaboration over developing and deploying technology. Post-2020, all countries should form part of an international cap-and-trade system and adhere to technological agreements.

Successful policy implementation requires an effective institutional structure. In the short run, the international community will need to build on the expertise of existing institutions, working through existing channels of international dialogue. Longer term, successful implementation rests on a new institutional framework, capable of drawing all parties into a single common process. New institutions should be based on common principles and applied with sufficient flexibility such that they manage:

- The development of global and distributed emissions targets timetables and milestones.
- A new and more extensive CDM mechanism.
- The establishment of the trading element of a cap-and-trade system.
- The creation of technological networks and institutional systems to supervise, monitor, and verify delivery against commitments.
- The emerging forestry carbon regime.
- The coordination and increased funding of advances in climate change–related science and low-carbon technology development.
- The development of an improved and joined-up understanding of the potential local risks from climate change and responses they develop.
- The development of a dispute resolution process.

Finally, the next few years present a great opportunity to lay the foundations of a new form of growth that can transform our economies and societies. As capital, input, and raw material costs fall in the current economic slowdown, publicly supported low-carbon development can both reduce risks for our planet and spark off a wave of new investment, which will create a more secure, cleaner, and attractive economy for all of us. In so doing, the developed world can demonstrate to the developing world that low-carbon growth is not only possible, but that it can also be a productive and efficient route to advance technological capacities and overcome world poverty.

SUMMARY AND CONCLUSION

In climate change policy, the conclusions that the costs of action are much less than the costs of inaction do not rest on any one particular modeling approach or assumption. Nor, for the most part, do they rely on this or that discount rate. An understanding of the risks shows that there is a clear case for strong and timely action, supported by well-designed economic policies. Formal attempts to quantify the case for action on climate change must reflect the full array of climate risks, as determined by the latest science, and use economics that consistently reflect these risks.

The good news is that most of these risks can be avoided at reasonable cost while promoting a cleaner, safer, more sustainable pattern of low-carbon growth and development. Several years on from the *Stern Review,* in the light of new developments in science, I would argue that this case has never been stronger.

Tackling climate change at both national and international levels requires leadership from the top levels of government. This policy clearly goes beyond the remit of environment ministers and should be at the heart of national economic policy. The risks and possible economic impacts of climate change, as well as appropriate public policy, affect all sectors. The strategy for a global deal needs to be one of strong, effective, and timely action to protect growth, support poverty reduction, and create new economic opportunities. A move to a carbon-constrained world will create opportunities for companies and economies that manage the transition by developing the technologies, institutions, and processes necessary to capture new markets. New markets in low-carbon, energy-efficient infrastructures to transform the economy represent a multibillion-dollar investment opportunity.

Delayed and piecemeal actions, relying on old technologies, will only exacerbate poverty, reduce growth, and curtail market choice. Delayed action will eventually be the antigrowth strategy. With wise policy and international collaboration and support for the application of new connected technologies, the world can find a way to a low-carbon, high-growth, resilient development path.

ACKNOWLEDGMENTS

Special thanks to following for their support and inspiration through the years: Simon Dietz, Chris Taylor, Professor Lord Nick Stern, Mattia Romani, Melinda Bohannon, Su-Lin Garbett-Shiels, Simon Reeve, Juan Carlos Castilla-Rubio, Jennifer Carr, Alex Bowen, Chris Kelly, Graham Floater, Bernice Lee, and Nick Bridge.

NOTES

1. In fact, capital markets are full of distortions. Hepburn (2006) and Dietz et al. (2007) argue that it is hard to find any markets that can reveal clear answers to the question "How do we as a generation value benefits to collective action to protect the climate for generations 100 or more years from now?" For a more detailed discussion, the reader is directed to Lord Stern's recent Ely lecture at the American Economic Association meetings (Stern 2008).
2. Professor Mohammed Dore of the Climate Change Lab at Brock University, St. Catharines, Ontario, puts this succinctly in *A Question of Fudge: Professor Nordhaus on Global Policy for Climate Change* when he remarks "it is strange how

the whole Cambridge welfare economics tradition—from Ramsay, de Graaff, to Mirrlees—has not made any difference to Nordhaus; he claims that climate change *is the mother of all public goods* and then forgets about the public good in his optimal policies!" (Dore 2009).

REFERENCES

Anand, S., and A. K. Sen. 2000. Human development and economic sustainability. *World Development* 28(12):2029–49.

Dietz, S., C. Hope, N. Stern, and D. Zenghelis. 2007. Reflections on the *Stern Review* (1): A robust case for strong action to reduce the risks of climate change. *World Economics* 8(1):121–168.

Dietz, S., D. Anderson, N. Stern, C. Taylor, and D. Zenghelis. 2007. Right for the right reasons: A final rejoinder on the *Stern Review*. *World Economics* 8(2):229–258.

Dore, M. 2009. A question of fudge: Professor Nordhaus on global policy for climate change. *World Economics* 10(1):91–106.

Ellis, J., and S. Kamel. 2007. Overcoming barriers to Clean Development Mechanism projects. OECD and UNEP/RISOE, May.

Eliasch, J. 2008. Climate change: Financing global forests. *The Eliasch Review*, October, UK Office of Climate Change.

Hepburn, C. 2006. Discounting climate change damages: Working note for the *Stern Review*. Mimeo, Oxford University, October, available from www.economics.ox.ac.uk/members/cameron.hepburn/research.html, accessed November 9, 2007.

Hoel, M., and T. Sterner (2007), Discounting and relative prices. *Climatic Change* 84(3/4):265–280.

International Energy Agency. 2008. *World energy outlook 2008*. Paris: International Energy Agency.

IPCC. 2007. *Fourth assessment report of the Intergovernmental Panel on Climate Change*. Cambridge, UK: Cambridge University Press.

Mirrlees, J. A., and N. H. Stern. 1972. Fairly good plans. *Journal of Economic Theory* 4:268–288.

Nordhaus, W. D. 2008. *A question of balance*. New Haven, CT: Yale University Press.

Pigou, A. C. 1932. *The economics of welfare*, 4th ed. London: Macmillan.

Ramsey, F.P. (1928), A mathematical theory of saving. *Economic Journal* 38 (December):543–59.

Sen, A., and B. Williams, eds. 1982. *Utilitarianism and beyond*. New York: Cambridge University Press.

Solow, R. M. 1974. The economics of resources or the resources of economics. *American Economic Review* 64(2):1–14.

Stern, N. 2007. *The economics of climate change: The Stern review*. Cambridge, UK: Cambridge University Press.

Stern, N. 2008. Key elements of a global deal on climate change. London School of Economics, April.

Stern, N. 2008. The economics of climate change, Richard T. Ely lecture. *American Economic Review: Papers & Proceedings*, Pittsburgh.

Sterner, T., and U. M. Persson. 2007. An even Sterner review: Introducing relative prices into the discounting debate. RFF Discussion Paper 07-37, Washington, DC: Resources for the Future, July.

CHAPTER 3

Climate Change Policy

What Investors Need to Know

David Gardiner

Mitigation of and adaptation to the threats of climate change provide a significant challenge to the world's policy makers. Investors should be concerned with climate policy because it provides information about risk and opportunity and directly impacts markets, prices, and demand for goods.

This chapter provides an overview of the international policies that frame the climate debate—the United Nations Framework Convention on Climate Change and the Kyoto Protocol; discusses the impacts of Kyoto on signatory countries; addresses some of the issues policy makers face in crafting domestic and international climate policy regimes; and provides an overview of some prospects for climate policy looking forward.

OVERVIEW OF POLICY AND POLICY DRIVERS

Scientists have reached a near-unanimous consensus on climate change over the past two decades. Their message is stark: the planet is warming and will continue to do so; we are already confronted with significant climate change impacts; and if we want to avert the more drastic consequences, we have to take action now. *Why* the climate is changing is important, as it informs the essential inquiry into *what* we are going to do to mitigate greenhouse gas (GHG) emissions and *how* we will adapt to those consequences that are now inescapable.

David Gardiner is president of David Gardiner & Associates.

In identifying and defining the problem, scientists inform the understanding and actions of policy makers. As the scientific community has become ever clearer in its predictions—"warming of the climate system is unequivocal," the Intergovernmental Panel on Climate Change (IPCC) stated in 2007, adding that it "could lead to some impacts that are abrupt or irreversible" (IPCC 2007)—and economists have become more accurate in their assessment of the costs of climate change, so too have policy makers begun to reach consensus that action of some sort is necessary to reduce the GHG emissions that cause global warming and to implement adaptation and mitigation measures. The scientists help policy makers to understand the risks and benefits associated with possible targets, solutions, and certain policies, establishing part of the context within which policy makers design measures to address climate change.

Science, of course, is not the only force shaping the context for climate policy. Energy security and economic concerns, for instance, have both become increasingly important drivers for climate change policy, as nations seek out not only clean but also stable and affordable energy supplies. Domestic politics also comes into play since climate policy will have repercussions on powerful interests and constituencies, and international political challenges arise as the global community attempts to structure some sort of coordinated policies across geopolitical boundaries.

Given the magnitude and complexity of the issue, the challenge for policy makers, unsurprisingly, has been to articulate just what those policies should be and how they should be implemented, especially across all countries with wildly diverse economic, social, and political circumstances. What are the appropriate GHG reduction targets, and on what timeline should reductions be mandated? How can policies be structured so they match and perhaps accelerate the time horizon required to develop and deploy transformative technologies? What is the role of market-based policies (such as cap-and-trade proposals), and what is the role of direct regulation (such as tailpipe emission standards)? Should GHGs from specific sectors be targeted, or should policies more appropriately be economy-wide? These are just a few of the questions that policy makers find themselves facing. Adding to the complexity of the challenge, policy makers cannot confront climate change in isolation; policies concerning energy, transportation, food, land use, and other areas all contribute to the success or failure of a given climate policy approach.

To some extent, the unique challenge of the climate change threat is simply its enormity. The sheer magnitude of the problem, however, may make climate change, unlike other environmental issues, uniquely suited to market-oriented solutions. Mitigating GHG emissions and adapting to climate change impacts is not a national challenge but a global one, and markets may have an easier time going global than regulation. Either way, the complexity of the climate change issue and the degree to which it is tied up with a range of other policy areas may require policy solutions unlike any in history.

WHY INVESTORS CARE ABOUT POLICY

For investors, these policy debates are both relevant and of immediate import, since inherent in the debates are both risk and opportunity for investors. On the one hand, climate policy presents clear risks. The suite of policy solutions with which policy makers respond to climate change will impose cleanup requirements and costs on some businesses. The choice of these tools, singly or in combination, will outline the rules within which businesses and investors will operate. Each approach has different impacts on markets, prices, and demand. Even the absence of any policy at all presents risks, as it makes the playing field uncertain for companies with significant emissions and imposes costs on those that experience the impacts of climate change, such as insurance companies. Investors, and the companies in which they invest, prefer policy certainty to policy uncertainty. As American Electric Power (AEP), one of the largest electric utilities in the United States, said in a 2004 report to its shareholders: "The central challenge the company faces is that of making decisions about large investments in long-lived assets in a setting of uncertain public policy and rapidly evolving technology" (AEP 2004).

On the other hand, where there is risk, there is opportunity, and climate change presents enormous opportunities for investors and the companies in which they invest. As governments begin to limit GHG emissions, some companies will come out as winners. Policy can help contribute to the development and deployment of new technologies (e.g., through policy support for particular low-carbon energy sources) or can drive game-changing shifts in the structure of the economy (through the imposition of a price on carbon). Indeed, the scale of change envisioned in the energy sector in the next 40 years is very significant as scientists call for emission reductions of 80 percent or more. As one well-known investor and entrepreneur said in 2007, "We've got to get the rules right so that this new frontier of opportunity is opened up to business enterprise" (Turner 2007).

Ultimately, the outcome of the climate policy debates is of great concern to investors. The risks posed by climate change, the opportunities it presents, and the ways in which society responds, both nationally and internationally, will directly impact investors and businesses.

Accordingly, this chapter aims to help investors understand the key historical and current policies dealing with climate change and some potential prospective policy issues looking forward. Understanding the climate policy universe is a prerequisite for understanding the risks and opportunities it might present for investors.

CLIMATE POLICY AND REGULATION: AN OVERVIEW

In 1992, recognizing climate change as a significant global issue that needed coordinated attention, the nations of the world negotiated the United Nations Framework Convention on Climate Change to establish a general framework for international cooperation. Five years later, the Kyoto Protocol to the Convention added more details and enforceability, committing developed nations—the largest historic emitters—to specific GHG emission reductions and establishing a series of market mechanisms to facilitate those reductions. The Kyoto Protocol, in turn, has shaped the domestic climate policy regimes of developed nations in a range of ways, with both successes and failures.

United Nations Framework Convention on Climate Change

In 1988, NASA scientist James Hansen testified before the U.S. Congress that climate change was real and was already happening. It was not the first time that scientists had publicly expressed concern about climate change, but it was a pivotal moment—the first time climate change became not just a cause for concern but breaking news. His statement helped propel the first legislation for control of emissions of GHGs.

Four years later, in June 1992, the United Nations Conference on Environment and Development (UNCED) in Rio de Janeiro, Brazil, adopted the United Nations Framework Convention on Climate Change (UNFCCC). The Convention, which came into force in March 1994, established an overall framework for intergovernmental cooperation to address climate change. "The ultimate objective of this Convention," reads the ratified text, "is to achieve ... stabilization of greenhouse gas concentrations in the atmosphere at a level that would prevent dangerous anthropogenic interference with the climate system. Such a level should be achieved within a time-frame sufficient to allow ecosystems to adapt naturally to climate change, to ensure that food production is not threatened and to enable economic development to proceed in a sustainable manner" (UNFCCC 1992). The Convention framed the policy debate in five key dimensions:

1. *Objective.* To prevent "dangerous" (though not "all") interference with the climate system "for the benefit of present and future generations of humankind."

2. *Scope.* International, yet requiring national action "on the basis of equity" in accordance with the "common but differentiated responsibilities and respective capabilities" of individual countries—a distinction between developing and developed countries.

3. *Time frame.* "Lack of full scientific certainty should not be used as a reason for postponing" action.

4. *Humanity.* The human dimensions of the challenge must be addressed (e.g., actions must "ensure that food production is not threatened").

5. *Economic.* Economic considerations are also critical ("economic development is essential for adopting measures to address climate change.... Parties should cooperate to promote a supportive and open international economic system that would lead to sustainable economic growth and development in all Parties, particularly developing country Parties, thus enabling them better to address the problems of climate change") (UNFCCC 1992).

With these five elements, the Convention laid out a climate policy framework that still shapes current policy.

The Kyoto Protocol

Under the Convention, governments agreed to share information on GHG emissions, policies, and best practices; to take national action on mitigating and reducing GHG emissions, including providing financial and technological support for developing countries; and to cooperate on preparing to adapt to the impacts of climate change. However, the Convention, as originally drafted, set no mandatory limits on GHG emissions and contained no enforcement provisions for inaction. Therefore, the Parties to the Convention (as countries that have ratified the treaty are known) agreed as well to several "protocols" (updates to the Convention itself).

The most significant of these is the Kyoto Protocol, which has become much better known than the Convention itself. Negotiated in 1997 in Kyoto, Japan, the Protocol set binding emissions reduction targets for 37 industrialized countries and the European community. Whereas the Convention *encouraged* Parties to stabilize GHG emissions, the Protocol *committed* them to doing so. On average, the targets amounted to a reduction of 5 percent against 1990 levels over five years—2008 through 2012—though each nation had a unique target.

While the Convention framed the climate policy debate, Kyoto set the stage for future discussion by emphasizing certain policy tools available for GHG reduction. Recognizing that the world's developed countries were (and are) primarily responsible for the concentration of greenhouse gases in the atmosphere as a result of their head start in industrial development, and reflecting the Convention's principle of common but differentiated responsibilities, Kyoto placed a heavier burden on those nations that had historically been large emitters due to early industrialization. The Protocol required these countries to meet their targets primarily through national measures—internal carbon reduction plans. However, the Protocol also created three international market-based mechanisms through which countries could reduce their emissions further:

1. Emissions trading
2. The Clean Development Mechanism (CDM)
3. Joint Implementation (JI)

These three mechanisms turned GHGs into a marketable commodity. Under the Protocol, so-called Annex I or Annex B Parties—the developed countries—accepted targets to reduce their emissions; these targets were expressed as levels of allowed emissions (or "assigned amounts") over the 2008–2012 commitment period. These became, in essence, a form of currency, denominated in assigned amount units (AAUs). Emissions trading, the first Kyoto mechanism, allows countries with emissions to spare—emissions "assigned" to them but not used—to sell this excess capacity to countries that are over their targets. Emissions units are a commodity.

The second Kyoto mechanism, the Clean Development Mechanism (CDM), allows countries to undertake emissions reduction projects in developing countries. In effect, the project-based trading under the CDM allowed trading between a country with an emissions target and one without a target. (This mechanism was designed to counter fears that meeting Kyoto commitments would be expensive for developed countries with both high targets and efficient national industries for whom further emissions reductions would be challenging. It was also a way to direct some financial flows from the developed world to the developing world.) These projects earn certified emission reduction credits (CERs), one of which is equivalent to one ton of carbon; removal units (RMUs) are one specific type of CER that is equivalent to one ton of carbon removed via land-use or forestry activities. CERs and RMUs, like AAUs, can be traded among countries. So, too, can the emissions reduction units, ERUs, generated by the third Kyoto mechanism, Joint Implementation (JI), whereby countries with reduction targets can invest in emissions reduction projects in other countries with targets as an alternative to meeting reductions domestically. Like the other mechanisms, JI allows countries to lower the costs of complying with Kyoto by allowing them to invest in reductions in countries where those reductions are cheaper and then apply credit for those reductions toward their own target.

While that may seem like a lot of acronyms, the three Kyoto mechanisms can basically be summarized as follows: CDM involves collaboration between developed and developing countries on emission reduction projects; JI involves similar collaboration between developed countries; and emissions trading is just the purchase of a developed country's extra emissions capacity by another developed country that is over its own assigned amount. Together, these three mechanisms created much of what is known as the "carbon market." As of 2008, the carbon market was worth $118 billion.

The market mechanisms are only the most high-profile part of the Kyoto package. The monitoring and verification aspects of the Protocol—whereby purchases and sales of carbon units are tracked in the Protocol's international GHG registry, reductions are verified to have actually occurred, and a compliance system ensures that countries meet their targets—are essential aspects of Kyoto as well.

Impact of Kyoto on Signatory Countries

Through its market-based system and its emphasis on monitoring and verification, the Kyoto Protocol has shaped the climate policy regimes of signatory nations. Unlike the Convention, Kyoto forced countries—especially the Annex I countries—to craft domestic climate policies. Among the Parties to the Protocol, there have been significant successes, significant failures, and significant differences among policy approaches.

European Union The European Union has generally been at the forefront of climate policy. The EU ratified the Kyoto Protocol in 2002; at the time, there were 15 members of the Union. The EU as a whole has an 8 percent reduction target, to be met by distributing different rates among member states. In order to meet this target, the EU in 2000 implemented the European Union Climate Change Programme (ECCP). Policies promoted by the ECCP include commitments to renewable energy, voluntary commitments by automakers to reduce carbon dioxide emissions, pollution control and permitting, aviation-industry emission reduction, and carbon capture and storage. The most important of the ECCP policies, however, is the European Union Emissions Trading Scheme (EU ETS), a cap-and-trade system covering emissions from six industries (energy, steel, cement, glass, brick-making, and paper/cardboard).

Like the Kyoto mechanisms, cap-and-trade systems are market-based carbon control tools. Under the EU ETS, which currently covers emissions from around 11,500 heavy emitters in the power and heat generation industries and in several energy-intensive industrial sectors, operators receive emissions allowances from their governments, giving them the right to emit a certain amount of GHGs each year. At the end of each year, plant operators are required to return the number of allowances equal to their actual emissions to their governments. Plants that do not use their entire allowance allocation trade their extra allowances—essentially, carbon permits—to others whose emissions exceed what they are allowed. Under the EU ETS, fines are imposed on member states that fail to meet their obligations. These started at €40/ton of carbon dioxide in 2005 and rose to €100/ton in 2008. Thus, the carbon allowances have a monetary value.

While the EU ETS is separate from the Kyoto market mechanisms—it is the European Union's primary domestic carbon reduction tool—it does recognize most credits generated by the JI and CDM mechanisms.

Japan   Japan approached the task of meeting its Kyoto obligation through a national implementation plan requiring GHG reduction targets for major economic sectors. A significant feature of this plan is an emissions trading program similar to that in place in Europe. Instituted in 2005, the Japan Voluntary Emissions Trading Scheme (JVETS) is voluntary, unlike the EU ETS. Participating companies pledge specific emissions cuts and receive allocations from the government. The JVETS is targeted—the government selects from the voluntary candidates based on the cost-effectiveness of potential reduction activity—and under the scheme, the government subsidizes the cost of emissions reduction equipment. In return, the companies commit to a certain reduction in emissions. As under the EU ETS, facilities have to return the number of allowances equivalent to their actual emissions to the government, and they can trade allowances in order to meet their targets. Also like the EU ETS, Kyoto mechanism credits (CERs, ERUs) can be used to meet emissions targets.

Covered sectors under the JVETS include the food and beverage, building, textile, paper and pulp, chemical, metal, and ceramic sectors. Notably absent is the energy industry.

Japan has also taken other, non-market-based measures to reduce emissions, including increased fuel economy standards for passenger vehicles, commercial vehicles, and aircraft and renewable energy standards for electricity produced. Japan also provides tax incentives for low-emission vehicle technologies and for energy efficiency.

Russia   Russia ratified the Kyoto Protocol in November 2004. The Protocol, as described above, limits the amount of GHGs a Party can emit relative to that country's 1990 emissions levels. For Russia, the target was to return emissions to 1990 levels. However, following the dissolution of the Soviet Union in 1991, the economies of most countries in the former Soviet Union collapsed, and Russia's GHG emissions fell to an estimated

30 percent below 1990 levels. In other words, even without taking action, Russia far exceeded its required reductions. This has resulted in what is known as "hot air"—allowances that can be traded without any efforts at emission reduction having taken place.

Russia does engage in carbon reduction projects under Kyoto, however, as it is widely recognized as the country with the largest potential for hosting projects under the JI mechanism.

Canada   Canada's target under Kyoto was to reduce emissions to 6 percent below 1990 levels. Canada ratified the treaty in 2002, but four years later the incoming Conservative government carried through with its campaign promise to implement a "made-in-Canada" solution to climate change. In 2006, Canada publicly proclaimed its support for the Asia-Pacific Partnership, an alternative to the Kyoto Protocol with strictly voluntary, rather than compulsory, emissions targets. The Environment Minister at the time, Rona Ambrose, urged support for the Partnership because it includes China and India, large emitters who are not bound by reduction obligations under Kyoto. (The other member countries of the Partnership are the United States, Australia, South Korea, and Japan.) Later that year, the government announced that it could not meet its Kyoto targets but advocated setting new ones; by early 2007, the government announced it would not attempt to meet its 6 percent goal.

Instead, the federal government under the Conservatives released a climate change strategy to set "intensity" targets for industry and other sources. Rather than the absolute reduction targets used in the Kyoto Protocol and other reduction programs, intensity targets require companies to reduce the amount of emissions released for every unit of a product, but do not require a reduction of emissions overall—even if a company's emissions intensity falls, its absolute emissions can rise if it increases production.

In lieu of federal action, some Canadian provinces are taking action individually. British Columbia, Quebec, Ontario, and Manitoba have joined the Western Climate Initiative (WCI), a collaboration that also includes seven American states. The WCI is setting up a regional cap-and-trade scheme to reduce emissions to 15 percent of 2005 levels by 2020.

United States   The United States is a signatory to the Kyoto Protocol but is the only industrialized nation not to have ratified it. Without ratification, the Protocol is nonbinding in the United States. In 1997, the Senate unanimously passed a resolution (95–0) stating that the United States should not be a signatory to any international climate agreement that did not include targets for developing nations. In 2001, President George W. Bush withdrew the United States' support for the Protocol.

The United States has been slow to act on climate policy on a national level. President Bush announced in 2005 that the United States would be party to the Asia-Pacific Partnership and abide by voluntary intensity reduction targets, but his administration was largely reluctant to push for any mandatory national climate policies. As discussed later in this chapter, though, the United States has pursued some other relevant policies (such as fuel economy standards) and has begun seriously debating climate legislation in Congress. In addition, President Bush's successor, Barack Obama, has indicated that addressing climate change will be among his top priorities.

In lieu of federal action on climate change, states and local municipalities across the United States have taken action independently. Nearly half of U.S. states are covered under regional cap-and-trade programs:

- Ten northeastern states have created a Regional Greenhouse Gas Initiative, under which they have committed to reduce emissions from the power sector by 10 percent by 2019 through implementation of a cap-and-trade mechanism.
- As noted, seven western states (and four western Canadian provinces) have joined together in the Western Climate Initiative, which aims to reduce emissions to 15 percent below 2005 levels by 2020 through a cap-and-trade system.
- Six midwestern states (and one Canadian province) agreed in 2007 to establish GHG reduction targets and develop a market-based, multisectoral cap-and-trade program.

Market-based approaches are not the only solutions being pursued. The mayors of more than 900 cities across America representing more than 80 million citizens have committed their communities to individually meeting the United States' Kyoto target of 7 percent below 1990 levels by 2012 through the U.S. Conference of Mayors Climate Protection Agreement. Policies pursued depend on the municipality but include renewable power generation, transportation and land-use planning changes, and energy efficiency improvements.

California has also implemented state legislation, Assembly Bill 32 of 2006, which requires the state to reduce GHG emissions to 1990 levels by 2020. AB 32 is comprehensive, covering every sector of California's economy. While emissions trading will be part of California's approach to meeting its target, AB 32 also includes building and appliance standards, renewable power mandates, energy efficiency policies, and land-use and transportation initiatives. (Senate Bill 375, passed in 2008, was the first law in the United States to explicitly link land-use planning and GHG emissions.) California also implemented tailpipe emissions standards stricter than those in place at the federal level, but those standards remain on hold due to the Bush administration's 2008 refusal to grant California a federal waiver under the Clean Air Act; the new Obama administration has pledged to reconsider that decision.

Besides state-level policies and regulations, states have also been acting to force climate action at the federal level. For instance, 12 states and several cities brought suit against the Environmental Protection Agency (EPA) to force it to regulate carbon dioxide and other GHGs from new motor vehicles as pollutants under the Clean Air Act. In 2007, the Supreme Court decided in favor of the plaintiffs in *Massachusetts v. EPA*, holding that the EPA does in fact have such authority under the Clean Air Act. In response, the EPA has begun drafting potential climate regulations under the Act, providing additional impetus to many in Congress to enact a law specifically designed to address climate change.

China China, as a non-Annex I Party to Kyoto, does not have a reduction target to meet. In fact, China often strenuously argues that the developed nations of the world must take the lead in carbon reductions given historical emissions patterns. It further argues that its emissions intensity is comparatively low, despite the fact that China has either already overtaken the United States as the leading global emitter of carbon dioxide or will soon do so (depending on which study one looks at).

Nevertheless, China does have several domestic climate change policies, among them strong fuel efficiency standards for automobiles and a national renewable energy standard. In 2007, China also formulated two climate policy documents to lay out the basic national strategy on climate change, which it aims to address in the context of sustainable economic development.

DESIGNING A DOMESTIC APPROACH

As the above examples indicate, the Kyoto Protocol has helped drive domestic climate action in many countries; even some countries without targets have taken action. Policy makers designing domestic climate change policies grapple with some challenging questions.

The Role of Market-Based Policies

There are two broad types of market-based policies that countries have used to reduce their domestic carbon emissions: cap-and-trade policies and carbon taxes.

In general terms, cap-and-trade policies, as outlined above, set limits ("caps") on the amount of GHGs that a given party (depending on the context, a plant, an industry sector, or a state) is allowed to emit. The government distributes emissions "permits" or "credits," the equivalent of one unit of carbon dioxide equivalent (usually one metric ton), among the various parties. Parties must have a credit for every unit of carbon emitted; if the party does not have enough credits to stay within the set cap, it must either reduce its emissions or obtain permits from other parties that have excess permits. Typically, the number of permits or credits issued every period (every year or every several years) declines; parties can usually "bank" credits from one period to the next, but over time the number of permits issued declines, resulting in an overall reduction in carbon emissions.

Carbon taxes are conceptually simpler; the government sets a price on GHG emissions and indicates which sectors will be covered by the tax. For every unit of carbon a covered entity emits, it must pay the tax, thus providing a financial incentive for parties to reduce carbon emissions.

While both of these policies have the same aim—GHG reduction—there is one crucial distinction: cap-and-trade systems provide certainty about the level of emissions reductions that will be achieved, but uncertainty about the financial costs of doing do; carbon taxes provide certainty about the cost of carbon reductions, but no guarantee of reductions achieved.

Cap-and-Trade

In designing a cap-and-trade system, some critical issues must be addressed.

Timeline and Targets   Policy makers first grapple with the timeline and target. This can be fairly controversial in itself; short time frames and higher targets raise the cost of compliance, while long time frames and lower targets risk "too little, too late" in confronting climate change. Under the EU ETS, the target is a 20 percent reduction below 1990 levels by 2020; while future targets have not been set, the European Council has affirmed that a target of 60 to 80 percent below 1990 levels by 2050 would be appropriate. The proposed but failed Lieberman-Warner climate bill in the United States in 2008 would have capped emissions at 19 percent below 2005 levels by 2020 and 71 percent below 2005 levels by 2050. (As we saw in Chapter 1, many scientists currently argue that a decrease of at least 60 to 80 percent below 1990 levels by 2050 is necessary to have a good chance of averting the most catastrophic consequences of climate change, and some aver that even steeper cuts are necessary.)

Sources and Sectors Covered   Another question for policy makers is which sectors to cover. Japan's JVETS scheme had no participants from energy-intensive sectors such as power, steel, and cement—major contributors of GHGs to the country's total. The 2008 Lieberman-Warner bill in the United States would have resulted in almost an economy-wide system. Policy makers also have to consider how to treat some tricky sectors under cap-and-trade plans, such as the transportation sector. Some approaches, such as the Lieberman-Warner bill, include energy processors (like refineries) under the cap to reduce emissions from fuels but use other policies to reduce emissions from the vehicles themselves. There has also been debate about whether and how to include emissions from deforestation in a cap-and-trade system.

Auction versus Allocation   By creating a cap-and-trade system, the government is creating a new asset class—carbon permits—with monetary value, so distributing them can be politically charged. There are two primary ways to distribute allowances among parties: via allocation, in which a government distributes credits to parties for free, or via auctioning, in which parties buy the amount of permits they think they will need. Both approaches encounter distributional challenges. If a government allocates the permits, it has to determine who gets how many, which can lead to a battle among various interests. Similarly controversial is the question of what to do with auction revenues, which can be used to purchase additional reductions, reinvested in energy efficiency and clean energy projects and technology, returned (at least in part) to consumers, or distributed to industries to help them transition to a low-carbon future. In practice, a hybrid allocation-auction system is often used, where permits are initially distributed via allocation but over time a greater and greater percentage are auctioned. The EU ETS, for example, distributed permits for free but will likely auction the majority of its permits in its third round (2013 on). The Lieberman-Warner bill in the United States would have allocated roughly 75 percent of permits in 2012 and auctioned 25 percent; by 2036, that ratio would have been reversed. Under either an auctioning or an allocation system, policy makers have to carefully gauge the right number of permits; too many permits and the carbon price will drop to nothing, which is basically what happened in the first round of the EU ETS.

Cost Containment versus Environmental Integrity   Policy makers also have to consider the costs to the parties subject to the cap-and-trade mechanism. Because cap-and-trade allows for certainty about the level of carbon reductions achieved, but no certainty about cost, policy makers are often urged to implement cost containment measures like "safety valves" (a ceiling price for carbon above which parties can opt out of the cap-and-trade system in order to reduce excessive expenses). While politically popular, such mechanisms risk distorting market signals, thereby weakening the emissions reduction capability of the system and delaying action.

Flexibility versus Policy Certainty   Companies and investors desire policy certainty; they want to know what the playing field is going to look like, and they make significant decisions based on that knowledge. But what if our scientific knowledge advances and we realize that much greater reductions are necessary in a shorter period of time? Policy makers need to consider how to balance the need for policy certainty with the need to adapt to changing circumstances, such as by deciding how long commitment periods will last, how far out to set targets, and so on.

Carbon Taxes

Although conceptually simpler to understand, carbon taxes still present policy challenges:

- *Cost.* The major difficulty for policy makers is choosing the tax level that will produce sufficient emissions reductions. Sweden enacted a carbon tax in 1991 equivalent to $100/ton, which has since risen to $150; the Swedish government estimates that the tax reduced emissions 20 to 25 percent below what they would have been otherwise (Johansson 2000). Great Britain has taken a different route and instituted a levy on the purchase of different commodities, ranging between 0.07 and 0.44 pence per kilowatt-hour for fuel sources and 0.12 pence for other taxable commodities (equal to less than one cent U.S.).
- *Exemptions.* Certain industries are often exempt under carbon taxes. In Great Britain, electricity produced from renewable sources or combined heat and power processes is exempt, as is electricity used for aluminum smelting. In Sweden, no tax is paid on fuels used for electricity generation, and the industrial sector pays only 50 percent of the tax. In Finland, fuels for hobby aviation and pleasure use were exempt until 2008, and biofuels used for heating are still so.
- *Political concerns.* Taxes can be politically unpalatable; policy makers therefore often choose to offset carbon taxes with decreases in other types of taxes. For instance, the Canadian province of British Columbia instituted a $10/ton carbon tax in 2008, which will rise to $30/ton by 2012, and the tax is to be offset by reductions in personal and business income taxes.

The Role of Direct Regulation

Market-based policies are beneficial, from a policy standpoint, because they allow regulated entities to pursue the cheapest emissions reductions. Correctly designed, they can provide broad cross-sectoral or economy-wide carbon reductions. There are times, however, when direct regulation can be more efficient and more effective, particularly for industries or other sources that are not easily reached via market-based policies.

Tailpipe standards are a good example. Unlike cap-and-trade programs, which most easily cover stationary sources of carbon (power plants, industrial factories, etc.), tailpipe standards regulate emissions from mobile sources. In the United States, transportation amounts to more than 30 percent of the country's emissions; thus, this type of regulation can have a significant impact on a country's climate mitigation policy. The Clean Air Act of 1970 provides the framework in the United States for regulating emissions from mobile sources such as cars and trucks. Beginning in 1970, it established nationwide air quality standards; it also allowed the state of California special permission to promulgate stricter standards. Other states have the choice of adopting either the national standards or the California standards, and several have followed California. In 2002, California enacted a law to regulate the GHG emissions of cars and trucks, requiring a 30 percent reduction by 2016, phased in gradually beginning with model year 2009. In order to implement these separate standards, California had to obtain a waiver from the U.S. Environmental Protection Agency; that request was denied in 2007 by the Bush administration, amid accusations of political interference, and the denial is currently being challenged in court. The new Obama administration has also pledged to review the decision to deny the waiver. As of this writing, Arizona, Connecticut, Maine, Maryland, Massachusetts, New Jersey, New Mexico, New York, Oregon, Pennsylvania, Rhode Island, Vermont, and Washington have indicated that they will adopt California's standards if a waiver is granted, and several others are poised to do so; this amounts to more than a third of the North American market.

There are many other possible types of GHG regulation as well. California's AB 32, mentioned earlier, contains numerous regulations designed to reduce GHG emissions. And the EPA, as noted earlier, is considering regulations to address climate change in response to the Supreme Court's 2007 decision in *Massachusetts v. EPA*.

THE IMPACT OF OTHER POLICIES

Climate policy, however, cannot be considered in isolation; other policies that are not climate-specific nonetheless can have a significant impact on a country's ability to reduce domestic carbon emissions.

Electricity Policy

The way a government handles electricity production, distribution, and consumption obviously has significant implications for climate policy; GHG emissions from the electricity sector make up approximately 40 percent of the total emissions in the United States, for example. Energy policies with carbon impacts include policies to stimulate energy efficiency programs by electric or gas utilities, policies to make appliances or buildings more energy efficient, and policies to mandate that a certain percentage of a municipality's, state's, or country's power comes from renewable sources. Indeed, in the United States, 29 states have already adopted Renewable Electricity Standards, which require electric utilities to sell increasing amounts of renewable energy, and 19 states have adopted Energy Efficiency Resource Standards (EERS), which require utilities to meet power needs through increasing amounts of energy efficiency.

Transportation Policy

Transportation policies represent a significant opportunity to reduce emissions; in the United States, transportation accounts for approximately one third of all emissions. Governments are adopting policies that regulate both GHG emissions and fuel economy from vehicles, directly regulate or change the fuels, and shift travel from cars, trucks, and airplanes to transit and rail. In the European Union, for example, the level of carbon emitted by new cars must be reduced to 130 grams of carbon dioxide per kilometer by 2012. In December 2008, the EU also adopted a new directive that mandates that each member state meet 10 percent of its fuel needs for all forms of transport with alternative fuels (including biofuels, hydrogen, and renewable electricity) by 2020.

Land-Use Policy

Land-use policy also affects efforts to address climate change. For instance, in 2008 California passed the first U.S. law explicitly linking land-use planning and climate policy; its SB 375 requires the 18 metropolitan planning organizations in the state to show that their planning scenarios will result in a reduction in GHG emissions, which they would achieve by denser development patterns and expanded use of driving alternatives, including walking, biking, and transit.

DESIGNING AN INTERNATIONAL APPROACH

Climate change is a global challenge, unlike perhaps any other in history. While domestic action is necessary in order to achieve the needed emissions reductions, concerted international effort is also essential. Regardless of where GHG emissions originate, the impact is global.

However, designing an international climate policy regime carries unique policy challenges. The developed world bears a disproportionate responsibility for historical carbon emissions due to early industrialization, yet the most significant effects of climate change will disproportionately impact the developing world, which has relatively less mitigation and adaptation capacity. And while the largest share of GHG emissions originated in developed countries, the current rate of emission of GHGs from the rapidly developing countries is rising as their economies expand.

Who Should Participate?

The Kyoto Protocol attempted to address these concerns through its focus on "common but differentiated" responsibilities—the developed (Annex I) countries agreed to mandatory carbon reduction targets, while the developing countries had no targets but shared in the common international responsibility to reduce emissions. In this way, Kyoto ensured near-universal international participation.

However, as noted earlier, there is one significant exception to the near-unanimous consensus on Kyoto; the United States, which signed the treaty in 1998, refused to ratify it on the grounds that developing nations such as China and India were not also subject to emissions reduction targets. The claim is not without merit; China likely surpassed the United States in 2008 as the world's leading emitter of carbon dioxide. China, however, argues that it has low per-capita emissions and that it bears little historical responsibility for the problem.

And so there is something of an impasse, with the United States not wanting to act without the rapidly developing economies, and with those countries wanting the United States to take the lead. At the heart of this impasse are concerns about competitiveness and development. The United States fears taking action that might give a competitive advantage to rapidly developing countries that lack carbon constraints, while China and India want to continue developing and to bring their people out of poverty.

These priorities are reflected in the policies these countries propose to address the impasse. Concern that strict carbon regulation might harm the competitiveness of exports has led several nations to propose carbon import tariffs. In January 2008, the European Commission proposed a tariff on goods from countries whose emissions policies did not match EU standards; the tariff would force companies that import products into the EU to buy emissions permits through the EU ETS to account for the imports' carbon emissions. The Liberal Party in Canada made such tariffs part of its 2008 election platform. However, where carbon import tariffs provide negative incentives for countries to enact domestic climate policy, climate technology transfer plans offer positive incentives. Developing countries are concerned that they will not have access to the funds or technologies needed to grow their economies and also combat climate change. Accordingly, they have insisted that financing and technology transfer be critical elements of any international climate regime.

What Should Each Country Do?

Given the "common but differentiated responsibilities and respective capabilities" of the developed and developing world, it can be challenging to devise actions and commitments that will address the climate crisis and yet will be acceptable to all UNFCCC Parties.

In terms of actions in the developed world, the main proposal thus far has essentially been a global cap-and-trade system, with each developed country having a specified level of emissions it must not exceed. There are other ideas out there as well. Japan, for instance, has suggested a sectoral approach that would set emissions reduction targets on industries across national boundaries.

As for the developing world, one way developing countries could participate would be to pursue a set of actions that would reduce emissions, without any particular target. Indeed, the Parties to the UNFCCC agreed in Bali in December 2007 to a basic road map for the post-Kyoto (post-2012) international accord; in that road map, developing countries agreed to pursue "nationally appropriate mitigation actions." There are also other ideas for ways the developing countries could contribute to the global mitigation effort. For instance, they could accept emissions intensity targets (tied to gross domestic product) or "action targets" (which require reductions in proportion to actual emissions), both of which would allow continued economic growth but would reduce emissions growth from what it otherwise might have been. Developing countries could also focus some of their efforts on protecting and enhancing their carbon sinks (e.g., forests), and mechanisms like CDM and carbon offsets might play a further role in tying those sinks into a global carbon market.

It is important to remember, however, that market mechanisms are not the only option. The Montreal Protocol on Substances that Deplete the Ozone Layer, which set timetables for phasing out the production of chemicals responsible for ozone depletion, provides an excellent example. That Protocol allowed developing countries more time to phase out the ozone-depleting substances, provided for periodic reevaluation of the science, and created a Multilateral Fund to help developing countries implement the requirements. Because of these mechanisms, the Montreal Protocol has been a major success. (It has also had a major climate impact—parties to the

Montreal Protocol agreed in 2007 to accelerate the phase-out schedule for hydrochlorofluorocarbons [HCFCs], which are potent GHGs, potentially yielding even more climate benefit than Kyoto.)

Regardless of the precise mechanisms chosen for an international climate regime, policy makers will have to recognize the linkages to (and potential competition with) other critical policy areas. As already noted, developing countries are principally focused on economic development. As of this writing, the world is in an international financial crisis and an international food crisis. Billions of people in the world live in poverty. Money spent to address climate change can take money away from these other policy priorities. However, the cost of climate inaction is much greater than the cost of action (the Stern Report estimates the cost of action at approximately 1 percent of annual global GDP and the minimum cost of inaction at 5 percent of global GDP), and the effects of unaddressed climate change would greatly exacerbate global poverty, the food crisis, and the other issues of survival and development that developing countries prioritize. Charting the right policy path through these issues is understandably challenging.

LOOKING FORWARD

In 1988, climate change was just beginning to come to prominence as an urgent global concern requiring action. Twenty years later, the international community has moved from studying the issue to encouraging action to requiring action. Individual nations have taken significant action. But climate scientists continue to warn that the impacts of climate change are happening more quickly than even the best models had predicted, suggesting that more serious action is needed more quickly. The policies outlined in this chapter, important though they are, are just the beginning.

Prospects for U.S. Action

The United States has been viewed as a laggard on climate action, at least on the national level. However, the environmental platform and political appointments of the new Obama administration indicate that the new president will reposition the United States to become a world leader in addressing climate change. President Obama has appointed experienced climate experts to the EPA and Department of Energy, as well as a well-respected climate scientist to be the Assistant Advisor to the President on Science and Technology and the former Clinton EPA administrator to report directly to the president and lead energy and climate change initiatives across the government. The

State Department also appointed an experienced climate negotiator to be the Special Envoy for Climate Change.

President Obama's platform calls for establishing mandatory climate targets and an economy-wide cap-and-trade program to reduce GHG emissions 80 percent by 2050; investing $150 billion in renewable energy over 10 years; committing to develop a "smart grid" and provide a 25 percent rebate for "smart grid" investments; and boosting energy efficiency.

The Senate took up a national cap-and-trade bill during the summer of 2008, and both chambers of Congress held numerous hearings on climate change. This interest and activity is expected to continue in 2009. And states and cities continue to take the lead in implementing strong climate policies and in pushing the federal government to take action.

Prospects for Further International Action

The timeline by which the Annex I countries are bound to meet their assigned reduction targets under Kyoto is 2012. Policy makers have been working to craft the outline of the post-2012 international climate policy regime, and there has been some progress. In December 2007, the Parties to the United Nations Framework Convention on Climate Change met in Bali, Indonesia, to hammer out a post-2012 roadmap; the four pillars of the roadmap that emerged were mitigation, adaptation, transfer of technology to developing countries, and financing. Bali also included "avoided deforestation" as a potential element of an international climate agreement. In December 2008, the next round of UNFCCC climate talks took place in Poznan, Poland, where countries approved financing for an Adaptation Fund to help developing countries adapt to climate change. In December 2009, the nations will meet in Copenhagen, Denmark, where they hope to adopt the post-2012 treaty. That is an incredibly ambitious timetable, particularly given that the new U.S. administration will be arriving in the middle of the process, but that is still the hope.

SUMMARY: IMPLICATIONS FOR INVESTORS

This chapter has outlined key historical and current policies dealing with climate change and has briefly touched upon some potential issues to consider as new policies are crafted. Given this broad policy overview, the central messages for investors are:

- The array of tools available to policy makers to confront the climate threat—including market-based and regulatory approaches in both domestic and international regimes—and the interplay between climate, energy, transportation, and land-use policies (among others) mean that there will never be one single uniform international climate response. There will be a large and varied set of policy approaches utilized by different actors.

- Each policy choice results in different risks and opportunities. Investors have a vested interest in climate policy because it will provide a relative level of certainty about the business environment in which they (and the companies in which they invest) will operate. Investors will need to be aware of the impacts of all of these policies, as each approach has different implications for markets, prices, and demand.
- Climate change, perhaps more than many other problems, is uniquely suited to international market-based solutions (though there will be critical non-market-based policies utilized as well). Sizeable markets already exist—the carbon market, as described earlier, is already worth billions of dollars—and these markets will continue to grow, again suggesting the enormous potential for investors who closely follow and weigh in on the policies that will move us into a low-carbon future.
- It can be daunting for investors who are used to operating in the world of finance and economics to try to keep track of all the nuance and detail in this policy web. However, the enormity of the challenges posed by climate change and the immense opportunities possible in a low-carbon economy mean that the policy structures devised to address climate change will directly affect investors.

ACKNOWLEDGMENTS

Sincere thanks go to Maja Gray, Dave Grossman, and Tamara Withers for their invaluable assistance in researching and reviewing this chapter.

REFERENCES

American Electric Power (AEP). 2004. *Shareholder resolution report: An assessment of AEP's actions to mitigate the economic impacts of emissions policies.* www.aep.com/environmental/reports/shareholderreport/docs/FullReport.pdf.

Intergovernmental Panel on Climate Change. 2007. *Climate change 2007: Synthesis report: Summary for policymakers.* www.ipcc.ch/pdf/assessment-report/ar4/syr/ar4_syr_spm.pdf.

Johansson, B. 2000. *Economic instruments in practice 1: Carbon tax in Sweden. Swedish Environmental Protection Agency.* www.oecd.org/dataoecd/25/0/2108273.pdf.

Turner, T. 2007. Quoted in *UN foundation's leadership challenges energy industry to look for opportunities in alternative energy economy.* February 7, UN Foundation. www.unfoundation.org/press-center/press-releases/2007/un-foundations-leadership.html.

United Nations Framework Convention on Climate Change. 1992. *Full text of the convention.* May 9, United Nations Framework Convention on Climate Change. http://unfccc.int/essential_background/convention/background/items/1349.php.

Climate Change and Institutional Investors: Key Strategic Issues

Risks and Their Impact on Institutional Investors

Mindy S. Lubber

The 2008 global financial crisis offers a cautionary tale about evaluating risks, with lessons that resonate far beyond subprime mortgages. Banks, investors, and regulators failed to recognize the risks associated with subprime mortgages: namely, the inevitable downturn in housing prices and the inability of low-income borrowers to repay debts. Today, companies and investors overlook the consequences of ever-rising carbon emissions created by a petroleum-based economy predicated on continuous growth.

Instead of standing by while climate risks actualize into financial hits, many institutional investors are beginning to reduce their exposure to subprime carbon assets. Complicating this task is the fact that climate risk is not equal opportunity: it is spread unevenly across companies, sectors, and regions, carving out a varied landscape that makes it difficult to map. For example, a U.S. utility and a European retailer face many distinct—as well as some overlapping—climate risks. But many institutional investors have global investment strategies, exposing them to a wide spectrum of climate risks.

Climate risk also presents multiple faces, encompassing regulatory, physical, competitive, and litigation risks. Whether and how companies move to address these risks will have profound implications for their future viability and for investors' portfolios. Companies need to first assess the climate risks they face, disclose material risk, and then minimize these risks. Ideally, companies will turn climate risks into opportunities, enhancing investment returns while simultaneously promoting the stable climate system upon which the global economy relies.

Mindy S. Lubber is the president of Ceres.

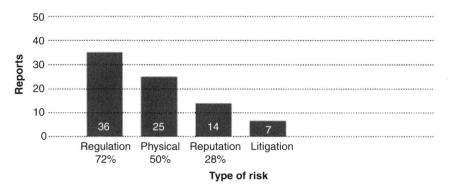

FIGURE 4.1   Risk types mentioned in reports.
Source: Barend van Bergen, Director of KPMG Sustainability, Figure 2.1: Risk types mentioned in reports in *Climate Changes Your Business: KPMG's Review of the business risks and economic impacts at sector level*, KPMG International, April 2008.

FOUR CATEGORIES OF CLIMATE RISK

Regulatory risk tops the list of the most significant climate risks cited in recent reports on climate change and investment, according to a 2008 KPMG report. The KPMG publication analyzed 50 reports by investment banks, business associations, insurance companies, nongovernmental organizations (NGOs), consultancies, rating agencies, and intergovernmental organizations that address the business risks and financial impacts of climate change. Almost three-quarters of the reports discuss regulation as a key climate risk (see Figure 4.1), and half mentioned physical risks. Fourteen percent of the reports discussed litigation risk.

Regulatory Risks

Regulation is a primary tool for governments to address the climate crisis. However, the lag in reaching national or global political consensus on climate change solutions has resulted in a patchwork of regulatory regimes.

For example, the European Union Emissions Trading Scheme (EU ETS) has regulated corporate carbon emissions since 2005. In the United States, the Regional Greenhouse Gas Initiative, the first mandatory, market-based effort in the United States to reduce greenhouse gas (GHG) emissions, initiated its first auctioning of emissions permits in 2008. Voluntary emissions markets such as the Chicago Climate Exchange are also in operation.

Yet the writing is on the wall for regulating carbon emissions in the United States, with strong administration and congressional interest in passing climate change legislation. Multiple climate change bills—most of which codify a carbon cap-and-trade system—were introduced in recent years, and one bill has emerged as the clear leader in the first half of 2009. Ultimately, climate regulation will span the globe, aligning government regulation with the reality of a "carbon-constrained" economy.

Of all immediate climate risks, regulatory risks are generally the most important to financial analysts because climate-related regulations have short-term impacts on investors in multiple sectors. Over a longer term, how well positioned a company is to operate in a carbon-regulated environment will significantly impact its future, as well as investors' portfolios. Investors who fail to prepare for the carbon-constrained economy not only face risks to their portfolios, but also risk losing out on substantial opportunities in the clean tech and green energy sectors, which will continue growing due to regulations capping carbon emissions and other incentives.

Regulatory Risk: Coal Regulatory risks related to coal-fired electric power plants are one of the most significant risks investors face. Coal is one of the most abundant traditional energy sources as well as the most carbon-intensive energy source, creating great tension among companies, policy makers, investors, and environmentalists over its future. While coal companies argue that so-called "clean coal" can solve carbon emissions problems, the underlying "carbon capture and storage" technology is unavailable on a commercial scale.

Nationally, eventual regulation of GHGs is fairly certain, although "program elements remain undecided, with major issues such as stringency of reductions and distribution of emissions allowances still undecided" (Schlissel 2008). If policy proposals trend away from lax measures such as grandfathering and free emissions allowances and toward the auctioning of allowances for all emissions, this trend will create significant regulatory risks for investors exposed to coal in their portfolios.

Regardless of national regulation, some state regulators are starting to reject coal plant investments as too risky, and some investment banks (as discussed later) are improving their risk management of their coal plant investments. In October 2007, for example, the Kansas Department of Health and Environment became the first government agency in the United States to "cite carbon dioxide emissions as the reason" for rejecting a proposed coal-fired electricity generating plant, withholding air permits due to concerns that GHG emissions threaten public health and the environment (Mufson 2007). In June 2008, a Georgia superior court judge ruled that a new coal plant needed to limit the amount of CO_2 it released, and in January 2009

Dynegy dissolved its joint venture for developing the Longleaf coal-fired power plant, citing "regulatory factors" among the barriers (Dynegy 2009; Zakai 2008).

Overall, in the last two-and-a-half years, 83 proposed coal plants in the United States have been denied permits by state regulators or voluntarily withdrawn (Warner 2009). While this trend is strong, counterexamples also exist. For example, in June 2008, Virginia's Air Pollution Control Board approved permits for a new coal-fired electricity plant in Wise County being built by Dominion Virginia Power (Fahrenthold 2008). However, environmentalist opposition continues, potentially posing risks to the project.

Physical Risks

Physical risks associated with climate change have the potential to affect every company, not just those subject to climate regulations. These risks include drought, flooding, sea level rise, wildfires, and severe weather events such as hurricanes.

Because physical risks are potentially devastating to portfolios, especially over the long term, every investor should be aware of these risks. However, because these risks have fewer short-term impacts, they don't show up on the radar screens of many investors. This is unfortunate because, if unaddressed, physical risks could be far more serious than regulatory risks for investors.

To adequately respond to long-term physical risks, investors should focus on the long-term sustainability of corporate profits. First, investors can engage with policy makers and corporations to ensure that government and corporate policies significantly reduce GHG emissions, which will help reduce physical climate risks. Investors can also explore deeper, structural changes in their relationships with portfolio managers—such as how often investment performance is evaluated—to deemphasize short-term profits and emphasize long-term, sustainable investing.

In the short term, recent examples of physical risks include the increased damage to offshore oil infrastructure by Hurricanes Katrina and Ike, and melting permafrost in the Arctic, which has required costly repairs to oil pipelines. Insurers have faced acute vulnerability to property damage and loss from sea level rise and severe weather events—risks shared across all sectors to varying degrees. Most sectors face risks of supply chain disruptions, the costs of which can affect consumer prices, and, in turn, investor profits.

Physical Risk: Drought   Water availability risks have received increasing attention from investors, and these risks are exacerbated by climate change.

For example, steam powers many electricity generating plants, including nuclear plants, and additional water is used to cool the steam for re-release into the environment (Lochbaum 2007). This makes nuclear and other steam-based energy generating systems vulnerable to drought, which is occurring with increasing frequency and duration due to climate change, according to a 2007 Intergovernmental Panel on Climate Change report (Sittler 2007). In January 2008, of the 104 operating nuclear power reactors in the United States, 24 were located in areas that faced severe droughts, raising concern that water levels of the lakes and rivers where they are located could dip below Nuclear Regulatory Commission minimum standards—or below water intake valves altogether (Weiss 2008).

In addition to impacting nuclear and fossil fuel–based electricity generating systems that rely on steam turbines, low water levels also affect hydroelectric facilities. Droughts resulting from climate change halved hydropower output in California and the southeastern United States in 2007 (*Gas Daily* 2008). Global warming also increasingly raises water temperatures too high to cool steam in both nuclear and fossil-fired plants, as occurred in France and Poland in the summer of 2006 (North American Electric Reliability Corporation [NERC] 2007).

Drought can force nuclear, fossil-fired, and hydro plant shutdowns, impacting investors by forcing utilities to buy replacement electricity on the wholesale market (*Gas Daily* 2008), where prices could run up to 10 times higher than the $5 to $7 per megawatt hour cost of nuclear power (Weiss 2008). Of course, replacement power may not be available from fossil-fired or hydro plants, given that they are vulnerable to some of the same climate change impacts as nuclear plants (*Gas Daily*).

Competitive Risks

Investors are increasingly aware that a company that manages the climate-related risks discussed here and *also* aggressively pursues climate-related opportunities may gain a competitive advantage. The risks posed to companies that ignore these growing opportunities are often called competitive risks. For example, utilities that are diversifying their energy mix, using co-generation and other clean technologies, and developing energy conservation and efficiency programs will be better positioned for the carbon-constrained economy than other utilities.

Competitive Risk: Utilities For investors in utilities, relative rates of carbon emissions provide a useful proxy for competitive risks, with bigger carbon emitters generally at greater risk. However, in the United States, the ultimate shape of the carbon-regulating regime remains up in the air, making the risks posed to particular companies difficult to quantify. A key issue is the percentage of carbon-emitting allowances that will be auctioned versus the amount distributed for free based on historical emissions. Coal companies are lobbying for free allowances to reduce their compliance costs, but the carbon price collapse in the first phase of the EU ETS, which provided an overabundance of free permits, advances a strong counterargument (Van Atten 2008). This uncertainty underscores the importance of investors factoring competitive climate risks into their assessment of corporate strategies.

Competitive Risk: Banks   The largest diversified banks, investment banks, and asset managers are subject to growing pressures from investors to reduce their climate risk, suggesting that climate risks in these industries may become material to investors. A number of these banks, which generally did little to address climate risk until the last three years, have announced significant programs to reduce their carbon footprint, invest in carbon trading, or take other steps to address climate risk.

The banks' relatively small carbon footprints from their operations are obscured by the outsized footprints of their lending and project finance portfolios. According to a recent Ceres/RiskMetrics Group report that evaluated climate governance practices at 40 of the world's largest banks, only 14 had adopted risk management policies or lending procedures that address climate change in a systematic way. Only Bank of America had set a specific target to reduce the rate of GHG emissions in its lending and had disclosed that it will use a $20 to $40 per ton cost of carbon in evaluating loans (Lubber 2008a). Only six banks were formally calculating carbon risk in their lending portfolios, and no banks had a policy to avoid investments in carbon-intensive projects such as new coal-fired power plants (Cogan 2008).

However, in 2008, Morgan Stanley, Citi, and JPMorgan Chase announced that any future lending for coal-fired power and other carbon-intensive projects would face increased scrutiny using their newly developed Carbon Principles (Odell 2008). Bank of America, Wells Fargo, and Credit Suisse later announced they will use the same principles in their lending practices (Herrera 2008).

Litigation Risks

Climate risk litigation appears to be following a similar trajectory to tobacco litigation, which increased over a number of years until it became a significant threat to the industry.

A total of approximately 100 climate-related lawsuits were filed in the United States through 2007, with the annual numbers trending markedly upward—from 6 suits in 2004 to 38 in 2007 (Gronewold 2008). While impacts on particular companies are difficult to predict, litigation thus far has been focused on companies responsible for large amounts of emissions, from either their operations or products.

There are several different types of climate litigation, however, and the risks are not limited to high-emitting sectors. For example, some lawsuits have challenged or sought government regulation of emissions. If successful, they could result in new regulatory risks for carbon-intensive or other impacted sectors.

Some lawsuits have directly targeted industry to make businesses curtail or pay damages for their GHG emissions, posing obvious income statement and balance sheet risks. In July 2005, eight state attorneys general sued the five largest U.S. electric utilities for being substantial contributors to the "public nuisance" of global warming. Plaintiffs sought an injunction forcing utilities to reduce their CO_2 emissions by a specified percentage each year; the suit was dismissed by the district court and is currently on appeal (*Connecticut v. American Electric Power* 2005; Gerrard 2007).

Similarly, in September 2006, the California attorney general filed a public nuisance suit against six automakers, arguing that they should be held responsible for some of the costs of combating climate change in California and for damage to the state's economy, environment, and health, which could run to hundreds of millions of dollars. This suit was also dismissed by the district court and is also currently on appeal (*California v. General Motors* 2007; Gerrard and Howe 2008).

Public nuisance suits like these are currently speculative risks for companies, but even if the likelihood of being found liable seems small, the potential liability is not. One expert has stated that damages in climate litigation "could reach the same level as in the tobacco and asbestos cases, and if anything they could be even higher" (Kanter 2007). According to the co-chair of the Sustainable Development, Ecosystems and Climate Change Committee of the American Bar Association, "the prospect of liability is a serious matter for people who understand climate change and take it seriously" (Choo 2006).

Other suits have focused on more specific activities like the construction of new coal-fired power plants. Cases have also been filed against companies for failing to include climate change as part of their environmental impact statements for new projects or developments.

One potential source of litigation is shareholders who may seek compensation from companies or boards of directors for failing, through negligence or mismanagement, to disclose to investors the material risks associated with climate change. Shareholders might sue a company with poor climate risk disclosure if, for instance, a climate-related incident causes the company's stock to drop. Shareholders could claim that the company had insufficient due diligence and did not communicate to shareholders the climate risks the company faced.

While no lawsuits focused on disclosure have been filed to date, the issue has drawn the attention of the New York attorney general. In September 2007, Attorney General Andrew Cuomo subpoenaed the executives of five major energy companies for information on whether their Securities and Exchange Commission (SEC) filings adequately described the companies' financial risks related to climate change. In 2008, he announced agreements with Xcel Energy and Dynegy in which the companies agreed to improve their disclosure of material climate risks in SEC filings, including information about regulatory risks, litigation risks, physical risks, and emissions management (Associated Press 2008; Confessore 2008).

The effect of these agreements on investors is not yet clear. In the power sector, the agreements are likely to result in improved disclosure by publicly traded companies. More broadly, the agreements could influence the SEC's position on climate disclosure, and therefore represent important first steps toward SEC guidance on mandatory climate risk disclosure applicable to other sectors.

INVESTOR ENGAGEMENTS WITH COMPANIES AND POLICY MAKERS ON CLIMATE RISK ASSESSMENT AND DISCLOSURE

This book is focused on how investment decisions can reduce climate risk while maintaining or improving investment returns. However, investor *actions*—particularly engagements with policy makers and corporations—also play a critical role. By encouraging companies to better assess, disclose, and respond to climate risk, investors can protect their portfolios better in the short term, before the shape of climate-related policies solidifies. These efforts are important for two reasons: the serious nature of the climate change threat and the nature of institutional investing today, in which many institutional investors effectively own large swaths of the economy.

The seriousness of the climate change threat has already led many institutional investors to engage with policy makers and companies. Today's climate crisis is global, long-term, persistent, multicaused, and potentially irreversible (Makower 2008). Carbon emitted in one city affects the local region while also exacerbating Siberian permafrost melt, sea level rise on South Pacific islands, desertification in China, and other climate impacts. For global investors, the potential negative effects of climate risk cannot be avoided by diversifying portfolios; engagement with policy makers is needed to address the risks.

Similarly, the enormous growth of institutional investment since the 1960s has resulted in mutual funds and public and private pension funds effectively owning companies in every sector of the economy. Indeed, portfolio performance for many large institutional investors depends more on the overall health of the economy than on any particular company (Hawley & Williams 2002.) On the issue of climate change, this leads investors to seek both reduced risks at individual companies and industry-wide changes that could positively affect both the global economy and investors' portfolios, such as increased use of renewables by power companies.

Climate Risk Assessment and Disclosure

During the last 5 to 10 years, steadily increasing numbers of institutional investors have urged publicly traded corporations to assess, disclose, and manage climate risks. Investors are also beginning to assess climate risk in their own portfolios by evaluating the capacity of their asset managers to analyze corporate climate risks (see sidebar). Investors have focused their efforts on assessment, disclosure, and mitigation by large corporations because of the large potential benefits to their portfolios.

Florida CFO Evaluates Its Asset Managers' Climate Risk Assessment Capacities

Florida CFO Alex Sink is pioneering strategies for assessing climate risk in the state's $24 billion in Treasury funds that she oversees as head of the Department of Financial Services. In November 2007, she announced "New Disclosure Requirements for Treasury Investment Managers" and asked managers to review the *Global Framework for Climate Risk Disclosure*, a standard for disclosure created by Investor Network on Climate Risk member pension funds and other organizations in October 2006 (*Pensions & Investments* 2008).

In September 2008, CFO Sink announced a robust semiannual review, with RiskMetrics Group evaluating and rating the Treasury's 20 external (and one internal) fund managers on their climate risk assessment capabilities. Researcher Bill Baue interviewed Florida's director of the Division of Treasury, Bruce Gillander, in October 2008 about the innovative assessment strategies the state is implementing.

Bill Baue: What prompted CFO Sink to evaluate the climate risk assessment capabilities of its fund managers?

Bruce Gillander: Our theory is that climate change risk will affect corporate profits going forward, so we think that corporations that are better prepared to deal with climate change are going to do much better financially.

Q: Will the scoring be used to prioritize retention of managers based on climate risk assessment performance?

A: Our primary evaluation will always be on financial performance and return on investment. The second point of review will be their use of climate change risk information to make their decisions on new purchases, since they didn't have the evaluation criteria until

May 2008. When they buy new securities, they have a choice between firms in the same sector; we're asking them to factor the climate change risk score in their decision. It will take time—a year, maybe two years—before they will have had a chance to incorporate climate risk into their new investment decisions and for their scores to have any real meaning.

Q: Will you engage with low-scoring managers to encourage them to use climate risk assessment more robustly?

A: Yes. Again, it probably will not be in the short term. It would probably be a couple of years into the program if we don't see any improvement. And really, the first set of scores is the benchmark, and we're hoping they will show gradual improvement as we get new scores every six months.

Q: You currently evaluate fixed-income managers. Are there plans for the state to apply similar evaluation of stock managers, and, if so, when?

A: CFO Sink is one of the three principals of the State Board of Administration buying equities for pension fund investments. The SBA has a new executive director coming in late October 2008, and I'm sure that's a question that will be posed to him, certainly by CFO Sink.

Q: Florida has some particular vulnerabilities to climate change, especially sea level rise and severe weather. To what degree do the state's vulnerabilities inspire this action?

A: Well, they certainly make it a more urgent policy initiative in the state of Florida. We've had two years with severe hurricane damage, and we are mostly very close to sea level. Most of our population lives within 10 miles of some coast.

Q: In sum, what are your hopes for the influence of this program on companies, asset managers, other institutional investors, and the environment?

A: As an institutional investor, our hope is that the asset managers of the world and obviously the corporations we invest in will adopt a deeper interest in climate risk. And we hope that other institutional investors will follow us in this program so we will have better corporate citizenship related to the environment.

Investor efforts to improve corporate climate risk disclosure fall into three main categories: shareholder resolutions leading to corporate engagements, disclosure mandated by the SEC or other securities regulators, and voluntary disclosure. While members of the Investor Network on Climate Risk (INCR) have asked the SEC for guidance on mandatory reporting of material climate risks, they have also encouraged companies to disclose voluntarily, which can lead to significant GHG mitigation efforts by companies (Baue 2004). (See Chapter 14 for a discussion of INCR.)

Shareholder Resolutions and Corporate Engagements

In the last 14 years, U.S. and Canadian institutional investors have filed over 275 climate change–related shareholder resolutions with publicly traded corporations. One hundred fifteen of those resolutions were withdrawn after successful negotiations, and these engagements have led to significant corporate progress addressing climate risks and capturing opportunities. The year 2008 was the most effective year on record for climate change-related resolutions. In the United States, 57 resolutions were filed, up from 43 the previous year, resulting in 25 negotiated withdrawals, up from 15 previously (Berridge 2008).

While investors are pushing for "wholesale," economy-wide regulations—such as mandatory climate disclosure and national emissions reduction legislation—shareholder resolutions remain one of the most important, effective tools for corporate engagement. They result in new practices at high-emitting companies and changes throughout these industries, as companies follow the best practices of corporate leaders.

Even when resolutions do not lead directly to corporate commitments, the cumulative effects of several years of dialogues can be effective. Resolutions and resulting dialogues with Centex, one of the largest single-family home builders in the United States, helped encourage the company to announce a new energy efficiency program the day before its 2008 annual meeting. The program will apply in all Centex homes and will boost energy efficiency up to 40 percent over a typical 10-year-old home, and up to 22 percent over new homes built to the 2006 International Energy Conservation Code (Centex 2008).

Mandatory Climate Risk Reporting

Investors have long argued that the U.S. Securities and Exchange Commission (SEC) is failing to enforce rules on its books requiring companies to disclose material environmental risks. In 2002, investors filed a petition with the SEC proposing a new rule to improve the Commission's material disclosure requirements with respect to financially significant environmental liabilities and to help ensure compliance with existing material financial disclosure requirements. The petition cited a 1998 Environmental Protection Agency (EPA) study and a 1993 U.S. Government Accountability Office (GAO) study finding underreporting of corporate environmental liabilities (Baue 2002).

Between 2003 and 2008, members of the Investor Network on Climate Risk wrote repeatedly to SEC chairmen and met with staff and commissioners, asking for SEC guidance on climate risk disclosure. In September 2007, 20 INCR members, including state treasurers, asset managers, and the nation's largest public pension funds, together with Ceres and the Environmental Defense Fund, submitted a *Petition for Interpretive Guidance on Climate Risk Disclosure* to the SEC (Rheannon 2007). The petition requested that "the Commission issue an interpretive release clarifying that material climate-related information must be included in corporate disclosures under existing law" (Ceres et al. 2008).

While the SEC has not yet issued an interpretive release, the petition has changed the climate disclosure landscape. Although mandatory climate risk disclosure has been an element of three leading Senate climate change bills for the past several years, the petition led to new state efforts to improve climate risk disclosure, like the New York attorney general's subpoenas of energy companies' disclosure. The petition has also affected the congressional response to climate risk disclosure, leading to a Senate hearing, a letter from two members of the Senate Banking Committee asking the SEC to respond to the petition, and language in an appropriations bill encouraging a response to the petition.

Investors have also begun to engage with other financial regulators, like the Financial Accounting Standards Board (FASB), and to deepen their engagement with the SEC. In 2008, investors submitted a letter to FASB regarding its draft rule on disclosure of "certain loss contingencies." While signatories applauded FASB for proposing stricter disclosure requirements, they recommended against limiting disclosure to contingencies expected to be resolved in one year, pointing out that climate risks, like tobacco liabilities, have a much longer time horizon (Lubber 2008b).

Investors have begun to deepen their engagement with the SEC by focusing on particular sectors (discussed below), and by seeking a role in specific SEC initiatives. Responding to an SEC request for comments on its Twenty-First Century Disclosure Initiative, INCR members submitted a letter emphasizing the need to include climate risk disclosure in the effort. The letter also noted the opportunity for the SEC to codify guidance on integrating material environmental, social, and governance (ESG) data into registrants' filings (Stausboll et al. 2008).

Voluntary Climate Risk Disclosure

As the financial materiality of climate risk becomes clearer, the case for mandatory climate risk reporting is strengthened. In the meantime, growing numbers of companies are reporting voluntarily—primarily in sustainability reports such as those using the Global Reporting Initiative's (GRI) Sustainability Reporting Guidelines, as well as in response to investor questionnaires like the Carbon Disclosure Project (CDP).

While investors typically welcome voluntary disclosure, they also note a number of problems, beginning with an ongoing concern: voluntary reporting is generally inconsistent from company to company, and hence not easily comparable. In addition, investors do not get the information they need from every publicly traded corporation because some companies do not respond

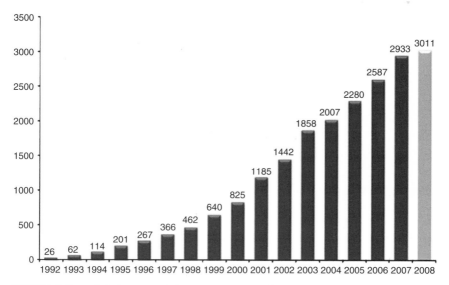

FIGURE 4.2 Global sustainability report output by year.
Source: © CorporateRegister.com.

to voluntary requests for climate risk disclosure. Without mandatory disclosure and mandatory U.S. limits to reduce CO_2 emissions, companies are not responding rapidly enough to climate risk to help avoid dangerous changes to our climate. Each company that forgoes disclosure is missing an important opportunity to thoroughly assess and reduce their climate risk. Finally, uneven disclosure prevents investors from understanding how companies are linking climate-related risks and opportunities to short- and long-term business strategies.

Nevertheless, the trend toward increased sustainability reporting has taken off and resulted in improved responses to climate change by companies. (See Chapter 15 for a discussion of corporate responsibility reporting.) The past 15 years has seen a 100-fold rise in sustainability reporting, from 26 such reports published in 1992 to over 2,900 in 2007, and topping 3,000 in 2008 (see Figure 4.2) (CorporateRegister.com 2008). A 2008 study of S&P 100 companies found that 86 have sustainability web sites, compared to 58 in mid-2005, and almost half issue sustainability reports (Sustainable Investment Research Analyst Network [SIRAN] 2008).

Strong investor pressure to disclose climate risk, from sources such as resolutions and corporate engagements, has resulted in improved climate disclosure in sustainability reports. One study found that two-thirds of FT Global 500 companies issued sustainability reports, of which 87 percent

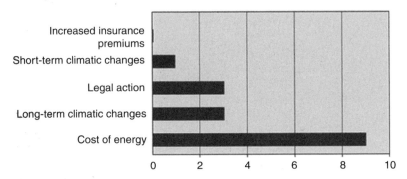

FIGURE 4.3   Number of companies that reported on risks arising from climate change.
Source: © KPMG Global Sustainability Services and the Global Reporting Initiative, 2007. Printed on recycled paper, 293_0707.

address climate change (Wayman 2008). Over three-quarters (78 percent) of these reporters publish quantitative GHG emissions data, an important element of any climate disclosure.

However, one study found that FT Global 500 companies' CDP responses demonstrate significant inconsistency in the protocols used for measuring emissions (Ethical Corporation Institute 2008). This study identified 34 different public protocols or guidelines being used by surveyed companies to report their GHG emissions, making it very difficult to compare performance between companies.

That said, voluntary reporting continues to yield growing disclosure rates, with 77 percent of the FT Global 500 companies and 64 percent of S&P 500 companies responding to CDP in 2008. Nevertheless, companies that engage in voluntary disclosure tend to focus more on upside opportunities (such as carbon trading) while downplaying risks (GRI & KPMG 2007). In one study, only 9 of 50 corporate sustainability reports discussed the increased cost of energy, the most often mentioned climate risk, while only three reporters mentioned long-term physical risks and three mentioned litigation risks (see Figure 4.3).

Sectoral and Regional Climate Risk Disclosure

Investors have long been aware that climate change poses specific sets of risks to different sectors, with high-emitting sectors like auto, electric power, and oil and gas facing the greatest risks. Therefore, investors have filed many resolutions with these high-emitting companies over the years. But sectors that are unlikely to be directly regulated under climate change legislation are receiving increasing attention, including banks that finance high-emitting projects; high energy users including retailers, restaurants, hotels, and real estate companies with large property portfolios; and manufacturers of major consumer brands.

In addition to filing resolutions, investors have begun to encourage government regulators to require climate risk disclosure by companies in high-risk sectors. In both the oil and gas and insurance sectors, investors have asked regulators to improve the disclosure of climate risks.

In 2007, at the request of oil and gas companies, the SEC began revising its reporting requirements for oil and gas reserves, an important issue that can affect the amount of reserves on company balance sheets to the tune of billions of dollars. In September 2008, the California Public Employees' Retirement System (CalPERS) and 14 additional investors and asset managers submitted a letter to the SEC regarding its proposed "Modernization of Oil and Gas Reporting Requirements," urging the Commission to require disclosure of the carbon intensity of petroleum reserves so investors can assess the climate risks embedded in reserve portfolios.

While investors have asked the SEC to issue generalized climate risk disclosure guidance for 6 years, this is the first request for Commission guidance on the climate risks facing a specific sector. The investor letter notes that "[o]il sand development releases three times as much global warming pollution in comparison to traditional oil extraction and refining," and so exposes these reserves to higher regulatory risks that should be disclosed.

In 2006, the National Association of Insurance Commissioners (NAIC) formed a Climate Change and Global Warming Task Force, spurred by the momentum of a few leading North American insurance companies that have been studying climate risk and its effect on insurers for several years and also following the lead of European insurers and reinsurers. Investors have filed resolutions with a number of insurers in recent years, registering concern that climate change has the potential to affect virtually all segments of the insurance business, including coverage of damage to property, crops, business interruptions, life, and health. The industry could also be affected because of climate risks embedded in the large investment portfolios managed by insurance companies.

In 2008, the NAIC task force published its third draft of a mandatory Climate Risk Disclosure Survey. The standard poses questions for insurers to answer in their annual disclosures to state insurance regulators, such as: "Has the company considered the impact of climate change on its investment portfolio?" and "Has it altered its investment strategy in response to these considerations?" (NAIC 2008). The survey, the first industry climate disclosure requirement in the world, passed the NAIC in

March 2009 and will be implemented beginning with the largest insurers in 2010 (NAIC 2009).

Investors are also closely watching regional climate disclosure initiatives, as the lack of climate action by the federal government has created a vacuum that states and regions in the United States are stepping in to fill with both voluntary and mandatory measures. The California Climate Action Registry established a voluntary regime for reporting and reducing GHG emissions, while the northeastern states launched the Regional Greenhouse Gas Initiative (RGGI) to auction mandatory emissions permits. Such regional variability leads to confusion and complexity for companies and investors alike, creating a mosaic of standards. However, these initiatives have created significant progress in climate risk assessment and disclosure in the United States, as well as emissions reductions, even in the absence of federal action.

INVESTOR ACTION ON CLIMATE RISK

As discussed in the sidebar "Florida CFO Evaluates Its Asset Managers' Climate Risk Assessment Capabilities," some investors have begun addressing climate risk in their portfolios by asking their investment managers to improve their capacity to assess climate risk. Environmental investment research providers facilitate such assessment, for example, by calculating the carbon footprints of investment portfolios.

An environmental investment research firm, Trucost, has developed a tool to measure the carbon footprint of investment portfolios—essentially a proxy for investor action on climate risk. In 2007, the firm calculated the carbon footprints of 185 U.K. investment funds and found significant variability: the most carbon-intensive fund's footprint was almost 10 times larger than that of the least carbon-exposed fund (Trucost 2007). In 2008, Trucost repeated the analysis on 100 portfolios in the 14 largest superannuation funds in Australia, and found a 36 percent difference between the largest and smallest carbon footprints of funds (Trucost 2008).

Institutional investors are starting to act to reduce their exposure to ESG risks such as climate risk. In 2008, Howard Pearce, head of environmental finance and pension fund management for the U.K. Environment Agency, dropped two fund managers, State Street Global Advisors and Capital International, in part because of the pension fund's drive to have its assets managed by signatories to the UN Principles for Responsible Investment (PRI) (UK Environment Agency 2008). PRI signatories, consistent with their fiduciary duties, commit to incorporating ESG issues into investment analysis and decision-making processes, as well as ownership policies and practices. In September 2008, Pearce replaced the fund managers with PRI signatories Impax Asset Management, RCM, and Generation Investment Management. Pearce "argued that appointing fund managers that consider risks such as climate change would produce better returns, in a way that was entirely consistent with his fiduciary duty" (Ethical Corporation 2008).

MARKET MELTDOWN: DOOM OR BOON TO GREEN ECONOMY?

The intersection of the market meltdown and the climate crisis represents the nexus of risk and opportunity for investors. As the economy began to collapse in the fall of 2008, a chorus of voices—from the *New York Times* to the *Wall Street Journal*—questioned whether the looming recession spelled doom for growth in renewable energy research and projects (Romm 2008). As renewable energy projects are capital intensive, there is some speculation that the credit crunch could strangle the burgeoning clean energy economy of the capital it needs to thrive. In other words, the deflation of toxic subprime assets could indirectly exacerbate climate risk.

However, others argue that a global recession creates a boon for the transition to a green economy. "Let us take the opportunity of the coincidence of the [financial] crisis and the deepening awareness of the great danger of unmanaged climate change: now is the time to lay the foundations for a world of low-carbon growth," said economist Nicholas Stern, whose 2006 review of the economics of climate change solidified the business case for climate mitigation and adaptation (Stern 2008).

A recent investment research study repeats the same message, noting that climate change investment can play a key role in recovering from the credit crisis and looming recession. "We believe that, when combined with energy security, climate change policies will play a role in government efforts to stimulate their economies in 2009," states report author Mark Fulton, head of climate change investment research at Deutsche Asset Management. "Governments now have an historic opportunity to define long-term regulatory frameworks to encourage private investment in climate change initiatives. Additional opportunity exists for governments to boost their economies by funding infrastructure projects that will serve to foster energy independence and climate-proof their economies" (DB Advisors 2008).

Institutional investors must insulate their portfolios against climate risk while also positioning their portfolios to seize climate opportunities. Failure to address climate risks and seize climate opportunities could, over a number of years, trigger a downward spiral in which dangerous climate change exacerbates problems with the economy. Conversely, protecting against climate risks and nurturing climate opportunities presents the possibility of fostering a shift to a clean-energy infrastructure that invigorates a green economy while simultaneously averting the climate crisis.

Institutional investors have a choice of which trend they want to support, and they have a key role to play in fueling their trajectory of choice.

ACKNOWLEDGMENTS

Research provided by Bill Baue, Jim Coburn, and Dave Grossman.

REFERENCES

Associated Press. 2008. *Dynegy to warn investors on risks of coal burning."* October 23. www.nytimes.com/2008/10/24/business/24dynegy.html?ref=business.

Baue, W. 2004. *Memo from: SRI analysts to: Companies—use GRI sustainability reporting platform!* SocialFunds.com, October 7. www.socialfunds.com/news/article.cgi/1535.html.

Baue, W. 2002. *SEC urged to strengthen rules governing corporate disclosure of environmental risks.* SocialFunds.com, August 21. www.socialfunds.com/news/article.cgi/911.html.

Berridge, R. 2008. *EthVest database analysis.* Ceres and Interfaith Center on Corporate Responsibility. www.iccr.org/ethvest.php.

Centex. 2008. *National homebuilder takes aim at energy efficiency: Centex to offer suite of energy efficiency features as standard."* July 8. http://phx.corporate-ir.net/phoenix.zhtml?c=112195&p=irol-newsArticle&ID=1172682&highlight=.

Ceres, Environmental Defense Fund, California Public Employees' Retirement System (CalPERS), *et al.* 2008. *Petition for interpretive guidance on climate risk disclosure.* September 18, 2007. www.incr.com//Document.Doc?id=187.

Choo, K. 2006. Feeling the heat: The growing debate over climate change takes on legal overtones." *ABA Journal,* July.

Cogan, D. 2008. *Corporate governance and climate change: The banking sector.* Ceres and RiskMetrics. January. www.ceres.org//Document.Doc?id=269.

Confessore, N. 2008. Xcel to disclose global warming risks. *New York Times,* August 27, 2008. www.nytimes.com/2008/08/28/business/28energy.html?_r=1&ei=5070&adxnnl=1&oref=slogin&emc=eta1&adxnnlx=1223560925-Ov3rHyxmDRlR0w3WRU4p1g.

Connecticut v. American Electric Power, 406 F.Supp.2d 265 (S.D.N.Y. 2005). http://ag.ca.gov/globalwarming/pdf/Connecticut_%20AEP_Decision_Dismiss_2005Sep152004July24.pdf.

CorporateRegister.com. 2008. *Global report output by year.* www.corporateregister.com.

DB Advisors. 2008. *Investing in climate change 2009: Necessity and opportunity in turbulent times.* Deutsche Bank Group. October. http://www.dbadvisors.com/deam/stat/globalResearch/climatechange_full_paper.pdf

Dynegy. 2009. *Dynegy announces dissolution of development joint venture.* January 2. http://investor.shareholder.com/dynegy/releasedetail.cfm?ReleaseID= 357286.

Ethical Corporation. 2008. *Ethical leaders of 2008.* December 1. www.ethicalcorp. com/content_print.asp? ContentID=6226.

Ethical Corporation Institute. 2008. *Only 62% of the FT 500 companies reporting emissions have carbon reduction targets.* CSRwire.com, July 2. www.csrwire. com/News/12547.html.

Fahrenthold, D. 2008. Dominion's coal-fired electric plant to advance. *Washington Post,* June 26. www.washingtonpost.com/wp-dyn/content/article/2008/ 06/25/AR2008062502798.html.

Gas Daily. 2008. Long-term droughts in California, Southeast seen as threat to grid stability. February 13. www.platts.com/Natural%20Gas/highlights/ 2008/ngp_iferc_021308.xml.

Gerrard, M. 2007. Survey of climate change litigation. *New York Law Journal,* September 28. www.abanet.org/abapubs/globalclimate/surveyoflitigation-. pdf.

Gerrard, M., and J. Howe. 2008. *Climate change litigation in the U.S.* Arnold & Porter LLP, December 11. www.climatecasechart.com/.

Global Reporting Initiative (GRI) and KPMG Global Sustainability Services. 2007. *Reporting the business implications of climate change in sustainability reports.* July 17. www.globalreporting.org/NR/rdonlyres/C451A32E-A046-493B-9C62-7020325F1E54/0/ClimateChange_GRI_KPMG07.pdf.

Gronewold, N. 2008. Lawyers see "growing legal storm" over emissions trading. *ClimateWire.* August 12. www.eenews.net/climatewire/2008/08/12/2/.

Hawley, J., and A. Williams. 2002. The universal owner's role in sustainable economic development. *Corporate Environmental Strategy* 9 (3): 284–291. www.sciencedirect.com/science?_ob=ArticleURL&_udi=B6VNW-45WGJX3-1&_user=10&_rdoc=1&_fmt=&_orig=search&_sort=d&view=c&_version= 1&_urlVersion=0&_userid=10&md5=c84fb5b1fa238098fecf2240a8776746.

Herrera, T. 2008. *The top ClimateBiz stories of 2008.* ClimateBiz.com. December 31. www.climatebiz.com/blog/2008/12/31/top-climatebiz-stories-2008?page=0%2C1.

Kanter, J. 2007. Fighting climate change, one lawsuit at a time. *International Herald Tribune,* August 15. www.iht.com/articles/2007/08/15/business/ greencol16.php.

Lochbaum, D. 2007. *Got water? Nuclear power plant cooling water needs.* Union of Concerned Scientists. December 4. www.ucsusa.org/assets/documents/ nuclear_power/20071204-ucs-brief-got-water.pdf.

Lubber, M. 2008a. Wall Street and the climate crisis. *Boston Globe,* September 29. www.boston.com/bostonglobe/editorial_opinion/oped/articles/2008/09/29/ wall_street_and_the_climate_crisis/.

Lubber, M. 2008b. Comment on Exposure Draft—Disclosure of Certain Loss Contingencies Amending FAS 5 and 141(R) – File Reference No. 1600-100. *Letter to Russell G. Golden, Technical Director, Financial Accounting Standards Board,* August 8. www.fasb.org/ocl/1600-100/52364.pdf.

Makower, J. 2008. *Strategies for the green economy: Opportunities and challenges in the new world of business* (pp. 13–14). New York: McGraw-Hill.

Mufson, S. 2007. Power plant rejected over carbon dioxide for first time. *Washington Post,* October 19. www.washingtonpost.com/wp-dyn/content/article/2007/10/18/AR2007101802452.html.

National Association of Insurance Commissioners (NAIC) Climate Change and Global Warming Task Force. 2008. *Climate risk disclosure proposal (draft 3).* August. www.naic.org/documents/committees_ex_climate_disclosure_proposal.pdf.

National Association of Insurance Commissioners (NAIC). 2009. *Insurance Regulators Adopt Climate Change Risk Disclosure: Requires Reporting of Risks, Responses.* March 17. http://www.naic.org/Releases/2009_docs/climate_change_risk_disclosure_adopted.htm.

North American Electric Reliability Corporation (NERC). 2007. *NERC 2007 long-term reliability assessment: 2007–2016.* Version 1.1, October 25. www.nerc.com/files/LTRA2007.pdf.

Odell, A. *Three major banks sign the carbon principles.* SocialFunds.com. February 13. www.socialfunds.com/news/article.cgi/article2468.html.

Pensions & Investments. 2008. "Florida's questionnaire for fund managers on climate change. March 3. www.pionline.com/apps/pbcs.dll/article?AID=/20080303/REG/397020869/1030/TOC.

Rheannon, F. 2007. *Investors, states, and activists petition Securities and Exchange Commission to mandate climate risk disclosure.* SocialFunds.com. October 8. www.socialfunds.com/news/article.cgi/2388.html.

Romm, J. 2008. Global recession? Must be time for the media's alternative-energy backlash. *ClimateProgress.* October 21. http://climateprogress.org/2008/10/21/global-recession-must-be-time-for-the-medias-alternative-energy-backlash/.

Schlissel, D. 2008. *Don't get burned: The risks of investing in new coal-fired generating facilities.* Interfaith Center on Corporate Responsibility. February 26. www.iccr.org/news/press_releases/pdf%20files/DontGetBurned08.pdf.

Sittler, M. 2007. Climate change panel foresees longer, dryer droughts. DroughtScape. *National Drought Mitigation Center,* Spring. www.drought.unl.edu/droughtscape/2007Spring/dsspring07-IPCC.htm.

Smith, J., Morreale, M., & Mariani, M., et al. 2008. Climate change disclosure: Moving towards a brave new world. *Capital Markets Law Journal* 3 (4): 469–485. *Capital Markets Law Journal* Advance Access, August 19, 2008.

Stausboll, A., Chiang, J., et al. 2008. Re: Comments on Roundtable on Modernizing the SEC's Disclosure System—File No. 4-567. *Letter to Florence E. Harmon, Acting Secretary, Securities and Exchange Commission.* October 22. www.ceres.org/Document.Doc?id=376.

Stern, N. 2008. Green routes to growth: Recession is the time to build a low-carbon future with the investment vital for economy and planet. *The Guardian.* October 23. www.guardian.co.uk/commentisfree/2008/oct/23/commentanddebate-energy-environment-climate-change.

Sustainable Investment Research Analyst Network (SIRAN). 2008. *Sustainability reporting by S&P 100 companies made major advances from 2005–2007.* July 17. www.siran.org/pdfs/SIRANPR20080717.pdf.

Trucost. 2007. *Carbon counts 2007: The Trucost carbon footprint ranking of UK investment funds.* July 16. www.trucost.com//CC2007.pdf.

Trucost. 2008. *Carbon counts—Australian superannuation investment managers 2008.* September 9. www.trucost.com/CarbonCounts-AIST.pdf.

Van Atten, C., Curry, T., & Saha, A. 2008. *Benchmarking air emissions of the 100 largest electric power producers in the United States.* Ceres, Natural Resources Defense Council, PSEG, PG&E Corp. May. www.ceres.org//Document. Doc?id=333.

Wayman, M. 2008. *The corporate climate communications report 2007: A study of climate change disclosures by the Global FT500.* CorporateRegister.com, February. www.corporateregister.com/pdf/CCCReport_07.pdf.

Weiss, M. 2008. *Drought could shut down nuclear power plants: Southeast water shortage a factor in huge cooling requirements.* Associated Press, January 23. www.msnbc.msn.com/id/22804065/.

Zakai, Y. 2008. *Controversial court ruling could affect permit plans for coal plants around the country.* CSRwire.com. July 2. http://vcr.csrwire.com/node/8986.

BIBLIOGRAPHY

Ackland, R. 2007. "Lawyers have been leaders in seeing green." *Sydney Morning Herald*, April 20. www.smh.com.au/news/opinion/lawyers-have-been-leaders-in-seeing-green/2007/04/19/1176696996408.html?page=fullpage.

Baue, B., and J. Cook. 2008. "Mutual funds and climate change: Opposition to climate change resolutions begins to thaw." Ceres, April. www.ceres.org//Document.Doc?id=322.

Boyle, R., et al. 2008. "Global trends in sustainable energy investment 2008: Analysis of trends and issues in the financing of renewable energy and energy efficiency." United Nations Environment Programme and New Energy Finance, June. www.unep.fr/energy/act/fin/sefi/Global_Trends__2008.pdf.

California Legislative Analyst's Office. 2006. "Cal facts 2006: California ranks among the world's top ten economies." www.lao.ca.gov/2006/cal_facts/2006_calfacts_econ.htm.

California Public Employees' Retirement System (CalPERS), California State Teachers' Retirement System (CalSTRS), Bennet Freeman, et al. 2008. Re: 33-8935, Modernization of the Oil and Gas Reporting Requirements. *Letter to Florence Harmon, Acting Secretary, Securities and Exchange Commission.* September 8. www.ceres.org//Page.aspx?pid=941&srcid=705.

California v. General Motors, et al. U.S. Dist. LEXIS 68547 (N.D. Cal. 2007.) www.pewclimate.org/judicial-analysis/ca-v-gm.

Connor, S. 2008. "Arctic thaw threatens Siberian permafrost." *The Independent*. June 14. www.independent.co.uk/environment/climate-change/arctic-thaw-threatens-siberian-permafrost-846951.html.

Emergency Economic Stabilization Act of 2008. http://thomas.loc.gov/cgi-bin/query/z?c110:H.R.1424.eas:.

Litz, F., and K. Zyla. 2008. "Federalism in the greenhouse: Defining a role for states in a federal cap-and-trade program." World Resources Institute, September. http://pdf.wri.org/federalism_in_the_greenhouse.pdf.

Monbiot, G. 2007. *Heat: How to stop the planet from burning* (pp. 20–42). Cambridge, MA: South End Press.

PricewaterhouseCoopers. 2008. "Carbon disclosure project report 2008: Global 500." September 23. www.cdproject.net/download.asp?file=67_329_143_CDP%20Global%20500%20Report%202008.pdf.

Rascoe, A. 2008. "Senate OKs energy tax breaks in bailout bill." Reuters. October 2. http://www.reuters.com/article/environmentNews/idU.S.TRE4905S120081002.

Sheppard, K. 2008. "Bail to the chief: House passes bailout plan with extensions for renewables, sends to Bush's desk." *Gristmill*. October 3. http://gristmill.grist.org/story/2008/10/3/104144/632/.

UK Environment Agency National Press Office. 2008. Press release: Green light to three new fund managers. September 8. www.environment-agency.gov.uk/news/86222.aspx?page=10&month=9&year=2008.

van Bergen, B. 2008. "Climate changes your business: KPMG's review of the business risks and economic impacts at sector level." KPMG Global Sustainability Services. April 2. http://kpmg.nl/Docs/Corporate_Site/Publicaties/Climate_Changes_Your_Business.pdf.

Warner, M. 2009. "Is America ready to quit coal?" *New York Times*, February 14. www.nytimes.com/2009/02/15/business/15coal.html.

CHAPTER 5

The Case for Climate Change as the Paramount Fiduciary Issue Facing Institutional Investors

Paul Q. Watchman

Climate change, like love in the words of the pop song, changes everything. Until very recently, many, if not most, institutional investors held the view that their discretion to invest was limited strictly by the need to maximize short-term financial returns for their beneficiaries (Freshfields Bruckhaus Deringer [Freshfields] 2005; Richardson 2008a). The principal justification put forward by Milton Friedman (1962, 1970) and others (Baue 2004a; Langbein and Posner 1980) for this conservative view of investment was the fiduciary duties owed by institutional investors to beneficiaries and by asset managers under the terms of investment mandates granted to them by institutional investors.

These fiduciary duties, it was stated correctly, were owed by law by institutional investors and, by delegation of their powers by such investors, to investment asset managers to both present and future beneficiaries (Flannigan 2004; Freshfields 2005; Megarry 1989; Richardson 2008a; Scanlan 2005). However, as history testifies, it can often be a short step from the statement of objective legal principles to their corruption into subjective political or economic dogma. That step was taken in a general attempt to fill ears with wax to deafen corporations, investors, and asset managers from hearing sweet seductive siren songs of values other than free enterprise financial values. Songs such as the "fundamentally subversive doctrine" (Friedman 1962, 1970) of social responsibility and later socially responsible

---

Paul Q. Watchman established Quayle Watchman Consulting.

investment (Ackerman and Bauer 1953; Prakash 1974), which if heard by businessmen or investors would wreck their enterprises and investments on the rocks and reefs of irresponsibility.

This orthodox view of the limiting effects of fiduciary duties is encapsulated notably and concisely by Friedman's well-known exhortation to business that it was the duty of businessmen to ignore socially responsible investments or any investments based on considerations other than what was mistakenly thought to be "purely" financial (short-term) monetary considerations as opposed to "nonfinancial" considerations, such as governance and sustainability of the business. Rather, it was their duty to adhere to a narrow fundamentalist principle that the business of business is business (Friedman, 1970).

This Friedmanite principle has been advanced in slightly different forms as providing guidance on fiduciary duties law to U.S. public sector pension funds (Freshfields 2005; Langbein 2005; Langbein and Posner 1980; Richardson 2008b) and other pension funds (U.S. Department of Labor [DOL] 1974, 2008) and, it must be accepted, has a large number of diehard adherents, at least among pension fund trustees, asset management advisors, and eminent academics in the United States (Ceres 2004; DOL, 2008; Freshfields 2005; Kinder 2008). These faithful followers of Friedman continue to hold the belief that the acceptance of any considerations other than what they would regard as purely financial considerations being taken into account in any form of business assessment, valuation, or investment is a breach of the fiduciary duties of institutional investors and asset managers—a belief held, it must be said, in the teeth of overwhelming evidence challenging their orthodox beliefs, including research by major financial institutions and international banks, such as UBS, HSBC, and Goldman Sachs (United Nations Environment Programme Finance Initiative [UNEP FI] 2004).

Notwithstanding the findings of these free enterprise institutions, which until recently have largely created great wealth for their investors and continue to share their values, the new Friedmanites persist in taking issue with the proposition that environmental, social, and governance (ESG) are valid investment considerations (DOL 2008). Indeed, some critics even fail to acknowledge that those who argue that ESG issues could be combined with other traditional financial considerations and accepted nonfinancial considerations to provide additional relevant investment information do not consider these ESG considerations as a substitute for other investment considerations (Baue 2004a; DOL 2008). Far less do these critics argue that ESG considerations can be employed or should be employed in any way to exclude assets, asset classes, or businesses from investment portfolios (Freshfields 2005; Principles of Responsible Investing [PRI] 2008; Richardson 2008b).

It is critical to understand that proponents of the inclusion of ESG in mainstream investment decision making also do not argue for exclusion or divestment; instead, they support engagement with investment businesses. (Freshfields 2005; PRI 2008). ESG considerations, they would suggest, are useful additions to the toolkit of the investor as they provide further valid information to assist in assessing and predicting investment and the value of investments more accurately.

THE FRESHFIELDS REPORT

In 2005, the shibboleth that fiduciary duties somehow prevented institutional investors from investing a cent in any way but to maximize short-term profits for beneficiaries was to be challenged by the United Nations Environment Programme Financial Initiative Report (UNEP FI), "A legal framework for the integration of environmental, social and governance issues into institutional investment" (Freshfields 2005). Perhaps because of its lengthy title, the report has come to be more commonly referred to by the investment community as the "Freshfields report." This title is carved from part of the name of the international law firm "Freshfields Bruckhaus Deringer," which was commissioned by the UNEP FI to carry out research pro bono public on the issue of ESG and investment decision making.

At the Freshfields London office, a core team of five lawyers was supported by over 30 legal and pension fund experts from Freshfields and other legal firms and academics with expertise in pensions fund and fiduciary duty issues. The Freshfields team spent six months reviewing legislation and case law on fiduciary duties and pension fund investment in nine countries and other legal jurisdictions. In consultation with the UNEP FI Asset Management Group, the team selected the following countries for review: Australia, Canada, France, Germany, Italy, Japan, Spain, the United Kingdom, and the United States of America. Where appropriate, an evaluation was carried of the laws of pension fund and investment laws of federal states and provinces, such as Manitoba, that proved surprisingly advanced in its understanding of the challenges facing pension fund trustees who wished to incorporate ESG into pension fund investment statements and mandates.

The authors of the Freshfields report (I was the senior author) came to very different conclusions than those who held the orthodox investment industry view that the fiduciary duties of institutional investors precluded investment that took account of ESG considerations:

> *Institutional investors who hide behind profit maximisation and the limits supposedly placed by their legal duties as fiduciaries do so at*

*their own peril. There is no legal bar to the integration of ESG con-
siderations into investment decision-making (provided the focus is
always on the beneficiaries best interests) and indeed failure to have
regard to such considerations where there is a proven link between
an ESG consideration and investment value may itself amount to
a breach of fiduciary duties by the pension fund trustee or on his
behalf by an asset manager. (Freshfields 2005)*

In essence, the Freshfields report had turned the orthodox view of the
fiduciary duties of institutional investors on its head. Why, it was asked,
should ESG considerations not be taken into account in investment decision making if it could be shown that there was some nexus between them
and value? For example, poor governance and lax regulation in the financial industries, human rights abuses in the mineral industries, pollution and
abstraction by industry of scarce drinking water resources in Africa and
India, displacement of indigenous people, environmental damage to rainforests and destruction of other carbon sinks in Asia and South America,
and global warming due to anthropogenic climate change have been demonstrated, over an extensive period of time, to damage the public reputation
and market brands of companies and industrial sectors with very real negative impacts to their bottom line, profitability, and solvency as goods or
businesses are boycotted or bypassed by consumers and customers (Baue
2004b; Joly 1999; *Just Pensions* 2001; Logan and Grossman 2006).

Reviewing the relationship on ESG investment considerations and fiduciary duties, Freshfields also suggested that, at the core, modern fiduciary
duties of a pension fund trustee would be found to include the following:

- Duty of loyalty
- Duty to act prudently
- Duty to act fairly
- Duty to seek advice

The contents and description of these modern fiduciary duties may be
expanded for the sake of clarity (Ceres 2004; Megarry 1989; Scanlan 2005).
Within the duty to act fairly, for example, a subsidiary duty lies to maintain
a balance between present and future beneficiaries. Equally, the duty of
loyalty suggests that it would be unlawful for pension trustees to take up
a moral, ethical, or political crusade on those grounds alone, whereas such
a "crusade" would be open to church charities and cause-related charities,
such as the Red Cross or cancer or other health charities. The duty to act
prudently includes the need to incorporate modern portfolio theory into
investment decision making, thereby spreading risk by holding assets in

businesses and different asset categories and to take a long-term view of investment returns rather than a short- or medium-term view. Finally, the duty to act prudently also requires pension fund trustees to be aware of their limitations as expert decision makers. Pension fund trustees are not required to be experts in investment, but pension fund trustees as representatives of asset owners and present and future fund beneficiaries are expected to seek independent advice on investment from appropriate experts, such as asset fund managers, accountants, and lawyers.

In addition to defining the nature of modern fiduciary duties, the Freshfields report is important for a number of other reasons. First, it criticized the view that *Cowan v. Scargill* ([1984] 3 WLR 50; Megarry 1989) in particular but also the other U.K. cases cited as justification for the conservative view of fiduciary duties taken by U.K. pension funds and institutional investment. Similarly, the report questioned whether the U.S. guidance on fiduciary duties was based soundly and fairly on objective interpretation of U.S. case law and legislation or was based on an unnecessarily restrictive view of the limits of the legality of fiduciary duties.

Second, the Freshfields report suggested that the concept of fiduciary duties should be regarded as organic, developing over time to reflect changes in society's view of the limits of those duties and investment practice, rather than being fixed in substance and content at a specific point in history. It was posited that the meaning of fiduciary duties given by the courts at any point in time did not necessarily coincide exactly or temporally with society's view of their meaning. The judiciary is, by design and inclination, a rather conservative traditional group of lawyers, not social reformers. A time lag in changes of society's view and that of the judiciary, therefore, is arguably both inevitable and desirable because of the far-reaching constitutional implications of judicial lawmaking. It is hardly surprising if the meaning of fiduciary duties acceptable to the courts does not, from time to time, coincide with changes in society's view of their meaning.

To illustrate this point, a number of cases on breach of fiduciary duties by trustees, albeit in each case other than *Cowan v. Scargill*, by a local authority trustee rather than a pension fund trustee or other institutional investor or pension fund, was examined. Each local authority trustee case turned on its own facts and circumstances at the point in time it was decided. However, in all three of these cases decided in the twentieth century (paying equal wages for women doing equivalent work to men, providing enhanced access to public transport for the disabled, and subsidizing the use of public transport), the local authority trustees were found to have acted *ultra vires* (beyond the powers).

The Freshfields report argues that it is very much open to question even in the absence of enabling legislation whether the use of these powers by local authorities would be regarded now as breaches of their fiduciary duties. Enhanced access to public transport, equality of treatment of men and women and discouraging the use of private transport by providing incentives to use public transport, are now central core values and hallmarks of a civilized society. It was stated by the authors of the Freshfields report, and therefore it follows in their view, that these decisions, given the development of fiduciary duties to reflect critical changes in social values, may not be regarded by a twenty-first-century judge as being beyond the limits of fiduciary duties.

UNITED NATIONS PRINCIPLES OF RESPONSIBLE INVESTMENT

Institutional investors are coming to recognize that ESG can affect the financial performance of investment portfolios and that to fulfill their fiduciary duties to their beneficiaries, pension fund trustees and other institutional investors may be required to take account of some of these considerations if they are to discharge their fiduciary duties lawfully (PRI 2008). While it is for the investor or asset manager to give weight to each relevant and material ESG consideration, ESG considerations, where appropriate, must be taken into account and given appropriate weight and must not be negated by giving the consideration no weight or wholly disproportionate weight. Until 2006, however, the rub was that there was no framework accepted by the investment industry to achieve these tasks. The United Nations Principles of Responsible Investment (PRI 2006) were developed by expert asset owners, asset managers, governmental and intergovernmental agencies, academics, and civil society to provide such a framework.

The PRI are:

Principle 1: We will incorporate ESG into investment analysis and decision-making processes.

Principle 2: We will be active owners and incorporate ESG issues into our ownership policies and practices.

Principle 3: We will seek appropriate disclosure of ESG issues by the entities in which we invest.

Principle 4: We will promote acceptance and implementation of the Principles within the investment industry.

Principle 5: We will work together to enhance our effectiveness in implementing the Principles.

Principle 6: We will each report on our activities and progress toward implementing the Principles.

The Freshfields report provided if not the legal foundations of the United Nations Principles of Responsible Investment, then a very important part of their supporting substructure. As can be seen, Principle 1, incorporation of ESG into investment analysis and decision-making processes and ownership policies and practices, and Principle 2, commitment to active ownership, are the governing Principles. The other four principles (disclosure, promotion, teamwork, and reporting) are subordinate or ancillary to those first two governing principles.

Evidence of the overwhelming success of the PRI is provided by the fact that by 2008 over 360 institutional asset owners, investment managers, and professional services partners representing over US$14 trillion had adopted the aspirational values of the PRI (PRI 2008). However important the PRI has been helping to mainstream ESG considerations, it must be acknowledged that there were other important factors that assisted the PRI in gaining market acceptance as quickly as they did. Legislative reform, including under U.K. pensions and company law legislation imposing requirements on pension fund trustees and company directors to consider and report on the environmental and social impacts of their investments and business activities (U.K. Pensions Regulations, 2005, and the Companies Act, 2006) was one important factor. However, there were also present a number of wide-ranging voluntary initiatives, which, together with PRI, influenced in different ways acceptance by the investment industry of ESG considerations as mainstream considerations and in doing so placed climate change center stage. These voluntary initiatives include, in no particular order of importance, the UN Global Compact, the Equator Principles, the Who Cares Wins Initiative, the London Principles, the Carbon Disclosure Programme, the Carbon Principles, and the Climate Principles. Even private-equity firms, thought of by some commentators, if rather unfairly, as the very antithesis of the responsible investor, have adopted a private-equity industry policy accepting ESG considerations as being relevant to investment decision making. As if to underline what to some may seem as taking irony a step too far, Kohlberg Kravis Roberts (KKR), the original barbarians at the gate, have become one of the PRI signatories, and in their recent dealings have begun to show a greater willingness than even Equator banks to respect and to include voluntarily carbon emissions limitations and climate change considerations in their investment decisions.

In addition to legislation and voluntary initiatives in the finance and other sectors, it must be acknowledged that from the outset of all ESG considerations, the environmental element had a much better chance of being understood and adopted by the investment industry in their decision making. First, of the three types of considerations, the soil of the investment industry was much more receptive to environmental considerations than the other two considerations because of familiarity of the investment industry with environmental issues. The profits of the investment sector, and in particular the insurance and reinsurance industry, had been affected adversely over many years by claims arising from environmental damage and pollution (e.g., severe weather damage arising from floods, hurricanes, tsunamis, and storms; other environmental matters such as asbestosis, leaking pipelines, oil and gas refinery explosions, chemical works fires, and plant and oil pollution following shipping disasters, such as the Exxon *Valdez*). Quite frankly, such familiarity made it easy for investors to assimilate environmental considerations into their decision making. Second, the financial industry was familiar with environmental considerations because of legislative and international law requirements since the 1970s for large-scale projects that have a material environmental impact to be subjected to environmental assessment. To meet the need for advice on environmental assessment, an army of corporate and financial risk managers, environmental consultants, and lawyers was marshaled. Later, to address an absence of consistency of approach among environmental practitioners, environmental practices, procedures, and protocols were developed by industry bodies and quality assurance bodies.

From dealing with environmental claims arising from severe weather conditions or evaluating the potential environmental impacts, risks, and value of projects, it was a relatively short step to accepting the relevance of environmental considerations to investment decisions generally.

Climate change, therefore, was to develop throughout the twentieth century as one of the most important environmental considerations to become primus inter pares of environmental considerations. However, by the first decade of the twenty-first century, the scientific and political consensus had emerged that climate change was anthropogenic and that action needed to be taken to address the problem—a problem that politicians would come to describe as a greater threat to mankind than international terrorism (Beckett 2007; Blair 2004; King 2004; Schwartz and Randall 2003). Almost from nowhere climate change had transmogrified into the leading threat to economic and social survival. Climate change thus became the leading social issue of its time and the steps to be taken to mitigate or adapt to climate change had become of great media, government, and public interest (Watchman 2008). Not a day passes when climate change issues are not reported by newspapers, radio, and television. No longer was the popular image of Armageddon, alien invasion, or meteorites smashing into the planet but something much more prosaic: the weather. Hitherto, the stock image of the only subject of English conversation post-Austen was the weather, but climate change was weather with a sting in its tail. The traditional presentation of the English and weather is rather obsessive and very Pooterish. Gentle Englishmen with little imagination or reading are portrayed as discussing endlessly and to little point whether it would rain before the second innings of cricket would begin. However, in Al Gore's documentary, *An Inconvenient Truth*, and films such as *The Day After Tomorrow*, the weather is not seen to be a fickle beneficence of the gods but rather is seen as the villain of the piece.

Climate change is a pale horseman, causing floods, hurricanes, and snowstorms, bringing economies and societies to a halt, washing out fields of crops, and drowning large tracts of land due to increases in the levels of the seas or rivers. To a lesser degree, business is also affected, for example, carbon allocation schemes that push the costs of carbon reduction onto capital expenditure or general expenditure and may even threaten the survival of some forms of business, such as traditional coal-burning energy generation plants, and transport infrastructures (e.g., as ports, roads and airports), upon which business depends.

THE CHALLENGE OF CLIMATE CHANGE

> *Climate change presents a unique challenge for economics: it is the greatest and widest ranging market failure ever seen. (Stern 2007)*

Of all ESG considerations, it has been suggested that one consideration alone stands head and shoulders above other ESG considerations: climate change. Its preeminence among ESG considerations arises not only because of its relevance to investment decision making but also its general relevance to business, society, and the survival of the planet (Abbasi 2006; Lovelock and Rapley 2007). The changes in global climate in recent years have created and are creating many challenges for business, government, and mankind. The melting of polar ice caps and mountain glaciers; the rise in sea and river levels and consequent flooding; the increased severity and frequency of hurricanes, ice storms, and heat waves; land slippage; and the changing pattern of monsoon rains are only a few of the examples of climate change. The loss of Pacific islands and large tracts of land in Asia, Europe, and America submerged by increases in the levels of the sea and rivers are other possible consequences of climate change. The past few years alone have witnessed the deaths of innocent people and the destruction of homes, buildings, and other property by wildfires in the United States and Australia; tsunamis, storms, and hurricanes of great severity beating against the Asian shores and the southern states of the United States; disputes over scarce water resources in Africa and Asia; and the displacement of indigenous groups of people from, among other regions, Sudan. Indeed, our language has been extended, if not enriched, by the adoption and use of climate change terms. Terms such as *climate refugees* and *environmental migrants* and *the first climate change war* are used to describe, respectively, those displaced from their homes by extremes of weather conditions; their impacts on natural resources, crops, and wildlife; and the Sudan conflict (Watchman 2008).

All these phenomena and occurrences can and do have profound effects for business and investment as well as communities. A way to illustrate this point is to examine risks and opportunities presented by climate change to some of those areas of business, and hence of potential investment or divestment, which have been most affected by climate change.

BUSINESS RISKS AND OPPORTUNITIES

There are a number of ways that the effects of climate change on business may be viewed. One perspective is risk (Carbon Trust 2005; Cogan 2006; Munich Re 2008; Swiss Re 1994, 2002, 2003, 2007). First, there are those businesses at risk from climate change directly: farmers awaiting monsoon rains in the Indian subcontinent, indigenous people and others displaced from their land because of the scarcity of water, heat waves, or increases in sea or river levels. Second, there are those businesses at risk from climate change as a consequence of businesses in the first group of business being affected directly. Insurance and reinsurance companies are perhaps the best example of those businesses indirectly affected by climate change. Third, there is the group of businesses affected by climate change neither directly nor indirectly but at risk from the need of the businesses to adapt to or to mitigate the effects of climate change. The transport and energy industries are examples of businesses materially impacted to their detriment by adaption to or mitigation of the climate change effects.

Another climate change taxonomy of companies may be suggested. First, there are those groups of businesses that see climate change not as a risk but as providing an opportunity. For example, climate change gives rise to opportunities to create new goods and services, including developing climate change litigation services and financing green businesses (California Public Employees Retirement System [CalPERS] 2007; Ceres and Calvert 2007; PRI 2008). GE's Ecomagination program and KKR and the Environmental Defense Fund's partnership (Primedia 2009) are examples of businesses in this group.

Second, there are those businesses that are created to grasp the opportunities of climate change adaption and mitigation. Businesses such as Acclimatise, Climate Change Exchange, Airtricity, and carbon trading firms that give advice on alternative energy projects and managing carbon credits in the European Union and other carbon emissions trading schemes would be examples in this group. Of course, some businesses will view climate change and carbon regulation as both a risk and an opportunity. Oil and gas companies, for example, may seek to develop alternative energy projects or create less pollution by reducing the carbon content of fuels, and automobile manufacturers may attempt to design smaller, greener cars with lower carbon emissions and greater efficiency.

For others, however, the perspective may be limited to a risk perspective as the risks associated with climate change will appear to outweigh the opportunities. Consider the future of air freight. Goods such as flowers from Africa, which in recent times have been shipped by air, may be transported by ship in refrigerated containers in the future because of the smaller environmental footprint created by shipping rather than air freighting such goods.

ENERGY AND INSURANCE AND REINSURANCE

To understand further the proposition that climate change is the prime ESG consideration and why it and its effects should be taken into account by institutional investors, we will focus on how climate change has affected two of those industries that arguably have been most negatively impacted by climate change. Out of a short list of five industries—energy (including oil and gas, electricity, alternative energy, and the extractive industries), automobile manufacture, insurance and reinsurance, transport, and commercial property—we will focus on the one industry most affected directly by climate change—energy—and the one industry most affected indirectly by climate change—insurance and reinsurance.

This section offers an assessment of how the energy industry and the insurance and reinsurance industry have responded to climate change. It includes an evaluation of the attempts of each industry to adapt to or mitigate the effects of climate change and how climate change in each case is a relevant ESG consideration because of the potential impact on investment decision making.

Energy

Before entering into an assessment of the energy industry, it may be helpful to explain the reasons for wrapping the energy banner around the oil and gas and essentially the coal extractive industries. First, a number of energy companies hold interests in more than one sector. These interests may not be interlocking but they are closely aligned, and action in one area, for example, the use of bio fuels or development of liquefied petroleum gas (LPG) or carbon capture and storage (CCS) plants and pipelines and storage facilities, impacts other areas of the energy industry, such as coal extraction or the burning of fossil fuels. Second, a number of energy companies have broad interests in natural resources and minerals as well as electricity and gas generation and transmission. Finally, energy companies, to varying degrees, have invested in alternative or renewal energy, carbon sinks, clean coal, CCS, and other energy forms or responses to climate change. However, the influence of climate change is not only on the energy companies themselves but the financial institutions that lend to them (Allianz Group 2005) and private-equity houses that acquire energy companies or energy companies' shares and other assets. KKR's bid in 2008 for TXU's power stations development is an example of the extended nature of climate change considerations in the energy industry. In the process of acquiring the TXU power station portfolio in the United States, KKR demonstrated a greater degree of sophistication in their dealings with the nongovernmental community than had been evinced by mainstream international Equator banks. KKR also showed a willingness to consider material changes to the development of TXU's portfolio of coal-fired power stations and, in particular, the impact of carbon-emitting coal-fired power stations on climate change, than other parties involved in the transaction had shown prior to KKR's intervention.

The energy industry is highly diverse, ranging from companies still dependent to a large degree on burning coal and other fossil fuels to companies that have well-diversified energy portfolios. For example, France is heavily dependent on nuclear power, and the French energy company Electricite de France (EdF) acquired British Energy, which runs nuclear generation in the United Kingdom. Scottish and Southern received the bonus of the legacy of Scottish Hydro's extensive hydroelectric scheme in the highlands of Scotland. German energy company EoN and Spanish energy company Ibredrola are world leaders in wind farm development.

Companies less fortunately placed include mineral companies that have large coal reserves, which may be impacted adversely by the implementation of Kyoto Principles and carbon emission limiting schemes, such as the EU Environmental Trading Scheme. Friends of the Earth has challenged the adequacy of Xstrata's disclosure of the impact of carbon restraint in statements made to the U.K. Stock Exchange (Mansley 2003).

There is now a record number of climate change resolutions being lodged against energy companies in the United States (Ceres 2008; Friends of the Earth 2006), and of the nine U.S. "climate watch" companies identified by investors because of their poor climate change response, seven were energy companies: Southern (electricity); Massey Energy and Consul Energy (coal); and Ultra Petroleum, Exxon Mobil, Chevron, and Canadian Natural Resources (oil and gas) (Sustainable Business 2009).

The destruction of and extensive damage to oil and gas platforms by U.S. hurricanes such as Katrina have already been mentioned as examples of the energy industry's vulnerability to climate change. However, it is not only energy extraction that is impacted directly by climate change but also the energy generation and transmission. The closing of nuclear plants during the 2003 heat wave and the longer times taken to replace or restring power transmission lines due to ice storms are both examples of the direct effects of climate change on energy companies and their profits.

Energy companies also have been affected by attempts to mitigate climate change. Coal-fired and other fossil fuel power generation plants have been targeted by leading EU member states, such as the United Kingdom and Germany, in their carbon reduction National Allocation Plan, to shoulder larger carbon reductions proportionately to their carbon footprint than other industry sectors (Spieth and Hamer 2005). Whether this has been done because energy companies may be able to pass increased costs attributable to carbon emission mitigation measures to the consumer, the economic externalities of needing to reduce carbon emissions for these energy generation companies may necessitate their buying carbon credits on the market, spending large sums of money on plant improvement, or, in some cases, mothballing plants or plant closure.

Equally threatening is carbon adaption. Automobile manufacturers seek to develop cars that are more efficient in their use of fuel and emissions to the atmosphere or that run on alternative fuel sources, such as hydrogen or electricity. Commercial and residential property owners and occupiers seek to source their energy from green suppliers or to reduce their energy consumption by the use of roof and wall insulation or to employ alternative sources of fuel supply, such as solar panels. Businesses discourage air transport in favor of videoconferencing and encourage public transport and cycling to work over private transport and the car. The social and political pressures on businesses to disclose the carbon footprints of business and wherever possible to reduce them, together with these other climate change considerations, has led to a fall in demand for coal and other fossil fuels. In investment terms, while some oil and gas and electricity generation and transmission companies, like Exxon Mobil and Chevron Texaco, remain among the largest and most profitable companies in the world, consideration must be given as to how well they are prepared to face the challenges of climate change in the twenty-first century (Logan and Grossman 2006).

Insurance and Reinsurance

> *The insurance sector has a key role in helping to mitigate the effects of climate change by providing financial indemnification, compensation and relief against climate change events and by developing products and solutions that can support emerging GHG and renewable energy markets. (Swiss Re 2007)*

The four watershed events for the insurance industry have been the great
dust bowl in the 1930s, urban riots in the 1960s and 1970s, global terrorism
in the last part of the twentieth century, and, from the 1990s onward, climate
change (Banana Skins 2007). Banana Skins Poll of Top Insurance Risks 2007
placed natural catastrophes and climate change as the second and fourth top
insurance risks. To put these findings of the relative severity of risks from
natural catastrophes and climate change in the context of other insurance
risks, Banana Skins rated terrorism, pollution, and asbestos as the 18th,
21st, and 33rd greatest risks.

The losses incurred by the insurance and reinsurance industry due to
climate change have been awesome (Association of British Insurers [ABI]
2004; Munich Re 2008; Swiss Re 2007; West Landesbank 2004). To give
some idea of the extent of those losses, it may be sufficient to look at
some of the losses incurred by insurance and reinsurance companies in
the United States recently due to hurricane damage. Six of the 10 most
expensive hurricanes in U.S. history—Katrina, Rita, Charley, Ivan, Frances,
and Jeanne—occurred within a two-year period from 2004 to 2005. In
Hurricanes Katrina and Rita alone, 113 oil platforms were lost completely
and 52 were extensively damaged (Minerals Management Service [MMS]
2006).

Extreme weather brings many risks for insurers and reinsurers: earth-
quakes, tsunamis, heat waves, substantial changes in global and polar ice,
ice storms, land slippage, extreme variability in temperatures, significant
rises in sea levels and river levels and incidence of flood, increased intensity
of storms, and the incidence of severe droughts. The extreme heat wave in
Europe in 2003 resulted in the closure or curtailment of power generation
in nuclear and other thermal power stations in France, Germany, Romania,
and Croatia as river water used for cooling plants became too warm, with
consequent business disruption and interruption costs.

In assessing the extent of losses sustained by insurers and reinsurers—
and in determining the impact of those losses—much depends on the elas-
ticity of demand for insurance or reinsurance products on offer and the
insurer's or reinsurer's time scale for individual cases. For example, the elas-
ticity of demand for real property insurance among the retired and poor
living on fixed and reducing incomes in states such as Florida and Louisiana
will be limited. Where payment of insurance premiums or increased insurance premiums cannot be afforded, some property owners may decide to
take the risk and to self-insure. It may not be possible for insurers or reinsurers to recover severe weather losses in the future by premium increases,
and some insurers may be exposed for only one year until the delivery of a
turnkey project, whereas a reinsurer's time horizon may be as much as 50
to 100 years. Equally, an insurer or reinsurer of a turnkey project may be

exposed for a limited number of years, whereas reinsurers may use a 50- to 100-year time horizon to calculate risk and return.

Climate change risks and losses, however, may be weighed against the opportunities to develop and invest in new product lines and to invest in green technology and carbon projects and companies that climate change has brought to the insurance and reinsurance industry. Insurance and reinsurance companies, such as AIG, Swiss Re, and Munich Re, are investors in their own right, investing billions of dollars in stocks and shares, bonds, and other assets each year with varying degrees of success. In addition, new insurance product lines developed as a result of climate change would include floodplain insurance, sea defense insurance, weather derivatives, green buildings insurance, energy savings insurance, renewable energy project insurance, carbon emissions trading insurance, insurance of Clean Development Mechanism (CDM) and Joint Implementation (JI) projects, pay-as-you-drive insurance, natural resources damage insurance, and CCS insurance. Climate change has presented insurance companies with the opportunity to invest directly or indirectly in renewable energy, clean technology, green funds, and CDM and JI projects and perhaps to consider more critically investment in those industries. Insurance companies also have the opportunity to invest in those industries and sectors most directly affected by climate change or climate change mitigation or adaptation or, as with the insurance and reinsurance industries, indirectly affected by the impacts of global warming on the businesses and property of the insured.

CASE STUDY: Insurance in a World with Climate Change

By C. Shawn Bengtson, PhD, CFA

The global insurance industry generated $4.1 trillion in 2007 premiums, primarily in property and casualty lines. However, the life/health lines dwarf the rest of the industry in total assets, with over $3 trillion—2.3 times those of property and casualty insurers. Climatic change poses significant risk to these bulwarks.

Operations in Climate-Affected Insurance

Mitigating downside risk traditionally includes a level of risk transfer through insurance. Climate change impacts are commonly associated with an increase in extreme weather-related events. When catastrophe strikes, property risk and risk to life, recovery, and return to "preloss state" are primary concerns. Actuarially, increases in claim costs (restoring property or risks to health) and costs of insurance (risks to life) result in premium increases.

C. Shawn Bengtson is senior portfolio manager of the securities department of Woodmen of the World Life Insurance Society.

Problematic to the valuation process is the impact of utter devastation on restoring an insured to preloss state (e.g., the 2004 Indian Ocean tsunami). Historically available policies such as earthquake or hurricane and business interruption insurance continue to be successfully marketed; however, the policies rarely insure to replacement value. Severe policy limits have replaced the generous social benefits that once existed. Higher benefit packages are prohibitively expensive, due to expected increases in the number and severity of future losses directly resulting from global warming.

Believing that the existence of insurance immunizes the portfolio from loss is analogous to ignoring counterparty risk in a derivatives market. Although insurance is in place, there is no guarantee that preevent valuations can be achieved once loss is sustained. A product solution to cost-prohibitive insurance is to mete the benefits into affordable pieces. For example, Swiss Re is offering climate change litigation protection for directors, similar in concept to an errors-and-omissions policy. As other product solutions emerge, policy costs will be affected by additional currently unforeseen risks resulting from climate change.

An example of those unforeseen risks may be those that are associated with infrastructure losses. Such costs are rarely considered in existing policy pricing. Independent property and asset valuation assumes that a viable infrastructure exists. Maintenance costs are assumed to exist, but are not part of property replacement claim costs. Significant climate changes could shorten the expected life span of infrastructure.

Risk management innovations exist: Consider Florida's 2004 and 2005 hurricane seasons. Many property and casualty insurers no longer underwrite policies in some Florida zip codes. The state established the Florida Hurricane Relief Fund (now known as the Florida Disaster Recovery Fund), a 501c(3) soliciting donations for Floridians who are victims of natural disasters. The fund takes the role of primary insurer when insurance is not otherwise in place, or takes the role of excess-share reinsurance when losses exceed the claims-paying ability of a primary insurer. As we now recognize, the losses sustained in Florida for that two-year period of time were too great for the private insurance industry to provide a viable long-term solution.

As catastrophic events (increased storms, broader paths of destruction due to hurricane-force winds, earthquake and other events) leading to catastrophic property damage move inland, additional structural standards will likely be adopted. An effective risk management program includes insurance; however, insurers cannot be solely responsible for having the capacity to appropriately restore every insured to his/her preloss state.

Insurers recognize this fact and have actively participated in developing broader market solutions. Catastrophe bonds, for example, permit underwriting risks to be bundled through geographically diversified risk exposure. The bundle is securitized and then made available to bond investors, who generally are not otherwise directly subject to global catastrophic property loss risk.

Investment Management in a Climate-Affected Insurance Environment

The insurance industry mitigates some risks in a climate-changed world, certainly, but also is a significant source of institutional investment. The industry's balance sheet assets are largely comprised of investments. Insurers survive through managing investment assets against their liabilities. Asset-liability matching does not occur on a one-year basis; rather, investment strategies include laddering portfolio assets and using duration and convexity tactics to mitigate interest rate and other cash flow risks.

Climate change affects investments by affecting risk, and risk must now be evaluated and modeled differently—not separately, but *differently*—because of climate risks. Climate risk goes beyond company-specific risks to include portfolio risk implications. In every aspect of risk management and mitigation, time becomes a consideration. Financial performance may be manageable near term, but downward pressure is exerted over time on equity prices due to reduced profitability and on bond prices due to increased default risk. Property values are subject to decline due to increased probability of devastation. Infrastructure costs rise because maintenance must be applied more frequently; in the worst scenario, maintenance is replaced with redesign and rebuild.

Existing controls for portfolio risk include asset-liability management (ALM) and risk-based capital (RBC) assessment. ALM involves stressing forecasts under various interest rate scenarios. At year-end 2008, RBC requirements are applied by asset class, ratings, diversification, and portfolio size. Additional risk-based capital may be assessed against an asset portfolio's largest concentration, for oversized portfolio weightings. Since RBC risks are categorized separately by asset risk, interest rate risk, operations risk, and other business risk, covariance between risk categories may exist. Capital calculations permit reducing the total risk-based capital assessment by covariance. However, climate change may materially impact the volatility of any asset portfolio—independent of interest rate risk, capital structure or diversification—due to portfolio composition. Asset volatility is accurately viewed as risk, so RBC is likely to increase, constraining capital further when organizations are seeking to free capital.

Most ALM models are ill equipped for forecasting event impact to the asset portfolio; modifications are generally incorporated using a constant default rate applied uniformly across the modeling period. Climate change risks and their associated impacts may require stress testing the asset cash flow stream itself. Stress tests might expose new portfolio risks in a climate-changed world; some of those tests could include imposing catastrophic event scenarios on the portfolio, structural and environmental assessments affected by temperature change, or regional events that cause a long-term business interruption.

New risk reporting may result from such testing. Engineering reports should be more common. Environmental assessments should become a mandatory part of an investment's documentation package, even though property is not being securitized. Summary statements of climate change impacts should ultimately become part of every financial filing, with stress tests attached as supporting documentation.

Increased reporting requirements inevitably lead to greater model sophistication. As models emerge to accommodate portfolio stresses (rather than individual asset level), new opportunities exist to securitize various aspects of that portfolio. Conceptually, any asset portfolio could be collateralized and tranched to appeal to the investment horizons of multiple market participants.

Woodmen of the World's awareness of climate risks is reflected in the management of its general account assets. For example, Woodmen commonly includes engineering and environmental assessments in its loan underwriting and security-level due diligence and requests management business resumption plans. Long-term investments in utilities securities are curbed by the proportion of coal-burning facilities and company location, both largely impacted by political risks resulting from a changing climate.

Ultimately, climate change will force changes to investment selection, evaluation, reporting, legislation, and risk management. The real question is the length of time these changes take to manifest themselves. Until reporting and assessment requirements are imposed through regulation or illustrative recommendations, insurance asset managers may need to reconsider existing risk controls. An existing portfolio construction approach may unduly expose the company to financial demise due to the accumulation of climate change impacts—redefining the measurement of portfolio diversification.

CONCLUSIONS

Traveling from Kyoto to Copenhagen has taken more than a decade. There have been bumps and disappointments along the way, most notably the failure of the Bush administrations to accept the need to provide global leadership in addressing the manifold challenges of climate change. However, through dogged persistence, there now appears to be genuine consensus about the cause and threat of climate change across a range of academic disciplines, governments, and civil society. Naturally, there will be different perspectives on climate change and global warming within such a wide group of diverse interests, which may range from scientists to conservationists to politicians concerned mostly with winning elections by appealing to the green vote.

Yet, it is important that politicians, scientists, economists, accountants, and even lawyers acknowledge the importance of climate change and begin to collectively discuss its consequences for their respective disciplines and to share information, research, and conclusions. Doing so will help deal with a genuine global threat to business, civilization, humanity, and the survival of the planet itself (Lovelock and Rapley 2007).

It may be difficult to see from this distance as Al Gore picks up the glittering prizes for his work on climate change, but for a long period of time he plowed a very lonely and unrewarding furrow. Seldom has a global issue been so identified with one politician, but Al Gore, shorn of the vice presidential comforts of Air Force Two, took his message of the risks of climate change all over the world, and in particular to small towns and cities in hinterlands of the United States, until global warming and the need for global action were accepted as "an inconvenient truth." Unprecedented storms, floods, and hurricanes in the United States devastating New Orleans and other great American cities no doubt helped give Gore the appearance of an Old Testament prophet, but, as was Moses, Gore is a prophet who deserves a little luck and decent props.

Not only has climate change now become a core global political issue, it is also recognized now as an issue that offers both immense risks and rewards. It is not surprising that government action is colored by where a state places climate change in the continuum between risk and opportunity. For example, the Obama administration appears to view climate change as more of an opportunity than a threat and has begun to invest large sums of money on green technology. The member states of the European Union, however, generally have tended to focus on mitigating risk and have developed a European Union Emissions Trading Scheme. That is not to say, of course, that any one government exclusively reacts to risk, whereas another reacts only to opportunity. Governments and state or multistate financial institutions, such as export credit agencies, the European Investment Bank, and the European Bank of Reconstruction and Development, both invest in opportunities presented by climate change and take steps to mitigate climate change risks. Governments now also accept the need to set hard targets rather than mouthing platitudes. For example, the United Kingdom has committed to a carbon dioxide reduction of 80 percent by 2050.

Eminent climate change scientists such as King (2008), Lovelock and Rapley (2007), and the members of the Intergovernmental Panel on Climate Change (IPCC 2007) have concluded that global warming is unequivocal and that most of the increase in global warming is anthropogenic (IPCC). This view was accepted by COP 13 in Bali (the 13th session of the Conference of the Parties and the 3rd session of the Meeting of the Parties to the Kyoto Protocol [CMP 3], Nusa Bali, September 3–14, 2007), which has provided the road map for negotiating international agreement on climate change measures for the period after 2012.

ESG considerations are now mainstream investment considerations partly because the significant number and standing of the signatories to the PRI have, in effect, created a de facto market standard from which it is impossible to deviate or to ignore because of the market dominance of the signatories of PRI. If the world's largest pension funds and asset management companies are PRI signatories and adhere to them, it is difficult for other institutional investors or asset managers not to join them.

The question of whether an institutional investor or asset manager has breached fiduciary duties ultimately is one of fact and circumstances and for the courts to decide on the evidence and arguments placed before it at the court hearing. However, while it should be emphasized that it is for institutional investors and their asset managers to decide what is the appropriate weight to give to climate change and other investment considerations, a failure to take account of climate change where climate change is a relevant consideration would appear in the light of reasons given in this chapter and elsewhere in this book to be so perverse, illogical, and absurd that it is difficult to foresee how a court properly advised could conclude otherwise than such failure's amounting to a breach of fiduciary duties. Institutional investors have to have the best interests of their beneficiaries at the heart of their fiduciary duties. They cannot, therefore, ignore climate change in their investment decision making without peril. A failure to do so may be a breach of several fiduciary duties, in particular the duty of prudence and the duty of fairness to present and future beneficiaries.

While climate change has emerged as the leading ESG investment consideration, it is critical that investors understand the interrelationship among the three considerations. What might be viewed as an environmental issue from one perspective might be regarded as a social or governance issue from another perspective, as one man's human rights issue may be the consequence of climate change or global warming.

For example, the mass exodus of climate change refugees from sub-Saharan Africa; the delay or advancement of the monsoon season, damaging the rice harvest that might cause hunger and starvation of Indian villagers; the deaths of children and elderly people due to the closure of hospitals because of the abnormally high temperature of coolant waters serving nuclear and fossil fuel energy plants due to a heat wave in Europe; and armed conflicts between states or groups in respect of scarce water resources due to mountain glacier melting in the Himalayas or Andes are all consequences of climate change.

The integration of climate change and ESG considerations makes the time ripe to assess the need for greater understanding of social and governance issues among the institutional investment community and its advisors and their close relationship with climate change issues. Environmental risks for investors, for example, may be quite small, but when a food shortage occurs in a failing state where bribery and corruption are rife, then the combined risk arising from the combination of environmental and social and governance considerations may make investment undesirable and divestment desirable. This, it is submitted, is where action is most needed now by the institutional investors and their advisors. Climate change, if properly understood, may be the catalyst for the development of broader fiduciary duties. The world is now in or moving into a global recession. This recession, and some of its causes such as the payment of bonuses based on short-term rewards for short-term investment goals and the ensuing credit crunch, may lend support to the argument for fundamental reform of the investment industry. Paradox, like hubris, has long been a feature of the financial sector, and it may be again for climate change and the institutional investment sector.

As economic activity slows down, fewer greenhouse gases should be emitted into the atmosphere. Governments appear prepared to spend significant sums of money on alternative or renewal energy, clean energy plants, green technology, and infrastructure, and, at the same time, ease the world out of recession. Institutional investors can also take the opportunity to promote better governance by banks and other corporations that acted outrageously in rewarding their directors and staff with grotesque bonuses for misusing financial instruments, such as derivatives, in a Friedmanesque pursuit of short-term profits that resulted in the recession and consequent credit crunch.

At the same time, the scientific community in Copenhagen reported in March 2009 that the impacts of climate change and global warming are now much worse than was thought by the scientific community, even as recently as a year before (Kintisch 2009). The polar ice caps are melting more quickly and temperatures are increasing more rapidly than scientists believed would happen, and governments are not acting quickly enough to bring the increase in temperature below the critical two degrees Celsius. It is critical now that institutional investors and their asset manager advisors come to appreciate that their fiduciary duties toward their beneficiaries are not a limitation but an enabling checklist for the unwary. Waking up to the investment potential of climate change and climate change mitigation and adaptation, and understanding their nature and likely impacts at microeconomic and macroeconomic levels, may result in a radical revival of investment portfolios. However, a failure to do so may result in a breach of fiduciary duties or a loss of profit from the many opportunities for investment climate change offers.

REFERENCES

Abbasi, D. R. 2006. *Americans and climate change: Closing the gap between science and action.* New Haven: Yale School of Forestry and Environmental Studies.

Ackerman, R. W., and R. A. Bauer. 1953. *Corporate social responsiveness.* Reston, VA: Reston.

Allianz Group. 2005. *Climate change and the financial sector.* June.

Association of British Insurers (ABI). 2004. *A changing climate for insurance.*

Banana Skins. 2007. Poll of top insurance risks.

Baue, W. 2004a. Fiduciary duty, undivided loyalty, and socially responsible investment performance. *Sustainable Investment News*, October 1. www.socialfunds.com/news/article.cgi/1530.html.

Baue, W. 2004b. Moving from the business case for SRI and CSR to the fiduciary case. *Sustainability Investment News*, February 19. www.socialfunds.com/news/article.cgi/article1346.html.

Beckett, M. 2007. *The case for climate security.* London: Royal United Services Institute.

Blair, A. 2004, The environment and the urgent issue of climate change. *Guardian*, September 15.

California Public Employees Retirement System (CalPERS). 2007. Climate disclosure: Measuring financial risks and opportunities. Sacramento: Author.

Carbon Trust. 2005. *A climate for change: A trustee's guide to understanding and addressing climate risk.* London: Author.

Ceres. 2004. *Sustainability and risk: Climate change and fiduciary duty for the twenty first century trustee.* Boston: Author.

Ceres. 2008. *Corporate governance and climate change: The banking sector.* Boston: Author.

Ceres and Calvert Investment. 2007. *Climate risk disclosure by the S&P 500.* Boston: Author.

Cogan, D. G. 2006. *Corporate governance and climate change.* Boston: Ceres.

Flannigan, R. 2004. The boundaries of fiduciary accountability. *Canada Bar Review* 83:35.

Freshfields Bruckhaus Deringer. 2005. *A legal framework for the integration of environmental, social and governance issues into institutional investment.* London: United National Environmental Programme Financial Initiative.

Friedman, M. 1962. *Capitalism and freedom.* Chicago: University of Chicago Press.

Friedman, Milton. 1970. The social responsibility of business is to increase its profits. *New York Magazine,* September 13.

Friends of the Earth (2006). *Fifth survey, climate change disclosure in SEC filings.* Washington, DC: Author.

Gore, A. 2006. *An inconvenient truth.* New York: Rodale Press.

Intergovernmental Panel on Climate Change (IPCC). 2007. *Fourth assessment report of the Intergovernmental Panel on Climate Change.* Cambridge, UK: Cambridge University Press.

Joly, C. 1999. Ethical demand and requirements in investment management." *Business Ethics: A European Review* 2(4): 199–212.

Just Pensions. 2001. Socially responsible investment and international development: A guide for trustees and fund managers. *Just Pensions,* May.

Kinder, P. 2008. *"Rigid rule" on economically targeted investment: New ERIS regulations on a plans' "economic interests."* The KLD Blog, November 25.

King, D. 2004. Climate change science: Adapt, mitigate or ignore?" *Science* 303 (January 9):176–177.

King, D., and G. Walker. 2008. *The hot topic.* London: Bloomsbury.

Kintisch, E. 2009. Projections of climate change go from bad to worse. *Science* 323 (March 2009):1546–1547.

Langbein, J., and R. Posner. 1980. Social investing and the law of trusts. *79 Michigan Law Review* 72.

Langbein, J. 2005. Questioning the trust-law duty of loyalty: Sole interest or best interest? *Yale Law Journal* 114:929.

Logan, A., and D. Grossman. 2006. *ExxonMobil's corporate governance on climate change.* Boston: Ceres.

Lovelock, J. E., and C. G. Rapley. 2007. Ocean pipes could help the Earth cure itself. *Nature,* September 27.

Mansley, M. 2003. *Open disclosure: Sustainability and the listing regime.* Friends of the Earth, February. www.foe.co.uk/resource/reports/open_disclosure.pdf.

Megarry, R. E. 1989. Investing pension funds: The Mineworkers' case. In T. G. Youdan (ed.), *Equity, fiduciaries and trusts.* Scarborough, Ontario: Carswell.

Minerals Management Service (MMS). 2006. *MMS updates Hurricane Katrina and Rita damage.* Minerals Management Service press release (Number 3486). May 1. www.mms.gov/ooc/press/2006/press0501.htm.

Munich Re. 2008. *Catastrophe report.* Munich: Author.

Prakash, S. 1974. *Up against the corporate wall: Modern corporations and social issues of the seventies.* Englewood Cliffs, NJ: Prentice-Hall.

Primedia. 2009. *KKR and EDF partnership helps companies save over $16 million while reducing emissions and waste.* Primedia press release, February 18. www.primedia.com/News/PressRelease/KKR_EDF.aspx.

Principles of Responsible Investing (PRI). 2006. *Principles of responsible investment.* New York: United Nations Environmental Programme Financial Institution.

Principles of Responsible Investing (PRI). 2008. *PRI report on progress 2008.* Paris: United Nations Environmental Programme Finance Initiative.

Richardson, B. J. 2008a. *Socially responsible investment law: Regulating the unseen polluters.* Oxford: Oxford University Press.

Richardson, B. J. 2008b. Putting ethics into environmental law: Fiduciary duties for ethical investment. *Osgoode Hall Law Journal* 46:243–291.

Scanlan, C. 2005. *Socially responsible investment: A guide for pension schemes and charities.* London: Key Haven.

Schwartz, P., and D. Randall. 2003. *An abrupt climate change scenario and its implications for United States national security.* Environmental Defense Fund. October. http://www.edf.org/documents/3566_AbruptClimateChange.pdf

Spieth, W., and M. Hamer. 2005. The implementation of the EU ETS directive in the EU member states in the light of the German national allocation plan. *Journal of European Environmental and Planning Law.*

Stern, N. 2007. *The economics of climate change: The Stern review.* Cambridge, UK: Cambridge University Press.

SustainableBusiness.com. 2009. *Insurance regulators adopt climate change disclosure.* SustainableBusiness.com, March 20. www.sustainablebusiness.com/index.cfm/go/news.display/id/17840.

Swiss Re. 1994. *Global Warming, Elements of Risk.* Zurich: Author.

Swiss Re 2002. *Opportunities and risks of climate change.* Zurich: Author.

Swiss Re. 2003. *Health effects of climate change.* Zurich: Author.

Swiss Re. 2007. *Global Climate Change: Swiss Re'sPrespective.* Zurich: Author.

United Nations Environment Programme Finance Initiative (UNEP FI). 2004. *The materiality of social, environmental and corporate government issues to equity pricing.* www.unepfi.org/work_streams/investment/materiality/mat1/index.html.

UNEP FI and Mercer Investment Consulting. 2007. *Demystifying responsible investment performance: A review of key academic and broker research on ESG factors.* Geneva: UNEP FI.

U.S. Department of Labor (DOL). 2008. Interpretative bulletin relating to investing in economically targeted investments. *Federal Register*, October 17.

Watchman, P. Q. 2008. *Climate change: A guide to carbon law and practice*. London: Globe Business Publishing.

West Landesbank. 2004. *Insurance and stability: Playing with fire*. Dusseldorf: Author.

BIBLIOGRAPHY

Acclimatise. 2009. *Climate finance business and community*. Acclimatise. www. acclimatise.uk.com/resources/reports.

Bowen, H. 1953. *Towards social responsibilities for the businessman*. New York: Harper & Row.

Bray, C., M. Colley, and R. Connell. 2007. *The credit risk impacts of a changing climate*. Barclays Environmental Risk Management and Acclimatise.

Carbon Principles. 2008. *Carbon principles*. http://carbonprinciples.org/.

The Climate Group. 2008. *The climate principles*. London: Author. www. theclimategroup.org/assets/resources/TCP_English.pdf.

Cogan, D. G. 2007. *Corporate governance and climate change: The banking sector*. Boston: Ceres.

Equator Principles Secretariat. 2008. *Equator principles*. www.equator-principles. com/principles.shtml.

Gerrard, M. B. (ed.). 2007. *Global climate change and U.S. law*. Chicago: American Bar Association.

Mercer Investment Consulting and Carbon Disclosure Project. 2009. *Carbon disclosure project*. February 11. www.cdproject.net/reports.asp.

Munich Re. 2005. *Weather catastrophes and climate change: Is there still hope for us?* Munich Re. www.munichre.com/en/ts/geo_risks/climate_change_and_insurance/weather_catastrophes_and_climate_change_is_there_still_hope_for_us/default.aspx.

REO Research. 2007. *In the front line: The insurance industry's response to climate change*. F&C Investments, September. www.climatewise.org.uk/storage/1250/co_gsi_climate_change_insurance_report.pdf.

Richardson, B. J. 2007. Do the fiduciary duties of pension funds hinder socially responsible investment? *Banking and Finance Law Review* 22.

Scott, A. W. 1967. *The law of trusts*. New York: Little Brown.

Swiss Re. 2009. *Natural catastrophes and man-made disasters 2008*. Zurich: Swiss Re.

Turner, A. 2008. *The economics of climate change*. February. www.ourplanet.com/imgversn/edge/edge_07_08/Lord Adair Turner.pdf.

UNEP FI and UK Social Investment Forum. 2007. *Responsible investment in focus: How leading pension funds are meeting the challenge*. Geneva: UNEP FI.

United Nations. n.d. *UN global compact*. New York: United Nations. www. unglobalcompact.org/.

U.S. Department of Labor (DOL). 1974. The Employee Retirement Income Security Act of 1974. Public Law 93-406, 88 Stat. 829, enacted September 2, 1974.

U.S. Department of Labor (DOL). 2000. *Meeting your fiduciary duties*. March 24. www.dol.gov/ebsa/publications/fiduciaryresponsibility.html.

CHAPTER 6

SRI or Not SRI?

Matthew J. Kiernan, PhD

There will be a large creation and re-distribution of shareholder value in the transition to a low carbon economy—there will be winners and losers at sector level, and within sectors at company level. The winners are more likely to be those businesses that take the time to understand and address this complex area.
—Tom DeLay, *Chief Executive,* "The Carbon Trust,"
Climate Change and Shareholder Value Report

For at least the next 5–10 years, climate change will affect sectors and stocks as a complex and enduring economic force. This should lead to market 'inefficiencies' and hence significant alpha opportunities for good stock pickers.
—Deutsche Asset Management

The arrival of Barack Obama in the White House brings the promise of tectonic policy changes in America's (and therefore the world's) energy and environmental policy and practice. Among other environmental initiatives, we have already seen a dramatic acceleration in U.S. regulatory, fiscal, and public investment initiatives to combat and mitigate climate change. President Obama and his environmental/energy "dream team" of Steven Chu, Carol Browner, Nancy Sutley, John Holdren, and Lisa Jackson are taking full advantage of a once-in-a-generation opportunity to align economic

---

Matthew J. Kiernan is co-head of the Sustainability Solutions Group at RiskMetrics Group.

development and job creation, energy security, environmental quality, and global competitiveness.

Reinforcing the major policy changes being driven from the White House will be the recent ascension of Henry Waxman to the chairman-ship of the powerful Energy and Commerce Committee of the House of Representatives. Mr. Waxman's views on energy, climate, and environment are completely aligned with those of Mr. Obama, and represent a 180-degree shift from those of his 30-year predecessor as chairman, the Detroit-based John Dingell. So, not only is there a new "climate sheriff" in the White House, but another one in the House of Representatives as well.

One of the first concrete opportunities that Team Obama had to demon-strate its new priorities and "green" credentials was the U.S. stimulus pack-age, touted by some as the "Green New Deal," and signed into law by the President in February of 2009. The legislation is formally known as the American Recovery and Reinvestment Act. While, by definition, the green component of the new package will be primarily driven by *public* funds (as much as $110 billion worth or more, depending on who's doing the counting), the team would be well advised to recall that it has another, even more potent weapon available to it: the mobilization of the enormous power of the *capital markets* to catalyze substantial changes in the business models, strategies, and performance of major corporations. These compa-nies, collectively, represent a major cause of the global climate challenge—as well as a significant proportion of its potential solution. As the providers of companies' "financial oxygen supply," the financial markets are in a perennially strong position to affect corporate behavior and strategy, and do so frequently. So far, however, insofar as issues such as climate change and environmental sustainability are concerned, this enormous power and potential have gone almost entirely unused.

Few environmental issues have created as real, significant, and widespread financial threats and opportunities for investors as climate change. International policy responses aimed at cutting greenhouse gas (GHG) emissions, together with both the competitive and direct physical impacts of climate change, will all require investors and money managers to take a much closer look at how their portfolios might be affected by company carbon risks and opportunities.

It should also be remembered that climate change is symptomatic of an even broader set of secular economic, political, and social change drivers that, taken together, could be accurately described as an eco-industrial revolution. Those changes collectively add up to nothing less than a global industrial restructuring; the entire basis of companies' competitive advantage is shifting, with an ever-higher premium being placed on companies' performance and strategic positioning on environmental and other "sustainability" issues. Climate change just happens to be—by light years—the *biggest* of those issues.

At least eight major global megatrends are driving this eco-industrial revolution:

- Dramatically increased levels of public and consumer concern and expectations for companies' performance on climate change (and other sustainability issues), turbocharged by unprecedented levels of information transparency with which to assess it.
- Accelerating natural resource degradation, scarcity, and constraints, driven to a significant extent by the explosive pace of industrial development, population growth, and urbanization, especially in rapidly growing and urbanizing emerging market economies such as those in Brazil, Russia, India, and China (the so-called "BRIC" countries).
- Accelerating economic interdependence internationally, so that economic, social, and political shocks occurring in any single region are more likely to reverberate globally.
- Tightening national, regional, and global regulatory requirements for stronger company performance and disclosure of "nontraditional" business and investment risks, notably climate change.
- The ongoing revolution in information and communications technologies (the Internet, YouTube, Facebook, webcasts, bloggers, etc.), which has enabled and accelerated the emergence of a stakeholder-driven competitive environment for companies with unprecedented transparency and, therefore, business risk.
- Growing pressures from international nongovernmental organizations (NGOs), armed with unprecedented financial and technical resources, credibility, access to company information, and global communications capabilities with which to disseminate their analysis and viewpoints.
- A substantial reinterpretation and broadening of the purview of legitimate fiduciary responsibility to include companies' performance on sustainability issues. Since there is now growing and incontrovertible evidence that superior overall environmental performance can in fact improve the risk level, profitability, and stock performance of publicly traded companies (Bauer, Derwall, Guenster, and Koedijk 2005; Gluck and Becker 2005), fiduciaries can now be seen to be derelict in their duties if they do not consider climate-driven risks and opportunities where they may be material.
- An institutional investor base that is increasingly sensitized to environmental and social (ES) issues, newly equipped with better information, and both willing and able to act on their concerns.

Going forward, investors and other fiduciaries would be well advised to assess their portfolios for "carbon risk," for at least four compelling reasons:

1. There is a growing body of evidence showing that superior performance in managing climate risk is a useful proxy for superior, more strategic corporate management, and therefore for superior financial performance and shareholder value-creation (Innovest Strategic Value Advisors [Innovest] 2007).
2. The considerable variations in carbon risk, performance, and strategic positioning among same-sector industry competitors are currently not transparent to, nor well understood by, mainstream Wall Street and city analysts. As a result, climate-driven risks and value potential remain—for the present, at least—almost entirely hidden from view.
3. In the longer term, the outperformance potential of companies with superior carbon management will become even *greater* as the capital markets become more fully sensitized to the financial and competitive consequences of climate change considerations.
4. There is strong evidence of dramatic increases in the level of institutional investor concern—and active intervention—concerning climate change issues and their investee companies.

The last of these four trends is perhaps best exemplified by the formation of at least four different groups of institutional investors: the Carbon Disclosure Project (CDP), the Investor Network on Climate Risk (INCR), the Institutional Investor Group on Climate Change (IIGCC), and the Investor Group on Climate Change (IGCC). The former is a global coalition of over 420 institutional investors, with combined assets of over $57 trillion; the second comprises over 50 U.S. institutional investors; INCR signatories include a number of U.S. state treasurers, as well as several leading labor funds with over $4 trillion in combined assets. The third organization includes over 35 of the leading institutional investors in Europe, and the fourth is a collaboration among concerned investors in Australia and New Zealand.

Unfortunately, however, one of the most potent weapons in the arsenal of those who would combat and mitigate the impacts of climate change has scarcely even been deployed to date. *That weapon is the mobilization of the enormous power of the capital markets to catalyze substantial changes in the strategies, business models, and performance of the major corporations that, collectively, represent a major cause of the problem—as well as a significant proportion of its potential solution.* As the providers of companies' "financial oxygen supply," the financial markets are in a perennially strong position to affect corporate behavior and strategy, and do so frequently. So far, however, insofar as climate change is concerned, this enormous power and potential has gone almost entirely unused.

A MISSED OPPORTUNITY

With over $57 *trillion* worth of institutional investor assets now *ostensibly* concerned about climate change as an *investment* risk, this represents a lost opportunity of epic proportions.[1] Perhaps somewhat controversially, I would *exclude* proxy voting and "engagement" with corporate emitters by investors from my personal definition of effective, substantive investor intervention. While there have indeed been some successes attributable to those techniques, Exxon Mobil's recent intransigence in the face of a *40 percent* shareholder protest over its climate strategy (or, more accurately, the lack thereof) is in fact far more typical. The fact that the protest was spearheaded by the descendants of the very family that *founded* the precursor of Exxon Mobil—the Rockefellers—only made their apparent refusal to act that much more poignant. No, personally, I set the bar for investor action much higher: it requires making concrete investment decisions at least somewhat differently than had they never even *heard* about climate change in the first place!

The "inconvenient truth" here is the fact that, while *awareness* of investment risk from climate change has clearly increased exponentially over the past five years, real *action* on the ground has been extremely limited. In my view, company engagement and shareholder activism are all well and good, but investors aren't paying *real* attention unless and until their actual investment decisions and *choices* are affected—and *known* by the corporates to be affected.

Even those rare investors who *have* changed their behavior seem to have trained 100 percent of their attention and efforts on only 1 percent of the opportunity set—*at most*. The overwhelming preponderance of "climate-savvy" investment has been concentrated in just two areas: clean tech (primarily in the private-equity space, but also with some smallish listed securities) and the carbon markets. While those two areas are undoubtedly both worthwhile and redolent with opportunity, for the typical institutional investor, those two asset classes represent no more than 1 percent of their total investment holdings. Yet, the last time I checked, *precisely the same regulatory, scientific, geopolitical, demographic, and competitive megatrends that are creating both the clean tech and carbon trading opportunities are also bearing down on the other 99 percent of the capital spectrum as well*. Ultimately, climate change will have just as much impact on the competitive prospects of mega-cap companies such as Royal Dutch Shell and Exxon Mobil as it will on those of a micro-cap, privately held, pure play fuel cell or wind turbine manufacturer. In my view, investors worldwide should start reflecting that fact in their concrete investment decisions across the *full* spectrum of their asset classes; today, almost none do so.

AFTER AWARENESS, THEN WHAT?

We have argued that, while institutional investor *awareness* of climate risk has increased dramatically (e.g., CDP), only a tiny handful have actually moved beyond rhetoric and shareholder resolutions to take concrete *investment* action—namely, incorporating climate risk considerations directly and systematically into their actual stock selection and portfolio construction processes. It is at that level—where the "rubber meets the road"—that investors can send the strongest signal to companies, produce significantly changed company behavior, and, most importantly, improve their own long-term, risk-adjusted financial returns. All things being remotely equal, any company capable of managing the complexities—and opportunity—of carbon risk better than its competitors will almost certainly turn out to be better managed and more "future-proof" overall. In short, it will simply be a better long-term bet for investors.

Historically, however, institutional investors—even foundations active in combating climate change on the program side—have been exceedingly slow to respond to climate risk. *We would currently estimate that, as a group, far less than 0.1 percent of the CDP signatories' $57 trillion in assets is currently invested in any investment strategy that explicitly and systematically takes climate risk into account.* (In the interest of full disclosure, Innovest Strategic Value Advisors, the firm the author formerly headed, wrote each of the first five annual global reports for the CDP.)

Historically, there have been at least four reasons for this intellectual and institutional inertia:

1. Investment professionals have long believed unquestioningly (albeit erroneously) that company resources devoted to environmental issues are either wasteful or even actually injurious to their competitive and financial performance, and therefore to those of their *investor* as well.
2. As a direct result, money managers, pension fund consultants, and even pension fund trustees have historically regarded explicitly addressing environmental factors in their investment strategies as incompatible with the proper discharge of their fiduciary responsibilities.

3. Until recently, there has been a dearth of robust, credible research evidence and analytical tools linking companies' environmental performance directly with their financial performance (Innovest 2007).
4. Climate change tends to be, quite inappropriately, lumped together with the litany of "socially responsible" investment issues, and thereby trivialized and delegitimized in the eyes of "professional," returns-driven investors.

This last point is of particular significance.

SRI OR NOT SRI?

There has been a highly unfortunate tendency for professional investors to confuse and conflate contemporary, sophisticated *sustainability* investment approaches with traditional "socially responsible" investment, or SRI. The differences between the two approaches are, in fact, quite stark and fundamental. Traditional SRI is primarily driven by *values*: investors view companies primarily through the prism of their own personal values: for or against contraception, alcohol, genetically modified foods, pornography, animal testing, company involvement in Sudan or Myanmar, and so on. Historically, financial returns have been only a secondary consideration; expressing one's personal values through investment practices has been paramount.

As a direct result, despite its considerable commercial success among individual investors, SRI has engendered considerable skepticism and rejection from *mainstream* institutional investors, for at least four reasons:

1. Values-based company exclusions, by definition, narrow investors' opportunity sets, and for nonfinancial reasons. In some cases, entire industry sectors such as mining and oil and gas are prohibited as having unacceptably high environmental or social impacts. Modern portfolio theory teaches us that this must inevitably lower the financial returns of the portfolio, increase the risks, or possibly both.
2. The research methodologies supporting SRI analysis, most of them heavily reliant on company responses to questionnaires, and generally devoid of serious competitive industry analysis and company-level financial analysis, are nowhere near sufficiently robust for returns-seeking investors and other fiduciaries.
3. There is a widespread (but largely erroneous) belief that traditional SRI funds have had outstandingly poor performance records.
4. Few public institutions are eager to set foot on the slippery slope of ethical exclusions. Drawing the line on what is and is not "ethical"

company behavior can be a highly contentious, no-win situation when, as is almost always the case, the beneficiaries are a heterogeneous group with widely diverging views. Moreover, many if not most of the ethical issues confronting investors are of a distinctly *gray* hue; the real world seems to contain an inconveniently small number of purely black-and-white situations.

Sustainability investing, however, is a different kettle of fish entirely; it is *all* about investment risk and return. Sustainability investors, including those taking climate change seriously, use the analysis of companies' performance and positioning on ES and other "nontraditional" issues as proxies and leading indicators to help identify better managed, more nimble, agile companies. Instead of expressing personal values, sustainability investing seeks to generate *financial* value through the application of a more robust, comprehensive, and forward-looking analytical framework. Let us be clear: environmentally aware investing is not the same as sustainability investing; it is, at least in our view, a subset thereof. We would define it as an investment approach that seeks improved risk-adjusted returns by addressing material environmental *and* social issues, *in addition to* more traditional investment factors. One good example of a "social" factor would be the quality of a company's ability to recruit, retain, and motivate superior talent.

Unfortunately, the penetration of mainstream investment thought and practice by sustainability considerations has been *severely* retarded by widespread confusion between these two fundamentally different philosophies and approaches. Of the two, SRI has a much longer pedigree and, as a result, sustainability investing (SI) has had enormous difficulty escaping the suspicion and even opprobrium created in the mainstream over the years by traditional SRI approaches and philosophies.

This has been unfortunate in the extreme, because both styles do share a common concern over issues such as climate change. What has separated the two camps has been their objectives, methodologies, and investment results. I can confirm from direct personal experience that, at the mere mention of the word *environment* or *social,* 99 percent of chief investment officers default to the presumption that one is about to discuss traditional (and therefore presumably underperforming) SRI funds. Their eyes then promptly either glaze over or actually roll back in their heads. This tendency is, unfortunately, particularly prevalent in North America.

The unfortunate result of this confusion has been that, for the most part, institutional investors have contented themselves with generalized hand-wringing at an abstract, "30,000-foot" level. Even their proxy voting and "engagement" activities with companies have been severely constrained by a lack of robust, granular analysis at the company level. With *same-sector* risk exposures varying by 30 times or more, generalizations are of precious little use to investors. Figure 6.1 illustrates graphically the enormous risk variations that exist among *same-sector* companies (and we use the term *risk* here as *net* risk, a function of the four major pillars of Innovest's Carbon Beta™ model, which we discuss later).

As we can see, in some cases 50 percent or more of individual companies' earnings before interest, taxes, depreciation, and amortization (EBITDA) could be threatened by climate risk. In order for investors to really take climate change seriously, however, at least four things will be required:

1. Compelling evidence that integrating climate risk analysis *can* in fact enhance risk-adjusted financial performance—in short, a robust *investment* case.
2. Compelling evidence that the variance in net climate risk exposure among companies is sufficiently large to warrant investor attention (see Figure 6.1).
3. A comprehensive and sophisticated analytical framework for assessing relative and absolute climate risk.
4. High-quality, financially oriented, company-specific information and analysis—an increasingly important information advantage and therefore potential source of alpha.

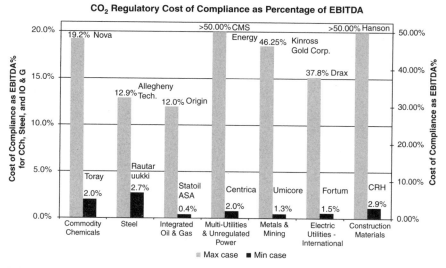

FIGURE 6.1 Risk exposures and costs vary widely, both between and within industry sectors.
Source: Innovest Strategic Value Advisors.

THE LIMITATIONS OF CURRENT APPROACHES

In my own view, investors cannot and should not make the all-too-common mistake of simply relying on companies' voluntary public disclosures about their "carbon footprint" as the sole or even primary basis for stock selection and portfolio construction. There are at least four major reasons for this:

1. At this stage, disclosed information remains notoriously unreliable, inconsistently reported, both across companies and over time, and generally not validated by independent third parties.
2. In today's regulatory environment—and for the foreseeable future—CO_2 emissions have very different regulatory, financial, and strategic implications depending on where they occur. All emissions are *not* created equal. Carbon footprints need to be analyzed on a granular, geographically sensitive basis.
3. Emerging empirical research suggests that the relationship between companies' carbon footprints and their financial performance is frequently a counterintuitive one—that is, contrary to widespread belief, the larger the carbon footprint, the *better* the financial returns often turns out to be.
4. Emissions data alone—if it is available at all, and even where it has been analyzed properly—provides less than 25 percent of the information a sophisticated investor requires.

   In order to investigate this relationship between disclosed information and financial performance in greater depth, our firm conducted a number of statistical tests. We compared the share price performance of CDP-based "disclosure leaders" to those of "disclosure laggards."[2] The results in the performance graph in Figure 6.2 should be somewhat unsettling for those placing undue reliance on purely disclosure-based analysis: *there was essentially no difference between the financial performance of disclosure leaders and laggards.* Simply put, it would appear that, whatever its other merits, *publicly disclosed information alone is an insufficient basis for achieving superior investment returns; the performance difference between the best and worst disclosures was essentially zero:*
   We believe that much more comprehensive and robust carbon finance models and analysis are required. At a minimum, such an analytical model should address the following factors.

   - Companies' overall carbon footprint or *potential* risk exposure, adjusted to reflect differing regulatory circumstances in different countries and regions.

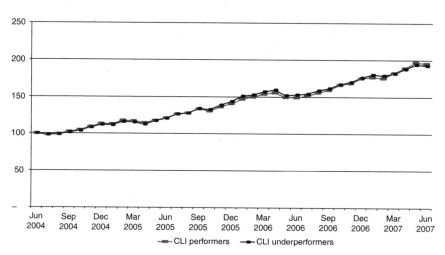

FIGURE 6.2 Comparison of total return performance of CDP disclosure leaders and laggards.
Source: Innovest Strategic Value Advisors.

- Their ability to manage and reduce that risk exposure.
- Their ability to recognize and seize climate-driven opportunities on the upside.
- Their rate of improvement or regression.

The *good* news, however, is that using a more sophisticated, multifactor analytical model, it appears that investors *can* in fact generate superior risk-adjusted financial performance.

LEVERAGING CARBON BETA IN THE EQUITY SPACE

International companies rated as top "carbon performers" using Innovest's, four-factor Carbon Beta model surpassed the return of companies rated as below average in the five year period from May 2004 to May 2009 *by an annualized rate of return of 2.66 percent* (a cumulative total return of 40.54 percent compared to 27.49 percent) (Innovest 2009). This is shown in Figure 6.3.

Not unexpectedly, these overall results mask a number of significant regional variations. As we would anticipate, the "carbon beta premium" was greatest in Europe. See Figure 6.4.

Similarly, overall results also tend to obscure the very different return premiums that exist in different industry sectors. As one would expect, the

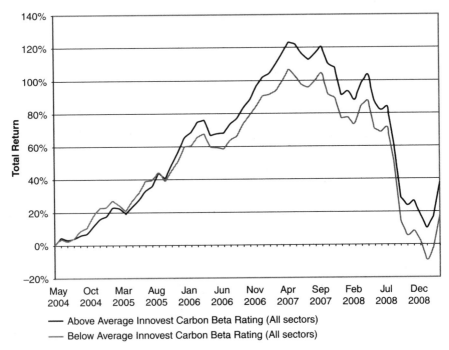

FIGURE 6.3   Carbon Beta results: leaders vs. laggards (globally).
Source: Innovest Strategic Value Advisors.

"carbon beta premium" is well above average in such high-impact sectors as electric utilities. See Figure 6.5.

As regulatory and competitive pressures for climate change mitigation and adaptation become even stronger, there is every reason to expect this "carbon beta premium" to become even larger going forward.

Carbon Beta in the Fixed-Income Markets

One concrete illustration of the purely financial benefits of factoring in climate risk exposure is provided by the recent performance of a live, "climate risk-adjusted" bond index jointly created by Innovest Strategic Value Advisors and JP Morgan. The index was built upon JP Morgan's conventional high-grade U.S. bond index (the JULI), with the weights "tilted" modestly to account for differences in the various companies' *net* exposures to climate risk. The thesis was that, unless companies' weights in the index had been adjusted by a sophisticated assessment of their net climate risk, the index could not possibly be "telling the whole truth" to investors.

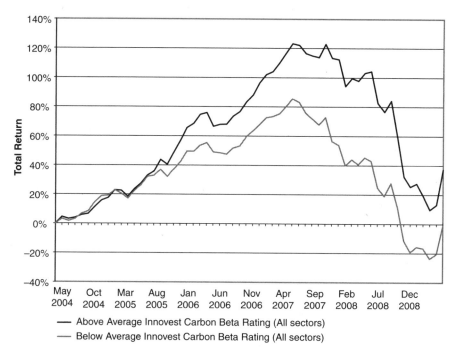

FIGURE 6.4 Carbon Beta results: leaders vs. laggards (Europe).
Source: Innovest Strategic Value Advisors.

After more than three years of live performance to date, this "carbon beta bond index" has generated over 120 basis points (1.2%) of annualized outperformance, despite having only a modest "tilt" toward carbon risk factors. See Figure 6.6.

THE CARBON BETA MODEL ITSELF

But what precisely constitutes a robust analysis of companies' climate risk exposure? At our firm, Innovest Strategic Value Advisors, we believe that there are at least four major dimensions:

1. Carbon risk exposure
2. Carbon management strategy
3. Strategic carbon profit opportunities
4. Improvement trend

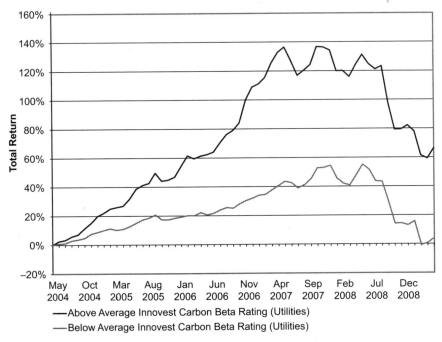

FIGURE 6.5 Corporate carbon risk and performance: utilities sector.
Source: Innovest Strategic Value Advisors.

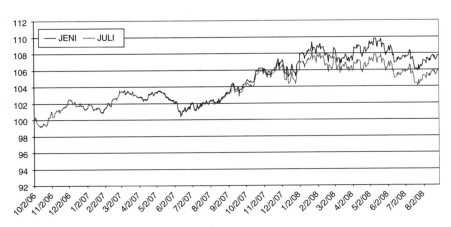

FIGURE 6.6 JENI Carbon Beta Index vs. JULI Liquid.
Source: Innovest Strategic Value Advisors.

In addition, each of these four dimensions can be influenced (to greater or lesser degrees) by four sets of factors:

1. Industry sector exposures
2. Geographic risk exposures
3. Company-specific factors
4. "Carbon financials"—the likeliest scenarios for an individual company's cost of compliance with current and/or emerging regulatory requirements

Let us examine each of these four factors in turn.

Industry Sector Exposures

In order to identify those industry sectors that are the most exposed to climate change risks and opportunities, Innovest has developed a three-stage approach to rate the specific risks of sectors along their entire value chain: upstream, internal, and downstream. A composite climate change (CC) intensity factor is derived from the three different categories of "carbon intensities":

1. *Climate Change Direct Intensity.* This indicator (1 = lowest exposure, 5 = highest exposure) captures the sector's exposure to carbon regulations and constraints (offsets, capping, bubbles, energy taxes, and other regulatory instruments). The CCDI is directly proportional to a sector's direct emission of CO_2 and other GHGs, that is, the emissions from its own physical assets (e.g., steel making is a strong emitter of CO_2 inherent in the coking process).
2. *Climate Change Indirect Intensity.* This indicator (1 = lowest exposure, 5 = highest exposure) captures the sector's sensitivity to increases and volatility in upstream energy costs as a result of a carbon-constrained economy. The CCII is directly proportional to a sector's consumption of electricity and other inputs that have caused substantial carbon emissions from their production or extraction, that is, the emissions from the suppliers' assets (e.g., aluminum making requires a large amount of electricity, which in turn may produce large emissions of CO_2 if reliant upon fossil fuels).
3. *Climate Change Demand Sensitivity.* This indicator (1 = lowest exposure, 5 = highest exposure) captures the sector's market sensitivity to climate change drivers. High-sensitivity sectors include sectors producing goods that engender large GHG emissions during the life use of these goods (e.g., oil and gas products, cars and trucks), those sectors whose

invested assets can contribute to high or low carbon emissions (e.g., the finance and insurance industries), as well as those sectors having strong carbon-related opportunities on the upside (e.g., energy generation technology manufacturers).

4. *Combined Climate Change Intensity.* This indicator (1 = lowest exposure, 5 = highest exposure) is a weighted average of the three indicators above, and reflect our firm's judgment as to the relative risk exposure of the sectors along the entire value chain.

Applying that methodology to the 60+ industries of the MSCI Global Industry Classification Standard (GICS), Innovest has determined the most exposed industries, that is, those industry sectors that have the highest average carbon exposure in terms of potential impact to net earnings, as well as those sectors offering the highest differential of exposure (Table 6.1).

Geographic Risk Exposures

The second key determinant of a company's carbon risk is the geographic distribution of its carbon assets and liabilities.

TABLE 6.1 Six of the Highest Carbon Impact Sectors

GICS Code	Industry	CC Direct Intensity (In-house)	CC Indirect Intensity (Upstream)	CC Demand Sensitivity (Downstream)	CCC Intensity (Combined Intensity)
551010	Electric Utilities	5	4	5	4.9
151020/ 201020	Construction Materials and Building Products	5	4	3	4.3
101020	Oil, Gas & Consumable Fuels	4	3	5	4.2
151050	Paper & Forest Products	4	3	5	4.2
151010	Chemicals	5	5	2	4.1
201050/ 201060	Machinery & Industrial Conglomerates	4	3	4	3.9

Source: Innovest Strategic Value Advisors.

Because of regional differences in approaches to Kyoto and other regulatory requirements, as well as the natural variations in climate conditions, the geographic distribution of a firm's operations and markets is a critical determinant of carbon risk. Investors heavily exposed to GHG-intensive sectors in regions aggressively pursuing emissions reductions—the European Union (EU), Japan, parts of the United States, and several provinces in Canada—will clearly face greater carbon finance risks than those with more carbon-diversified portfolios. (See Figure 6.7.) However, the threat of climate-related litigation hangs over U.S. emitters much more than probably any other.

To assist in carbon risk profiling, Innovest has developed the concept of the Weighted Average Country Carbon Reduction Target (WACCRT^{TM}), which represents the aggregate extent of emissions reductions over the full range of a firm's industrial activities. For example, Honda's emission-generating operations are divided among Europe (5.6 percent), the United States (18 percent), Canada, (ca. 5 percent), Japan (14.6 percent) and rest of world (57.5 percent). Under Kyoto, the regional emissions reduction requirements (below 1990 levels) are as follows: EU (8 percent), United States (0 percent—not a signatory), Canada (6 percent), rest of world (0 percent). Based on the WACCRT model, the company's overall emissions reduction obligations could be: Europe (5.6 percent × 8 percent) + United States (1.8 percent × 0 percent) + Canada (5 percent × 6 percent) + Japan (14.6 percent × 6 percent) + rest of world (57.5 percent × 0 percent) = 1.62 percent.

Note that companies are very likely to have increased emissions since 1990 if pursuing a "business as usual" course, though some may have started

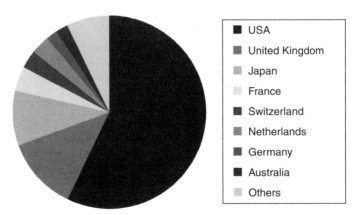

FIGURE 6.7 Geographic exposures of a typical global equity portfolio.
Source: Innovest Strategic Value Advisors.

mitigation efforts. Thus, the actual required reduction would be a much higher percentage than shown above.

Company-Specific Factors

Individual companies' financial exposures are essentially a function of eight company-specific key factors:

1. Energy intensity and source mix and consumption patterns.
2. Geographic locations of production facilities relative to specific regulatory and tax liabilities and compliance schedules in different countries.
3. Product mix—direct, indirect, and embedded carbon intensity.
4. Company-specific "marginal abatement" cost structures: Some companies can reduce emissions at much less cost than others.
5. Technology trajectory—level of progress that a company has already made in adapting/replacing its production technologies for a carbon-constrained environment.
6. Industry competitive dynamics—ability/inability of companies to pass on costs to consumers.
7. Company-specific risk management capability.
8. Ability to identify and capture upside and revenue opportunities, including new manufacturing cost efficiencies, new product/service opportunities, and emissions trading.

Carbon Finance: The Compliance Cost Model

The costs of adapting to the new environmental and regulatory carbon-constrained reality will become increasingly critical to the investment decision process. Innovest has developed a proprietary compliance cost model that estimates, as a percentage of EBITDA, the current or potential exposure a company has when complying with emissions' restricting regulations.

The elements comprising our firm's compliance cost model are:

- *Weighted Average Country Carbon Reduction Target.* The WACCRT refers to the expected emissions reduction targets according to applicable legislation where a company has relevant assets, domestically and internationally. In this sense, the metric shows a weighted average for the restrictions that a company faces in the countries and regions in which it operates during the mandated compliance period.
- *Industry discount rate.* The industry discount rate is calculated from the weighed average cost of capital (WACC) for each specific industry as of January 2008. This is calculated using the weighted average of the

cost of equity and after-tax cost of debt, weighted by the market values of equity and debt (using the cumulative market values for the entire sector for the weights).

- *Carbon cost.* Carbon cost is the weighed price for three different scenarios (expected, maximum, and minimum price) per emission allowances ($ per ton of CO_2 equivalent) in the European Union Emissions Trading Scheme (EU ETS) during a specific compliance period. In the case that this period extends beyond 2012, prices are estimated using available data for actual EU allowances prices from January 2005 to the present, and future prices from the present to 2012.

- *Net present value of meeting emissions reduction targets.* To calculate this figure we estimate the abatement compliance cost for each year of the commitment period. In the case of companies operating in countries that are Parties to Kyoto Protocol, the commitment period goes from 2008 to 2012. Previous commitment periods, as EU ETS Phase I, are taken into account on a case-by-case basis. For countries that currently have or are likely to have different climate change policies to the Kyoto Protocol, specific domestic legislation or possible scenarios are modeled by calculating the net present value of the compliance cost.

WHITHER CARBON FINANCE?

As we have seen, a robust and comprehensive assessment of companies' *net* risk exposure to climate change is not a trivial undertaking. We firmly believe, however, that it has already been demonstrated convincingly that the financial rewards for investors can be substantial. Perhaps even more important, the "carbon beta premium" is almost certain to become even larger going forward.

Clearly, the analytical model proposed here is only one of myriad possible approaches. The most *important* thing, however, is that *some* form of sophisticated carbon risk analysis actually be undertaken. In the absence of such analysis, investors are truly "flying blind"—largely if not entirely unaware of what can be same-sector risk differentials of *30 times* or more.

This is not, of course, to imply that investors should simply gravitate willy-nilly to the low-carbon company alternatives. Clearly, there are dozens of other investment factors at play. What I *do* insist upon, however, is this: Let's say that investors and fiduciaries normally examine 20 different investment factors and variables before making a final decision. Here's all I ask: they should find a way, somehow, to make room for a 21st—the selfsame factor that the World Economic Forum and others have rightly described as "the business issue of the millennium"—climate change!

To repeat a distinction that we attempted to make earlier, environmental investing is *not* the same thing as sustainability investing, although the two are frequently confused. The pursuit of "environmental alpha" is, by design, focused on a narrower range of "nontraditional" issues. By concentrating on more readily identifiable and quantifiable factors and indicators, environmental investing is arguably an even *more* compelling value proposition.

ACKNOWLEDGMENTS

The author gratefully acknowledges the technical contributions to this article of two senior colleagues at Innovest Strategic Value Advisors: Pierre Trevet and Mario Lopez-Alcala. Both were instrumental in the development of the Carbon Beta model, which is discussed at length in this chapter.

NOTES

1. The $57 trillion figure represents the combined assets under management of more than 400 institutional signatories to the Carbon Disclosure Project, arguably the largest investor collaboration ever assembled.
2. In technical terms, a parameter stability test was conducted to test for any financial performance differences. Regression coefficients were utilized with the two sets of time series. The results are valid at a confidence interval of 99 percent.

REFERENCES

Bauer, R., J. Derwall, N. Guenster, and K. Koedijk. 2005. The eco-efficiency premium puzzle in the U.S. equity market. *Financial Analysts Journal* 61(2):51–63.

Gluck, K., and Y. Becker. 2005. The impact of eco-efficiency alphas. *Journal of Asset Management* 5(4).

Innovest Strategic Value Advisors. 2007. *Carbon Beta and equity performance: An empirical study.* Innovest Strategic Value Advisors. http://www.riskmetrics.com/system/files/private/2007carbon_beta_equity_performance.pdf

BIBLIOGRAPHY

Kiernan, M. 2008. *Investing in a sustainable world.* New York: AMACOM.

Labatt, S., and R. White. 2007. *Carbon finance: The financial implications of climate change.* New York: Wiley.

Tang, K. 2009. *A guide to carbon finance.* London: Risk Books.

Environmental Alpha: The Investment Case for Climate-Related Strategies

Taxonomy of Environmental Investments

Angelo A. Calvello, PhD

A s should be clear by this point, climate change presents investors with both risks and opportunities. This chapter focuses specifically on the investment opportunities resulting from climate change, and in so doing seeks to provide the bridge between the critical background issues related to climate change and fiduciary responsibility and the specific categories of environmental investments and related implementation issues. Accordingly, the chapter covers four main topics:

- Drivers of returns
- Categories of environmental investments
- Challenges—risks and idiosyncrasies
- Benefits—environmental alpha

The definition of environmental investing is developing before our eyes, partly because of the newness of the inquiry and partly because of the nature of the subject matter itself. Some have chosen to define environmental investing by the content of the opportunity set. For example, Deustche Bank chooses to delineate the opportunity set:

> We then define the climate change investment universe to include all companies that provide any of a diverse range of goods and services that further mitigation or adaptation to climate change. We have identified four broad sectors: (i) Clean energy, (ii) Environmental

Angelo A. Calvello is the founder of Environmental Alpha.

resources management including agriculture and water, (iii) Energy and material efficiency, and (iv) Environmental services. (DB Advisors 2008)

The consulting firm Rogerscasey takes a similar but more circular approach:

Green investing generally targets environmentally oriented investments or those that are deemed to have a positive or neutral environmental impact. This can include:

> *Companies whose products and services are designed to have a positive contribution to the environment.*
>
> *Companies that take advantage of trends that reward mitigating environment-related risks in the future.*
>
> *Companies who, through their operational practices, work to reduce their overall environmental footprint. (Zraikat 2008)*

Both are certainly thoughtful definitions but environmental investing is simply too big and too dynamic to be defined by an ever-changing opportunity set. To avoid the limitations associated with such functional definitions, I have chosen to define environmental investing according to the drivers of return instead of content of the opportunity set. Environmental investing is *investing directly or indirectly in assets whose values are affected primarily by climate change.* This approach results in a slightly broader definition than others have offered.

This definition differs from others also because it fully recognizes that environmental investing is a type of thematic investing and its central theme that underlies environmental investing is climate change.

CLIMATE CHANGE: A CLUSTER CONCEPT

The climate change theme can be narrowly defined as follows:

The scientific community has reached a strong consensus regarding the science of global climate change. The world is undoubtedly warming. This warming is largely the result of emissions of carbon dioxide and other greenhouse gases from human activities including industrial processes, fossil fuel combustion, and changes in land use, such as deforestation. (Pew Center on Climate Change 2009)

To solve the climate change problem, we need to mitigate greenhouse gas (GHG) emissions to a specific level within a defined time period and adapt to the possible consequences of climate changes already in the pipeline. Mitigating and adapting to climate change will require changes in behavior and the development and dispersion of an evolving portfolio of market-based technologies. However, it is critical that readers understand that the climate change theme is actually much bigger than simply "global warming." In this book, climate change is a "cluster concept," encompassing both the previously mentioned narrow definition and such direct environmental topics and broader social issues as (Kiernan 2009):

Environmental Issue	Social Issues
Water quality	Energy supply
Air pollution	National security
Waste management	Human development
Deforestation and land degradation	Population growth
Chemical and toxic emissions	Global changes in demographics
Biodiversity loss	Poverty and income disparity
Depletion of the ozone layer	Public health
Quality of fisheries and oceans	Human rights
	Labor rights

A note from the German Advisory Council (2007) expresses this idea succinctly:

> ... *climate change could exacerbate existing environmental crises such as drought, water scarcity and soil degradation, intensify land-use conflicts and trigger further environmentally-induced migration. Rising global temperatures will jeopardize the bases of many people's livelihoods, especially in the developing regions, increase vulnerability to poverty and social deprivation, and thus put human security at risk. Particularly in weak and fragile states with poorly performing institutions and systems of government, climate change is also likely to overwhelm local capacities to adapt to changing environmental conditions and will thus reinforce the trend towards general instability that already exists in many societies and regions. In general it can be said that the greater the warming, the greater the security risks to be anticipated.*

The implication for investors is simple: Investors cannot reduce "climate change" to simply "global warming." Climate change is a socio-scientific phenomenon that transcends changes in global temperature. Investors need to think broadly about both the risks and opportunities associated with climate change.

Let's look at coal as a specific example of this idea of interrelatedness. Embedded in the discussion of coal as a power source are a multitude of interrelated issues. Table 7.1 juxtaposes some of the above issues with coal-related topics.

In order to fully understand the risks and opportunities arising from climate change and to properly assess and access the environmental investment universe, investors should understand the scientific definition while recognizing the cross-disciplinary meaning of climate change.

More importantly, investors should recognize that "climate change" is the concept chosen to express this cluster of topics because while each of the related topics represents a critical issue, only climate change comes with the exigency of time. Action certainly needs to be taken with regard to the other issues, but only climate change *demands* that timely action be taken if we are to avoid an ecological "tipping level ... a measure of the long-term climate forcing that humanity must aim to stay beneath to avoid large climate impacts." (Hansen et al. 2008) Said another way, where the other topics might come with a moral imperative, climate change comes with a temporal imperative.

It is this time driver that makes climate change the "mother of all themes." Investors cannot reduce "climate change" to simply "global warming." Climate change is a socio-scientific phenomenon that transcends changes in global temperature. Investors need to think broadly about both the risks and opportunities associated with climate change.

THE DRIVERS OF RETURNS

The climate change theme, like all other investment themes, is impacted by macroeconomic and broad sociopolitical issues, but the specific thematic factors affecting environmental investments—the drivers of returns—can be reduced to the following four climate change–related factors in Figure 7.1:

- Science
- Economics
- Policy and regulation
- Technology

Science

Professor Betts explains the science of climate change in detail in Chapter 1, so I will summarize it specifically as it relates to environmental investing. In brief, solar radiation passes through the clear atmosphere. Most of this

TABLE 7.1 Coal as an Example of Climate Change as a Cluster Concept

Issue	Relationship to Coal
GHG emission	"Coal plants represent the single biggest source (about one-third) of the U.S. share of CO_2 emissions—about the same as all of our cars, trucks, buses, trains, planes, and boats combined." (Massachusetts Institute of Technology [MIT] 2007)
Energy supply	Coal is the lowest-cost fossil fuel source for base-load electricity generation and, in contrast to oil and natural gas, is widely distributed around the world. MIT 2007)
National security	Countries with an abundance of coal are provided a certain amount of energy independence and security. The U.S. military certainly recognizes this: "Air Force officials said the [coal plants that convert domestic coal into a synthetic fuel] plants could help neutralize a national security threat by tapping into the country's abundant coal reserves." (Associated Press 2008)
Water quality Air pollution Chemical and toxic emissions Public health Depletion of the ozone layer Quality of fisheries and oceans	"The burning of coal releases more than 100 pollutants into the atmosphere. It is the largest source of sulfur dioxide emissions (which cause acid rain), the second largest source of nitrogen oxides (which contribute to smog and asthma attacks), and the largest source of fine soot particles (which contribute to thousands of premature deaths from heart and lung disease yearly). Coal plants are also the largest remaining source of human-generated mercury, which contaminates lakes and streams, the fish that live in them, and anyone who eats those fish." (Freese, Clemmer, and Nogee 2008)
Water quantity Quality of fisheries and oceans	"Cooling and scrubbing coal plants requires copious volumes of water. (A typical 500-megawatt coal-fired power plant draws about 2.2 billion gallons of water each year from nearby water bodies, such as lakes, rivers, or oceans, to create steam for turning its turbines. This is enough water to support a city of approximately 250,000 people.) Power plants in general are responsible for approximately 39 percent of U.S. freshwater withdrawals, second only to agricultural irrigation. While most of that water is returned to the source, the act of withdrawal kills fish, insect larvae, and other organisms, and aquatic ecosystems are further damaged by the return of water that is both hotter than when it was withdrawn and contains chlorine or biocides added to protect plant operations." (Freese et al. 2008)

(Continued)

TABLE 7.1 (*Continued*)

Issue	Relationship to Coal
Deforestation and land degradation Biodiversity loss Human rights	"Mountaintop removal mining in Appalachia permanently destroys mountains and adjacent valleys, has destroyed hundreds of thousands of acres of forests, and has buried more than 700 miles of some of the most biologically diverse streams in the country." (Freese et al. 2008)
Waste management Public health Human rights	"Coal mining and combustion both create wastes that must be disposed. Combustion results in more than 120 million tons of fly ash, bottom ash, boiler slag, and sludge from air pollution controls annually—roughly the same amount as all municipal solid waste disposed in U.S. landfills each year. Though uses have been found for some of this material, most of it goes into landfills and surface impoundments, from which mercury, lead, cadmium, arsenic, and other toxic constituents of this waste can leak out and contaminate water supplies." (Freese et al. 2008)
Labor rights	"Coal mining remains a dangerous occupation, with fatality rates at least five times higher than the average for all private industries." (Freese et al. 2008)

radiation is absorbed by the Earth's surface, warming the surface. Some solar radiation is reflected (reradiated) by the Earth and the atmosphere in the form of heat (infrared radiation). GHGs such as carbon dioxide (CO_2), nitrous oxide (N_2O), methane (CH_4), halogenated fluorocarbons (HCFCs), ozone (O_3), perfluorinated carbons (PFCs), hydroflourocarbons (HFCs), and water vapor absorb this infrared radiation and trap the heat in the atmosphere.

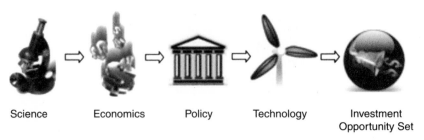

Science   Economics   Policy   Technology   Investment Opportunity Set

FIGURE 7.1 Drivers of return.

If the atmospheric concentrations of GHGs remain relatively stable, the amount of energy sent from the sun to the Earth's surface should be about the same as the amount of energy radiated back into space, leaving the temperature of the Earth relatively constant. If the amount of GHG concentrations (CO_2 equivalent—CO_2e) in the atmosphere increases, then we could likely expect average atmospheric temperatures to increase.

Here's where it gets interesting. Scientific inquiry shows that GHGs have been increasing, with CO_2e rising from 280 parts per million (ppm) in preindustrial times to about 380 ppm today (Intergovernmental Panel on Climate Change [IPCC] 2007).

In spite of what one former vice presidential candidate might say, there is strong evidence that this increase is the result of human activity; as Betts makes clear in Chapter 1, "we are certain that the observed increase in carbon dioxide in the atmosphere has resulted from a combination of burning fossil fuels and deforestation." The Nobel Prize–winning IPCC in its Fourth Assessment Report extends this conclusion to changes in temperature: "... most of the observed increase in globally averaged temperatures since the mid-twentieth century is very likely due to the observed increase in anthropogenic greenhouse gas concentrations" (IPCC 2007).

Why is this increase in temperature important? It's important because higher temperatures lead to adverse and irrevocable consequences like more severe weather events (drought, storms, etc.), melting of sea ice, loss of species, loss of glaciers, rise in sea levels, and changes in distribution of some disease vectors.

This scientific understanding of climate provides the opportunity to determine a response that might keep GHGs at "acceptable levels" and limit and possibly minimize such adverse impacts. This response is typically framed in terms of the stabilization of emissions or tipping levels (Hansen et al. 2008). The question becomes: at what level can we stabilize GHG emissions to limit the impact of climate change? Using a variety of models and data, the scientific community arrives at different conclusions. For example, numerous studies estimate that if we want to keep the increase in equilibrium temperature to 2 degrees Celsius or about 420 ppm, we need to reduce emissions worldwide by 80 percent by 2050; to limit GHGs to 450 ppm, we would have to reduce emissions worldwide by 50 percent from today's levels by 2050; and to limit GHGs to 500 ppm, we need to hold global emissions constant for 50 years (Pacala 2008).

But let's be clear about one thing. It's a foregone conclusion that we cannot simply stop GHGs and temperatures from rising. CO_2 has an atmospheric life of about 50 years (and probably four times that in the ocean), so there are climate change consequences "in the pipeline" that we cannot avoid. Also, the global economy will be powered by carbon-based technologies for years to come, especially given expected increases in global population and demographic changes occurring in the developing world.

Economics

While science provides us with scenario analyses, economic analysis allows us to calibrate the two sets of costs climate change imposes on society: the cost of adapting to climate change's expected consequences, like extreme weather events, changes in the hydrological patterns, rising sea levels, lower crop yield, and increased diseases; and the cost of mitigating GHGs to reach any emission target—"the costs of developing and deploying low-emissions and high-efficiency technologies and the costs to consumers of switching spending from emission-intensive to low-emission goods and services" (Stern 2007).

Policy

This understanding of costs allows governments and other organizations to create policies and regulations that set a clear price for carbon and provide incentives and structures to persuade businesses and individuals to reduce GHG emissions from both energy and nonenergy sources (mitigation) and prepare for changes in climate that are likely to occur regardless of our future actions (adaptation). The most obvious example of such policies is the Kyoto Protocol, which tried to impose regulations across a large number of industrialized or industrializing countries by providing a sense of how much carbon each could emit. From an investment perspective, it is critical that these policies be as unambiguous as possible, consistent across jurisdictions, long term, and enforceable, so capital could be committed with reasonable confidence.

Technology

Climate change policies will clearly result in changes in economic activity that could include:

- Reducing non–fossil fuel emissions, particularly from land use, agriculture, and fugitive emissions
- Reducing demand for emission-intensive goods and services
- Improving energy efficiency, by getting the same outputs from fewer inputs
- Switching to technologies that produce fewer emissions and lower the carbon intensity of production (Stern 2007)

In a word, "the policy elicits the technology. The interactions are fundamental." (Robert Socolow, quoted in Gertner 2008)

As should be obvious at this point in the book, GHGs "are produced by a wide range of activities in many sectors, so it is highly unlikely that any single technology will deliver the necessary emissions savings" (Stern 2007). Instead of a single silver technological bullet, our response to climate change will require a portfolio of technologies that mitigate GHG emissions (think of efficient lighting; improved insulation; more fuel efficient vehicles; technologies that use solar, wind, geothermal, and tidal and wave energy) and help us adapt to the consequences of climate change (such as desalination, digital controls, seawalls, and storm surge barriers). This portfolio of technologies gives rise to a broad array of environmental investment opportunities.

So when thinking about environmental investing, investors would do well to recognize the four drivers of returns shown in Figure 7.1 that produce environmental investment opportunities.

Other Characteristics

The other characteristics of the climate change theme are:

Long term. The nature of GHGs, scientific inquiry, policy development and implementation, research, development, demonstration, and deployment (RDD&D), the scale required for effective deployment of transformative and adaptive technologies—all of these and other features make climate change a long-term, multidecadal theme. As such, it transcends periodic economic cycles, which fits well with the investment horizon of many fiduciaries.

Secular change. Climate change is similar to other major themes (e.g., burgeoning global emerging middle class, information revolution) in that it represents a major transgenerational economic and societal change, one that cannot be reduced to simply a trend or fad.

Global. While mitigation and adaptation efforts will be local and the consequences of climate change will be experienced locally, climate change itself is a global, transnational phenomenon.

Nonconforming. "Climate change is also one of the most financially significant environmental issues facing investors today. While other environmental risks can be highly relevant to specific sectors, climate risk distinguishes itself through its widespread potential for impact on individual companies, across sectors and whole economies" (Mercer Investment Consulting 2005). Climate change cannot be reduced to a "sector play" because as an investment theme it cuts across:

Asset classes and sub–asset classes

Sectors

Geographies

Currencies

Public and private securities

Investment strategies and structures

Styles of investing

Investment approaches

This nonconforming attribute matches up quite well with the "universal owner" aspect of institutional investors.

CATEGORIES OF ENVIRONMENTAL INVESTMENTS

Environmental investments break down into five major categories:

1. Carbon
2. Financing solutions for land use, land-use change, and forestry (LULUCF)
3. Clean technology (clean tech)
4. Sustainable property
5. Water

This classification is selected in large part because it captures the relationship between the problem (GHG emissions) and the technologies and activities that attempt to mitigate GHG emissions or adapt to the consequences of climate change.

Carbon is a major category for several reasons. First, as we decarbonize economies, carbon will be the currency by which we measure the efficacy and success of market-based solutions. In a word, the price of carbon tells us what it costs to emit GHGs. Second, as will be discussed in detail in Chapter 9, a large global carbon market already exists, with over $118 billion in total transactions occurring in 2008, representing about 4 billion tonnes of carbon allowances (New Carbon Finance 2009). The notional value of this market already makes it one of the largest commodity markets in the world. This market is expected to grow in importance and size as the Obama administration implements its plans to put a price on carbon either through

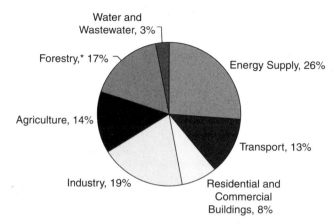

FIGURE 7.2 Global Anthropogenic GHG emissions by
sector (2004).
*Forestry includes deforestation.
Source: IPCC Assessment Report 4 (2007), Summary of
Policymakers: Figure SPM 3.

a carbon tax or a cap-and-trade system. This policy action will transform
carbon into what David Gardiner calls in Chapter 3 "a new asset class."

LULUCF, clean tech, and sustainable property qualify as major categories because they represent what Vinod Khosla (2009) calls the "relevant
scale solutions" that attempt to mitigate emissions from almost all of the
sources of global anthropogenic GHGs. (See Figure 7.2.)

We must mitigate GHG emissions from these sources (which include
power, industry, transportation, built property, and deforestation) if we
are to have any chance of achieving the policy targets. Moreover, beyond
sectorial GHG emissions, LULUCF, clean tech, and property have broad
societal, economic, and environmental impacts.

Consider forestry's broad impact:

> *In addition to regulating climate, forests provide a number of im-
> portant local services that can reduce communities' vulnerabilities
> to climate change. Forests are rich in biodiversity: they are home to
> the majority of terrestrial species. They regulate water flow; reduce
> runoff, erosion, siltation, and flooding; and provide food, medicine,
> building materials, fuelwood, and income sources for local commu-
> nities. These ecosystem services are critical to many rural and urban
> economies, provide environmental security, and are of heightened
> importance in adapting to climate change. The services provided by*

forest ecosystems can be thought of as the "natural insurance" that help buffer vulnerable communities against the negative impacts of climate change. Developing countries are projected to encounter some of the most severe impacts of climate change and are least able to cope. Losing forests could further destabilize societies that climate change may make vulnerable to political upheaval, migration, and conflict (Nicholas Institute for Environmental Policy Solutions 2009).

Similarly, property comprises 14.2 percent of the $10 trillion U.S. gross domestic product (GDP) (U.S. Department of Energy 2008) and that the property market's environmental footprint is substantial. Buildings in the United States account for:

- 39 percent of U.S. primary energy use
- 70 percent of U.S. consumption of electricity
- 12 percent of all potable water and 30 percent of raw materials use
- 30 percent of GHG emissions (Nelson 2007)

Water qualifies as a major investment category because it is a primordial and scarce resource and because "freshwater is one of the Earth's resources most jeopardized by changing climate" (Williamson and Schindler 2009). The possible impact of climate change on water quality and quantity will require broad adaptation in the form of new technologies. These technologies will present investors with a broad range of investment opportunities.

Additionally, all five categories share a common characteristic: All represent a nexus of climate change and investment opportunities, and more specifically, this nexus gives rise to a portfolio of *market-based, alpha-centric solutions* to the problem of climate change.

Returning to the categories themselves, each could be described more granularly. For example, clean tech could be broken down as in Table 7.2.

TABLE 7.2 Breakdown of Clean Technologies

Clean Tech/Electricity	Clean Tech/Alternative Fuels	Other Clean Technologies
Solar	Biodiesel	Nanotechnologies
Hydro	Ethanol	Battery storage
Geothermal	Cellulosic ethanol	Energy efficiency technologies
Wind	Hydrogen	Carbon sequestration

Source: Readey, Meek, Perencevich, and Hadzima 2007.

Additionally, there could be broad overlap among the major categories because technologies have applications in more than one category (e.g., environmental property investments would include clean technology investments; financing solutions for LULUCF are intrinsically linked to carbon markets). This could result in investment strategies that are "pure plays" (e.g., carbon-only funds) and broadly diversified and environmentally focused (like climate change equity funds).

CASE STUDY: Finding Environmental Alpha in All of the Usual Places

By Simon Webber

Climate change is an important issue that should be factored into virtually every investor's long-term investment strategy and asset allocation. This is particularly the case for pension funds that have liabilities over long periods, given our conviction that climate change will require a decades-long transformation process of the way the world does business and the economy is organized. The key, in our view, is selecting an actively managed strategy that provides genuine diversification and excess return yet captures the long-term implications of climate change.

The Schroder ISF[†] Global Climate Change Equity strategy is designed to achieve these goals by targeting a wide range of publicly traded companies in a variety of industries that either help mitigate the impact of climate change or help the world adapt to changes that are by now possibly irreversible (see table).

Selection of Industries Impacted by Mitigation and Adaptation

Mitigation

Mitigation Solutions	Examples	Industries Impacted
Energy efficiency	Reducing energy wastage	Autos, Electronic and Electrical Equipment, Industrial Engineering Materials
Shift from high-carbon to low-carbon fuels	Decreasing dependence on coal and oil	Natural Gas, Nuclear, Materials
Renewable technology	Carbon-free or low-carbon alternatives to fossil fuels	Wind, Solar, Hydro, Tidal, Geothermal, Fuel Cells, Biofuels
Carbon capture and sequestration	Physically removing or storing carbon dioxide	Oil and Gas, Industrial Engineering, Forestry

Simon Webber is portfolio manager at Schroders Global Climate Change Strategies.
[†]Schroder ISF refers to Schroder International Selection Fund, a Luxembourg-registered SICAV.

Adaptation

Areas Already Affected by Climate Change	Impact	Industries Impacted
Agriculture	Crop failure and lower yields from current strains	Biotechnology, Beverages, Food Producers
Forestry	Depletion of forests	Forestry and Paper
Water resources	Changing rainfall patterns and water becoming more scarce	Food Producers, Biotechnology, Beverages, Real Estate, Construction
Coastal	Rising sea levels and greater threat of flooding	Construction and Materials, Travel and Leisure
Ecosystems	Damage to fragile environments (e.g., rainforest, tundra)	Infrastructure, Travel and Leisure, Utilities

Using these general guidelines and proprietary fundamental research, we reviewed industries for the major implications of climate change and identified over 700 stocks globally, each with over $200 million in equity value, in which climate change has a significant positive impact on the investment case. We then select the 50 to 80 stocks in which we have the strongest fundamental convictions and think in aggregate will outperform global equities over time (i.e., generate alpha). Moreover, because the companies selected are involved in mitigating the effects of climate change or helping adapt to its possible impacts, they should have different return drivers than companies in other parts of the market. These characteristics should result in a portfolio that tends not to be highly correlated to more traditional equity portfolios.

Stock Selection and Portfolio Construction

In some cases, we invest in direct climate change stocks, such as companies that offer transformative clean technologies like manufacturers of wind turbines. In other cases, we try to identify companies that indirectly benefit from the climate change theme. For example, forestry is an area that has been bypassed by growth-seeking investors for many years. The principal markets for forestry products have been construction, furniture, and paper, but climate change developments will change this. Efforts to minimize further deforestation means timber supply will be restricted and its value will rise. Until fairly recently, biofuel production had been seen as the next generation of low-carbon fuel, but the true carbon benefit (as well as the ethics) of using different food crops in biofuel production has become the subject of heated debate. Timber, however, is renewable, pulls carbon dioxide from the atmosphere during its years of growth, and then returns the carbon when burnt. As a

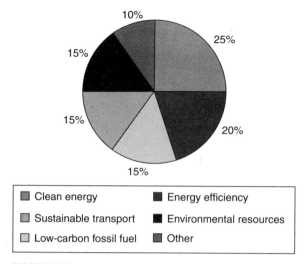

FIGURE 7.3   Sample allocation across investment themes.
Source: Schroders Investment Management. For Illustrative purposes only.

renewable fuel, several large utilities have already begun to replace coal with wood fuel in their existing power stations, boosting demand for forestry products.

Additionally, we believe value could be added by making active decisions on the optimal allocation across various sectors or regions. The chart in Figure 7.3 is an example of allocation among our main climate change investment categories.

However, it is critical to recognize that our investment universe is continually evolving as our understanding of the risks and opportunities increase and the related science, legislative and corporate policies, and technologies develop. This evolution provides us with new and additional opportunities to generate meaningful excess returns.

Conclusion

Investors are right to consider climate change as an investment theme that holds the promise of excess returns and genuine diversification. But they should realize that while environmental alpha might be found in some carbon funds and project finance deals, it could be in more familiar places—in the public equity markets—and through familiar vehicles—listed commingled funds—that provide many of the benefits institutional investors seek: pricing transparency, reasonable liquidity terms, custody, reporting and client service, and rigorous regulatory oversight.

THE RISKS AND CHALLENGES OF ENVIRONMENTAL INVESTING

Each major category of environmental investing will be discussed in detail in the following chapters so rather than continue to describe categories, I want to discuss certain risks, challenges, and benefits associated with environmental investments as a whole.

Risks

Environmental investments not only carry the same risks and challenges as more traditional investments (e.g., new, untested business, failure of business model, poor strategic planning, insufficient capitalization and financing, weak management capabilities, liquidity constraints, poor performance, etc.); they come with their own set of risks associated with the dynamic nature of the return drivers (science, economics, policy, and technology) mentioned above. Here are three examples of environmental-specific risks (adapted from Calvello 2009).

- EcoSecurities (LON:ECO), one of the largest developers and suppliers of emissions credits, puts together various environmental projects that generate carbon credits and sells to needy buyers. Each project must be independently certified to the United Nations as environmentally legitimate. However, in late 2007, the UN rejected several EcoSecurities projects on the ground that they would have been financially viable without the revenue from credits. This caused EcoSecurites to write off a portion of the credits it promised to deliver. The company's shares, traded on the London Stock Exchange's AIM index, dropped by more than 80 percent in the year since this announcement.
- In July 2008, the U.S. Court of Appeals, Washington D.C., vacated the Clean Air Interstate Rule. This unexpected decision shocked the emissions markets, sending prices into a tailspin, and introduced a level of complexity and uncertainty for other similarly planned cap-and-trade schemes. The price of sulfur dioxide allowances on the Chicago Climate Futures Exchange dropped from more than $300 per ton to about $100 per ton in two days. Prices have rebounded to about $215 per ton.
- In 2008, two MIT scientists made what one peer called "probably the most important single discovery of the century [for solar energy]" (Bullis 2008) and another described as "a major discovery with enormous implications for the future prosperity of humankind" (Trafton 2008). What did they do? Dan Nocera and Matthew Kanan appear to have

solved one of the key problems in making solar energy a dominant source of electricity: They developed a catalyst that can generate oxygen from a glass of water by using sunlight to split water molecules. If this breakthrough proves commercially viable at scale, it could render obsolete the current photovoltaic industry, allow solar power to truly compete with fossil fuels as a major energy source, and open the door to new environmentally friendly technologies, like at-home energy systems (thereby improving the lives of billions of people in developing countries) and highly efficient fuel cells to power electric cars. The move from the lab to the marketplace will create dozens if not hundreds of new investment opportunities (Calvello 2009).

The evolving nature of the science of climate change, changes in the economics of the responses; the expected consequences of climate change; the local, national, and international climate change policies; and technological advancements present investors with some rather substantial risks.

Idiosyncrasies

Beyond these risks, environmental investments come with the following four idiosyncratic conditions:

- Status quo
- Scale
- Size
- Speed

Investors need to understand these conditions in order to determine which technologies and investment strategies will prove to be winners.

Status Quo Breakthroughs like Nocera's and the potentially disruptive technologies they might produce generate a great deal of excitement and exuberance, especially among investors in environmental strategies, but it is critical that investors understand, in spite of the long-term promise of these technologies, that it will take a Herculean effort to displace existing technologies, especially in the power generation and transportation sectors (which account for the majority of anthropogenic GHG emissions). Current energy sources are deeply ingrained in our lives, our economy, and our history. Change comes slowly. Investors should realize that there has been one new large-scale energy source introduced in the past 100 years—nuclear power in 1957. In the entire history of energy consumption, there have been eight major energy sources, "if you add whale oil and animal fat (candles) to wood, coal, oil, natural gas, water, and uranium" (Tertzakian 2007). Let's look at just one energy source: coal.

Coal has been the foundation for the development of every advanced society. It is a base-load energy source that is:

- Relatively cheap—about a third of the price of other power sources given current pricing schemes.
- Abundant—by some estimates there are coal reserves to last between 300 and 3,000 years; it is mined commercially in over 50 countries and used commercially in over 70 countries (MIT 2007).
- Entrenched—coal accounts for about 52 percent of the United States' electricity generation, about 80 percent of China's electricity generation (and more than 50 percent of its industrial fuel utilization and about 60 percent of its chemical feedstocks), about 70 percent of India's total electricity generation, and about 40 percent of global electricity (MIT 2007).

There also is an existing infrastructure that supports coal's production (e.g., efficient mining activities), distribution (e.g., railroads), and use (there are over 600 operational coal plants in the United States). Additionally, there are large, established coal companies that collectively form a formal coal *industry* that benefits from government policies and incentives and is engaged in the environmental policy discussions taking place around the world.

These characteristics virtually ensure that the United States, India, China, and others will continue to burn their coal for years to come. In fact, the percentage of power derived from coal is expected to remain at similar levels over the next 30 years (World Coal Institute 2005). We only have to look to China, where coal output rose from 1. 30 billion tonnes in 2000 to 2.23 billion tonnes in 2005 and two new coal power plants are built weekly (MIT 2007).

Because of its long-term and widespread use and the processes used to mine and burn coal, coal accounts for about 80 percent of emissions from the power sector in the United States (Environmental Protection Agency and Department of Energy 2000). Beyond the GHG emissions, burning coal also produces other negative environmental consequences (see Table 7.1). For these reasons, we need to come up with technologies that will both mitigate its emissions and offer viable healthy alternatives. (Of course, it is always possible that governments will simply make the use of coal unaffordable or legislate away its use, but as of today, politicians find such choices completely unpalatable.)

Scale The enormity of the climate change problem and the entrenched nature of existing energy sources require scalable technological solutions. Ultimately, "alternative technologies must be judged on their impact at the scale of the conventional energy-generating operations they hope to replace" (Krupp and Horn 2008). Reducing emissions from transportation, power, built property, or land use requires *scalable* solutions. Let's continue with our coal example and the fact that for at least the next decade coal likely will continue to be a primary power source.

According to some thinkers, there might be a way to employ two existing technologies to arrive at a cleaner use of coal:

1. *Integrated gasification combined cycle (IGCC) systems.* Basically, IGCC systems burn coal more efficiently and are cleaner than conventional coal plants, typically producing 33 percent less nitrogen oxide, 75 percent less sulfur oxide, and almost no particulate emissions. IGCC also uses 30 to 40 percent less water than conventional plants (World Energy Council 2007).
2. *Carbon capture and storage (CCS).* CCS strips CO_2 emissions out of the exhaust steam from coal combustion and typically injects them into the earth's subsurface. The emissions are produced but they never enter the atmosphere.

Many believe that we will not be able to significantly reduce GHGs without the widespread use of these technologies (either separately or in combination). (Of course, others believe we will never hit the stabilization targets as long we continue to burn coal.) "The only realistic way to sharply curtail CO_2 emissions is to phase out coal use except where CO_2 is captured and sequestered" (Hansen et al. 2008). The challenge is deploying these technologies at a consequential scale. At present, there are only 160 IGCC facilities globally, and CCS remains in the demonstration stage, perhaps 10 years away from broad commercial deployment (World Coal Institute 2005).

Size With the challenge of scale comes the coequal challenge of size. Significant adoption of IGCC and CCS—two *existing* technologies—is an enormous global undertaking. (For example, the technology required for CCS has been used by the oil and gas industry for many years, but it has never been used at the scale required by the coal industry.) Bringing either or both of these two technologies to the approximately 600 coal-fired plants in the United States is a daunting undertaking. The challenge increases by an order of magnitude when the widespread use of coal globally is factored in—and the fact that many developing countries are large coal users.

The challenge is further compounded because simply retrofitting existing coal plants (and building more efficient new coal plants) is just the first step. CCS, for example, is site specific, meaning that not all coal plants have the proper geological structures nearby to store the now-liquid CO_2. A national pipeline system will be required to carry the CO_2 slurry to suitable sites. An MIT study (2007) estimates that if 60 percent of the CO_2 produced from U.S. coal-based power generation were to be captured and compressed into a liquid for geological sequestration, its volume would about equal the total U.S. oil consumption of 20 million barrels per day. The largest CCS project today stores about one million tons of CO_2 underground. By comparison, a single coal plant in Georgia emits 25 million tons of CO_2 (Krupp and Horn 2008).

One other point about size: To reach scalable proportions, developing and broadly deploying these transformative energy technologies and their supporting infrastructures will require a massive amount of investment capital. We're not talking about the information technology market, which by most estimates was about several hundred billion dollars in size. The energy market alone is a $6 trillion industry (and expected to double by 2030). Developing and dispersing transformative energy technologies will be far more capital intensive than required for information technology investments. As John Doerr points out, Electronic Arts was started with $2 million; Amazon was started with $8 million; Google was started with $25 million (Doerr 2008). Starting and nurturing an alternative energy firm like Bloom Energy requires greater capital (more than $250 million) and greater patience (at least five to seven years before the company achieves profitability) (Doerr 2007). In Chapter 8, Veronique Bugnion and Jurgen Weiss forecast that the capital cost of transforming just the U.S. power sector would be about $10 trillion, or about one year's GDP—and this does not include costs associated with managing changing water supplies.

CCS could prove to be quite expensive. The process requires a significant amount of energy, increasing fuel consumption by 10 to 40 percent. This makes CCS-equipped plants 30 to 60 percent more expensive than conventional technology (Llewellyn 2007). Add in the infrastructure costs, and the price of electricity generated from coal could increase by 50 to 80 percent. Costs certainly will increase as the optimal storage sites are taken up. While it is difficult to fix a total cost for CCS, Lord Oxburgh, former chairman of Shell, predicts that "the CCS sector will grow to become a $1,000bn industry and as big as the oil industry is today" (Byrne 2008). However, the benefit in this high price for CCS is that it will likely make other alternative energy sources more attractive.

Speed   With scale and size comes speed. It should be quite obvious that it will likely take years if not decades to deploy IGCC and CCS in scalable, functional ways that significantly reduce GHG emissions from coal. Perhaps more fundamentally, even before these technologies can be deployed, it will be necessary to create and ratify a post-Kyoto agreement and to design and implement the policies and regulations associated with the use of IGCC and CCS. The latter is in itself no trivial matter, as few of the basic policy issues associated with CCS—such as who is responsible for the CO_2 that will be stored underground for generations—have been addressed. "At present, there is no institutional framework to govern geological sequestration of CO_2 at large scale for a very long period of time" (MIT 2007). The absence of clear, long-term policies has had a direct impact on the development and deployment of these technologies. "IGCC plants in Illinois, Minnesota, Florida, and Colorado have been suspended or canceled, in several cases because the utilities were unwilling to invest in carbon dioxide–capture technology without greater regulatory certainty on both carbon limits and the requirements for underground storage" (Krupp and Horn 2008). It will take years to create and implement such a framework.

And we are not even talking about new disruptive technologies like Nocera's catalyst. The time to market will be much greater with disruptive technologies because displacing existing technologies will require the appropriate regulation and policy, extensive RDD&D (because we are still engineering the solutions; IGCC and CCS are in many ways market ready), new infrastructure, significant investment capital, and changes in behavior.

Even with these elements in place, there are still significant barriers to be overcome. First, replacing the extant fossil-fired power plants with clean sources of power is not a one-for-one replacement of existing capacity. More capacity will likely be needed because these clean sources typically have a lower capacity factor than traditional power generation due to the intermittency of the renewable resource used. Second, the time to commercial adoption will be substantial. Some, like Joseph Romm, an energy physicist and former Clinton Department of Energy official, point out that historically it takes about 25 years *after* the commercial introduction for a primary energy form to obtain a 1 percent market share. The first transition from scientific breakthrough to commercial introduction may itself take decades (Friedman 2008).

With regard specifically to CCS, Shell estimates that commercial deployment of this existing technology will begin sometime around 2020 and that by about 2030 one-fifth of all coal- and gas-fired power generation will be equipped with CCS. (Shell International BV 2008) CCS clearly illustrates what one writer described as new transformative technologies' "treacherous slog to commercialization" (Gertner 2008).

Attempts to adapt to the consequences of climate change—even with existing technologies—will likely require the same time frame, especially if the project schedule of the Great Man-Made River Project (GMMRP) in Libya is an indication.

Rapid urbanization, industrialization, and population growth coupled with improvements in the standard of living in Libya drastically increased the country's demand for potable water. Oil explorations in the 1950s and 1960s led to the discovery of vast amounts of water in aquifers underneath Libya's southern deserts. In 1984, the Libyan government began construction of the "Great Man-Made River Project" to extract and transport the water to the northern coastal belt through about 4,000 kilometers of prestressed concrete cylinder pipe. Once completed, the GMMRP is expected to deliver up to 6.5 million m^3 (1.7 billion gallons) of water per day. The increased water supply will drastically improve the lives of 5.6 million Libyans and allow for the cultivation of an additional 130,000 hectares of farmland.

After 25 years—and in spite of using readily available technologies and having the full support of the state, the GMMRP is only in the third stage of a five-stage plan, delivering over 4 million m^3 of water per day to Libya's two largest cities, with total costs in excess of $25 billion (Parsley and Liu 2009).

In sum, there are specific challenges associated with environmental investing. Displacing existing technologies will be difficult and costly and take a great deal of time in spite of the urgency associated with the climate change problem. Investors should be mindful of the four S's—status quo, scale, size, and speed—when assessing environmental investment opportunities.

THE BENEFITS OF ENVIRONMENTAL INVESTMENTS

Environmental investments are attractive for two main reasons: the promise of excess returns and, in some cases, genuine portfolio diversification.

The drivers of environmental investments are sufficiently fluid and complex that it is possible for some market participants to gain advantages that they could exploit to produce excess returns. As we said above, the science of climate change is evolving; our understanding of the economics is changing as science evolves; climate change policies are continuing to be debated and formed; existing technologies are finding new applications; and new technologies are continuously being developed and deployed.

We have only to look at ethanol as an example of the changing dynamics and resulting inefficiencies. By 2004, ethanol was a favorite environmental investment, with numerous initial public offerings on ethanol companies, and it was embraced by politicians and investors alike. It all ended badly. The turning point might have been April 2008, when a United Nations expert called biofuels a "crime against humanity" (Green blog 2008), or November 2008, when Goldman Sachs announced that its analysts would no longer cover listed ethanol companies because of their "dim future" (Moore 2008). Advances in scientific understanding (Fargione, Hill, Tilman, Polasky, and Hawthorne 2008), changes in policy and sentiment, and new technological developments (e.g., algae-based and cellulosic-ethanol technologies) all played a part in this change in affection.

The example illustrates that "in the case of ES [environmental and social] information and analysis, markets are woefully inefficient; information is by no means universally available, and it is certainly not widely considered credible or even potentially useful. Therefore, it cannot possibly be reflected in current stock prices" (Kiernan 2009).

The dynamic nature of the drivers of return gives rise to informational and regulatory inefficiencies that a skilled manager with access to capital could exploit to earn a return that is not attributable to market exposure. Additionally, these drivers make it possible for certain participants to control scarce resources, such as the prime locations for wind farms that will allow them to earn more favorable returns than participants controlling less optimal resources. Market participants with the right technology *and* access to such locations would be well positioned to earn an excess return.

This return derived from the deliberate application of manager skill to assets whose values are determined primarily by climate change–related factors is *environmental alpha*.

In this sense, environmental alpha is like other alphas—it is "the residual return that results from skilled active management—or luck—as opposed to the return that compensates investors for assuming downside risk" (Callin 2008). It just so happens that the manager is applying his or her skill in the environmental space to earn or generate alpha. Environmental alpha has two other common alpha attributes: First, it is subject to the immutable law of alpha (alpha is scarce, transitory, and capacity constrained) so the opportunities for creating environmental alpha will certainly diminish over time. But given the size of the climate change challenge and the variety and fluidity of the primary drivers of returns, we can expect new advantages to arise and this theme to play out for decades. However, because the source of the alpha arises out of the phenomenon of climate change, the potential for persistent alpha opportunities is great. "All Alpha factors will fade into the background eventually. ... Therefore, it becomes a question of how long the trend can last. Given the 40–50 year investment horizon and the size of the problem—$45 trillion of investment needed in energy markets alone—we believe that climate change will remain the source of identifiable Alpha for many years ahead" (DB Advisors 2008). Second, environmental alpha should have a low correlation with "the market."

The argument in favor of portfolio diversification is a familiar but poorly realized one primarily because few investors realize the high correlation of most of the investments in their portfolios. The diversification benefit offered by a portfolio of traditional investments is typically more perceived than real, as we learned in 2008 when the correlation of most assets went to 1.0. This is not surprising, as over 90 percent of a conventional portfolio's total volatility could be attributed to equity beta (Liebowitz and Bova 2007). So while there are genuine benefits to having a diversified portfolio of assets, finding investments that achieve these benefits is difficult. However, because of their structures and exposures, some environmental investments—especially those that do not invest in public equities—produce return streams that typically are not correlated with the performance of other strategies (such as those invested in stocks and bonds) held in an investor's portfolio. An example could be an environmental hedge fund that trades emission credits and other listed related instruments using a short-term technical trading model that has a correlation with the S&P 500 of −0.10 and with the JPM Bond Index of 0.04. (A counterexample would be Schroders ISF Global Climate Change Equity strategy, which, because of its exposure to global equities, tends to have a relatively high correlation to equity markets, especially over short periods of time.) The topic of correlation will be discussed in more detail in the following chapter.

Investments that offer genuine diversification are incredibly valuable to institutional investors because they can improve a portfolio's return per unit of risk. And insofar as environmental investments can provide such benefits, institutional investors should consider adding them to their portfolio.

CONCLUSION

Climate change presents institutional investors with both risks and investment opportunities. In order to properly understand both, investors need to understand that climate change, broadly defined, is the driver of returns for environmental investing. The fluidity and complexity of climate change result in a multitude of risks whose impacts and timing are difficult to measure. These uncertainties give rise to market inefficiencies that skilled managers could exploit to generate a return greater than what exposure to the market has to offer—what we've come to call *environmental alpha*.

Just as the risks associated with climate change impact every sector and economy, so, too, can the investment opportunities be found throughout the investment landscape. Investors must think broadly about environmental investing and refuse to succumb to the tyranny of the policy allocation. To do so is to place an unnecessary constraint on the investment process and to possibly limit fiduciary duty unnecessarily. Environmental investing does not belong in a single investment bucket in the portfolio; it should not be pigeonholed into a specific style of investing (e.g., negative screening), asset class, vehicle structure, or geography. While climate change is the dominant theme, environmental investing is a leitmotif that cuts across all investment decisions.

REFERENCES

Associated Press. 2008. "Air Force plans to switch fuel for coal." Military.com. March 22. www.military.com/NewsContent/0,13319,164531,00.html.

Bullis, K. 2008. Solar power breakthrough. *MIT Technology Review*, July 31.

Byrne, E. 2008. Dirtiest source of power aims to clean up its act. *Financial Times*, September 16. www.ft.com/cms/s/0/ac6f7174-8387-11dd-907e-00pre0077b07658.html.

Callin, S. 2008. *Portable alpha theory and practice.* New York: Wiley.

Calvello, A. 2009. Green trends and greenbacks. Portofolio.com. January 7. www.portfolio.com/news-markets/national-news/portfolio/2009/01/07/Environmental-Investing#page1.

DB Advisors. 2008. *Investing in climate change 2009.* Deutsche Bank Group. http://dbadvisors.com/climatechange.

Doerr, J. 2007. *Speed and scale.* Presented at the MIT energy conference, April 12.

Doerr, J. 2008. Where the money is. *Wall Street Journal*, March 24, R6.

Environmental Protection Agency and Department of Energy. 2000. *Carbon dioxide emissions from the generation of electric power in the United States.* Department of Energy. July. www.eia.doe.gov/cneaf/electricity/page/co2_report/co2report.html.

Fargione, J., J. Hill, D. Tilman, S. Polasky, and P. Hawthorne. 2008. Land clearing and the biofuel carbon debt. *Science* 319:1235–1238.

Freese, B., S. Clemmer, and A. Nogee. 2008. *Coal power in a warming world.* Cambridge, MA: Union of Concerned Scientists.

Friedman, T. 2008. *Hot, flat, and crowded.* New York: Straus and Giroux.

German Advisory Council on Global Change. 2007. *World in transition: Climate change as a security risk: A summary for policy makers.* London: Earthscan.

Green blog. 2008. www.green-blog.org/2008/04/30/biofuels-are-a-crime-against-humanity-says-un-official/.

Gertner, J. 2008. Capitalism to the rescue. *New York Times Magazine*, October 5, 82.

Hansen, J. 2008–2009. Tipping point: Perspective of a climatologist. *State of the Wild.* www.columbia.edu/~jeh1/2008/StateOfWild_20080428.pdf.

Hansen, J., M. Sato, P. Kharecha, D. Beerling, Robert Berner, Valerie Masson-Delmotte, et al. 2008. Target atmospheric CO_2: Where should humanity aim? *Open Atmospheric Science Journal* 2 (November). www.bentham.org/open/toascj/openaccess2.htm (accessed March 1, 2009).

Intergovernmental Panel on Climate Change. 2007. *Fourth assessment report to intergovernmental panel on climate change.* Boston: Author.

Khosla, V. 2009. *Session three: What are the economic opportunities on the pathway to a low carbon economy?* US Climate Action: A Global Perspective. Washington DC. March. http://dl.nmmstream.net/media/cgd/flash/030309panel3/mediaplayer.html

Kiernan, M. J. 2009. *Investing in a sustainable world.* New York: AMACOM.

Krupp, F., and M. Horn. 2008. *Earth: The sequel.* New York: Norton.

Liebowitz, M., and A. Bova. 2007. *Beta targeting: Tapping into the appeal of 130/30 active extension.* Morgan Stanley Research.

Llewellyn, J. 2007. *The business of climate change.* Lehman Brothers, February.

Massachusetts Institute of Technology (MIT). 2007. *The future of coal.* Cambridge, MA: Author.

Mercer Investment Consulting. 2005. *A climate for change: A trustee's guide to understanding and addressing climate risk.* The Carbon Trust. www.thecarbontrust.co.uk/trustees.

Mercer Investment Consulting. 2006. *Universal ownership: Exploring opportunities and challenges. Conference report.* Moraga: Saint Mary's College of California, April 10–11.

Moore, H. N. 2008. The death of ethanol: One thing Wall Street saw coming. *Wall Street Journal,* November 3, Deal Journal. Online edition. http://blogs.wsj.com/deals/2008/11/03/the-death-of-ethanol-one-thing-wall-street-saw-coming/.

Nelson, A. J. 2007. The greening of U.S. investment real estate—market fundamentals, prospects and opportunities. *RREEF Research* 57:3.

New Carbon Finance. 2009. Press release, January 8.

Nicholas Institute for Environmental Policy Solutions. 2009. *Forest and climate: The crucial role of forest carbon in combating climate change.* Duke University, March.

Pacala, S. 2008. *How much technology do we need to solve the climate problem?* Presented at the National Academy of Engineering regional meeting public symposium, May 2, Princeton University.

Parsley, R., and H. Liu. 2009. Unpublished manuscript. Used with permission. March 17.

Pew Center on Climate Change. *Global warming basic introduction.* www.pewclimate.org/global-warming-basics/about (accessed March 8, 2009).

Readey, A. M., D. Meek, J. Perencevich, and E. Hadzima. 2007. *Investing in clean energy and technology.* Boston: Cambridge Associates.

Shell International BV. 2008. Energy scenarios to 2050. www-static.shell.com/static/public/downloads/brochures/corporate_pkg/scenarios/shell_energy_scenarios_2050.pdf.

Stern, N. 2007. *The economics of climate change: The Stern Review.* Cambridge, U.K.: Cambridge University Press.

Tertzakian, P. 2007. *A thousand barrels a second.* New York: McGraw-Hill.

Trafton, A. 2008. "Major discovery" from MIT primed to unleash solar revolution. *MIT News,* July 31. http://web.mit.edu/newsoffice/2008/oxygen-0731.html.

U.S. Department of Energy. 2008. *Buildings energy data book.* November. http://buildingsdatabook.eren.doe.gov/docs%5CDataBooks%5C2008_BEDB_Updated.pdf.

Williamson, C., J. Saros, D. Schindler. 2009. Sentinels of change. *Science* 323 February 132009): 887–888; http://www.sciencemag.org/cgi/reprint/323/5916/887.pdf.

World Coal Institute. 2005. *The coal resource.* World Coal Institute, May.

World Energy Council. 2007. *2007 Survey of Energy Resource.* London: Author.

Zraikat, J. 2008. *Environment-friendly investing: The different shades of green.* Rogerscasey white paper, April.

BIBLIOGRAPHY

BP Statistical Review of World Energy 2009. 2009. British Petroleum. London. www.bp.com/liveassets/bp_internet/globalbp/globalbp_uk_english/reports_and_publications/statistical_energy_review_2008/STAGING/local_assets/2009_downloads/A_statistical_review_of_world_energy_full_report_2009.pdf.

CHAPTER 8

Investing in Climate Change

Mark Fulton and Bruce M. Kahn, PhD

Investing in climate change strategies gives an investor a concentrated exposure to a major economic force. Government regulations, economic and market trends, and the development of new technologies are acting in concert as drivers of adaptation to, and mitigation of, the impacts of climate change. The confluence of these factors has resulted in a broad and deep investment universe that not only takes advantage of these trends, but reflects a necessary shift in the organization of the global economy.

Chapter 7 described the five categories of environmental investing and explained some of its immediate characteristics. In this chapter, we extend this description to the economic and financial attributes of the climate change universe and discuss why environmental investing lends itself especially well to certain asset classes and opportunities. Within certain asset classes, the spectrum of investment is also a consideration when deciding how to deploy capital. By investing across many asset classes, including alternatives, a diversified portfolio may reduce overall portfolio volatility and correlation to the broad public markets. Including climate change sectors in an investment portfolio through proper asset allocation can improve the risk-return profile for investors, while giving them exposure to a transformation of the economy that has the potential to be on the level of the industrial revolution.

Mark Fulton is managing director at Deutsche Asset Management. Bruce M. Kahn is senior investment analyst at Deutsche Asset Management.

This chapter is divided into four major sections that discuss the considerations an investor should utilize when approaching an investment in climate change sectors.

1. Persistence of climate change as an identifiable alpha source
2. Fundamental attributes of climate change
3. Applying climate change to different asset classes
4. Benefits of including climate change in asset allocation

PERSISTENCE OF CLIMATE CHANGE AS AN IDENTIFIABLE SOURCE OF EXCESS RETURNS

We view climate change as a distinct economic and global theme and we expect the excess return profile of the climate change factor to resemble that of other global themes, that is, to outperform the market over a long period of time and then move generally back in line with the market as it matures and becomes fully priced in. This is shown in Figure 8.1.

We define persistence to mean a factor identifiable at a quantitative level that leads to distinct excess returns over the long term. In our paper, *Investing in Climate Change: An Asset Management Perspective* (Deutsche Asset Management 2007), we identified climate change as such a theme. The question is how long it will remain a distinct theme. In 2006 and 2007, climate change acted as a separate and identifiable theme, as evidenced by its excess returns relative to the broader market. During 2008, this corrected along with the broader fall in the markets, but we view this correction as having more to do with the credit crisis than a broader shift from secular to cyclical growth in climate change (Figure 8.2).

The climate change sectors recovered fairly well after the initial collapse in the fall of 2008 and, as Figure 8.3 shows, relative to the MSCI world have shown a 35 percent improved performance.

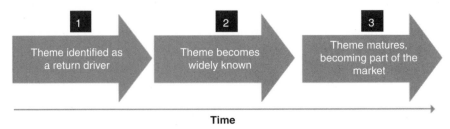

Time

FIGURE 8.1   The illustrative lifetime of an identifiable investment theme.
Source: DB Climate Change Advisors analysis, 2009.

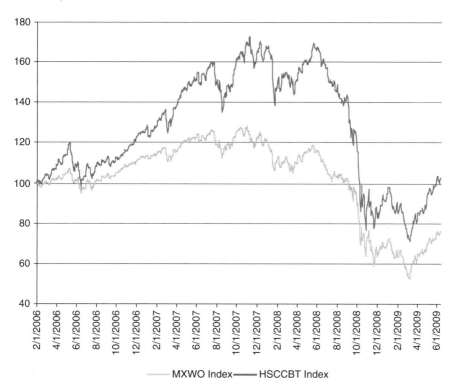

FIGURE 8.2 Performance of climate change sectors.
Source: Bloomberg, DB Climate Change Advisors, June 2009.

The question of when the theme begins to become the new "normal" and is submerged within the overall market will have different implications for different asset classes. Listed equity managers may find it more difficult at the far end of our time horizon to successfully pick stocks that can be appropriately classified within the climate change sector. This could happen if large, diversified companies become dominant and the notion of "renewable" fades into "energy." As discussed in Chapter 6, climate change could ultimately become simply a factor in an overall diversified portfolio, measured by carbon risk and carbon beta. Having said that, smaller, emerging pure plays representing innovative business models and technologies, are likely to remain for many years.

Indeed, for strategies focused on identifying, developing, and commercializing innovative new technologies, the potential for sustainable returns will remain. As the sector matures, technologies will become more complex, requiring ever more specialized expertise. Climate change venture capital and

FIGURE 8.3   Ratio of climate change index to the MSCI world.
Source: Bloomberg, DB Climate Change Advisors, June 2009.

private-equity investors should expect long-term persistence of the theme as it relates to their focus on innovative new technologies. The acceptance of the theme by the broader market only offers a more diverse end market into which they can sell technologies and companies. See Chapter 11 for additional commentary.

Infrastructure investment strategies are designed to provide stable returns to investors, and the scope of climate change will require enormous new infrastructure. The installation of new infrastructure projects will extend far into the future. It is possible that "climate change" aspects of infrastructure investment will merge into nearly every "normal" project. As such, they would no longer be identifiable in a separate sense. Overall, as a global economic megatrend, we believe that climate change will persist as an investment theme that requires specialized knowledge.

Government regulations are creating market conditions that are conducive to a global megatrend for decades to come. Governments have been stepping up the pace of legislation designed to help and support "green" industries. The burst of activity on this front by the Obama administration through the American Recovery and Reinvestment Act of 2009 is particularly visible and will provide welcome leadership and focus to similar efforts across the globe. At the same time, there has been a general ramping up of measures to counter climate change in countries as diverse as Greece, Britain, New Zealand, and China. We believe this trend toward greater regulation will provide crucial support to climate change industries during the current global economic downturn. Our research shows that, contrary to the widespread concern that recession would force governments to abandon initiatives on this front, governments have in fact been increasing their efforts. Between June 2008 and February 2009, we identified over 250 new governmental regulations that seek to mitigate and adapt to climate change. Coupled with increased government spending on key climate change initiatives as part of economic stimulus packages in countries such as the United States and the United Kingdom, this should provide a boost to "green" industries in contrast to other sectors of the global economy affected by recession. See Chapter 3 for additional commentary.

FUNDAMENTAL ATTRIBUTES OF THE CLIMATE CHANGE UNIVERSE

Looking at the climate change universe, sectors of the economy that give rise to significant investment opportunities often have low correlation to the broader economy. For example, annual growth in agriculture exhibits low correlation to gross domestic product (GDP) (Table 8.1). Second, the correlation of the annual growth in energy consumption with annual GDP is moderate and has declined over time (Table 8.2). The correlation of

TABLE 8.1 Economic Correlations of Climate Change Sectors

	1978–2006
Private Industries	0.99
Agriculture, forestry, fishing, and hunting	0.20
Farms	0.20
Forestry, fishing, and related activities	0.07
Utilities	0.32
Waste Management and Remediation Services	0.60

Source: DB Climate Change Advisors analysis, 2009.

TABLE 8.2 Climate Change Sector Correlations with Real GDP Growth

Correlations of Energy Consumption Growth (by Type) with Real GDP Growth

	Total Energy	Petroleum	Nat. Gas	Coal	Solar and Wind	Hydro and Geothermal	Nuclear	Wood and Waste	Total Renewables
1950–2007	0.78	0.69	0.48	0.57	n/a	−0.11	n/a	0.16	−0.05
1965–2007	0.73	0.69	0.36	0.59	n/a	−0.13	0.08	0.20	−0.07
1990–2007	0.51	0.75	0.02	0.47	−0.19	−0.07	−0.14	0.25	0.04

Correlations with Real GDP Growth

	Total Ag	Farms	Other
1948–2007	0.04	—	—
1978–2007	0.20	—	—
1978–2006	0.20	0.20	0.07
1985–2006	0.34	0.33	0.14
1990–2006	0.40	0.39	0.18

Source: DB Climate Change Advisors analysis, 2009.

renewable energy with GDP has been much lower. Water utilities also exhibit low correlation to real GDP growth. The economic attributes of key climate change sectors exhibit low correlation to the general economy.

APPLYING CLIMATE CHANGE TO DIFFERENT ASSET CLASSES

The climate change universe has different attributes that lend themselves to certain asset classes. The risk-return profile as well as the investment time horizon varies for each asset class. In our paper, *Investing in Climate Change: An Asset Management Perspective* (2007), we represented the opportunities in terms of risk and return and environmental focus, as shown in Figure 8.4.

In Figure 8.5, key asset classes are associated with a set of climate change attributes to match their suitability. Investment attributes provide background for different asset classes and climate change sectors offer investment opportunities across all stages of the investment spectrum from venture capital through to listed equities. It is deep knowledge of investment attributes that provides investors with an information advantage.

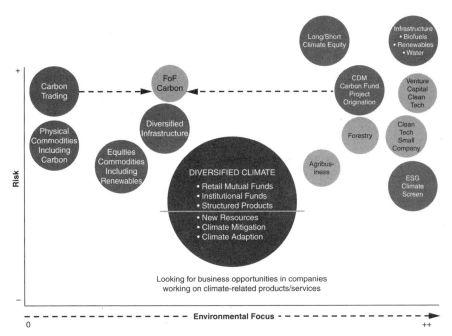

FIGURE 8.4 Specific investment strategies for climate change.
Source: DB Climate Change Advisors analysis, 2009.

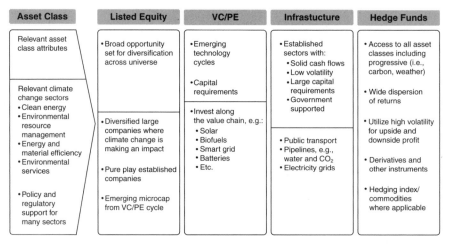

FIGURE 8.5   The climate change universe asset class fit.
Source: DB Climate Change Advisors analysis, 2009.

Listed Equities

Listed equities offer investment opportunities in established and new companies, a broad range of sectors and market capitalizations, and are for the most part highly liquid. In the DWS Climate Change Alpha Pool, which is the global pool of investable stocks used by the DWS Climate Change mutual funds, we have identified and tracked over 2,000 companies that fall within the scope of climate change related themes. In terms of alpha generation, we have already looked at the 2006–2008 bull market, where climate change generated outperformance. The outperformance by the climate change universe indicated that markets were responding to the broader economic demand of adapting to and mitigating against climate change, and this was a source of excess return. This was confirmed in an independent study by MSCI Barra (Chia et al. 2008), where they found that a "green" factor does exist in portfolio returns. In 2008, about 40 percent of the excess returns generated by climate change in 2006–2007 were lost on the downside. However, the regulatory support, along with the longer-term need for the products and services responding to climate change, indicates to us that as the dust settles, even with a period of weaker energy prices, climate change investments have the potential to outperform.

Opportunities for Hedge Funds

Within the universe of listed equities described previously, there has been a wide dispersion of returns, which is a measure of volatility. A wider dispersion indicates both a riskier investment as well as one that offers the potential for absolute higher returns. In the case of climate change–related listed equities, the dispersion of returns, as well as volatility, offers the opportunity for certain investment strategies, such as hedge funds, to utilize a variety of techniques or investment approaches beyond a long-only framework (e.g., short selling and derivatives) to capture outsized returns.

Measuring Carbon's Role in Portfolios

In addition to investing in mitigation and adaptation companies, integrating climate change parameters such as carbon risk into the overall investment process in listed equities has emerged as a new opportunity. Investors in listed equities can assess the degree to which portfolios are subject to climate change risk by addressing carbon intensity of different industry sector exposures, individual company risk positioning, and carbon financials (e.g., the costs of compliance). Going further and explaining the "risk" scale of the equation, investors could enhance climate change investments by including carbon leaders in the portfolio and avoiding or shorting carbon laggards. An example is given in Chapter 6, where Matthew Kiernan of Innovest Strategic Value Advisors explains this effect and significant outperformance by carbon leaders.

Private Capital (Private Equity/Venture Capital)

Private equity (PE) and venture capital (VC) have other attributes that are attractive for climate change investors. First, this asset class is the first sector to pick up emerging technology cycles. Venture capitalists typically invest in innovations around specific technologies, and they ultimately seek to be invested in disruptive technologies that can change whole industries. For example, many venture capitalists have been investing in cellulosic biofuel technologies and thin-film solar and smart grid technologies. As the technology emerges, PE investors step in and provide expansion capital in order for the start-up companies to take their technologies to market. See also Chapter 11.

The Financial/Investment Spectrum

Company stage, investment style, and investment attributes are also associated with the amount of capital deployed. Moving across the capital spectrum from infrastructure project finance to VC, the risk profile increases, thereby increasing the required return profile. Each investment style requires different amounts of capital and will be influenced by regulations and market dynamics. It is interesting to note that in the clean technology

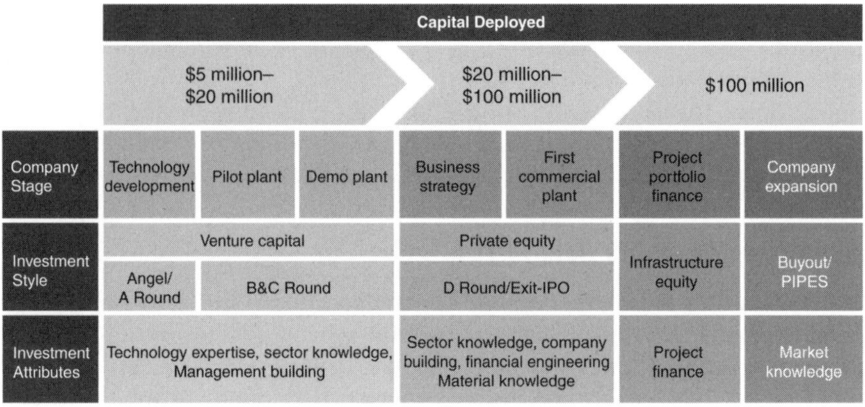

	Capital Deployed						
	$5 million–$20 million			$20 million–$100 million		$100 million	
Company Stage	Technology development	Pilot plant	Demo plant	Business strategy	First commercial plant	Project portfolio finance	Company expansion
Investment Style	Venture capital			Private equity		Infrastructure equity	Buyout/PIPES
	Angel/A Round	B&C Round		D Round/Exit-IPO			
Investment Attributes	Technology expertise, sector knowledge, Management building			Sector knowledge, company building, financial engineering Material knowledge		Project finance	Market knowledge

FIGURE 8.6   The investment spectrum for the private market climate change universe.
Source: Hudson Clean Energy Partners.

PE/VC/infrastructure space, there is much debate about potential blurring around the "D" round (Figure 8.6). VC investors still see plenty of risk at the first commercial plant stage and believe that high returns should reflect this level of risk. PE investors see this as a more mature phase with some risk. Increasingly, infrastructure/project finance investors are looking to derisk this phase as much as possible. However, the recent credit crisis is likely to put equity, where it is available, in the driver's seat.

PE/VC investments in the climate change universe are attractive for institutional and private investors. A study by the Cleantech Venture Network (2005) suggests that a hypothetical portfolio of North American clean tech companies in the period of 1987 through 2003 would have returned an estimated $6.2\times$ invested capital, and European markets have reported similar expansion in clean tech investing. Adding PE/VC to a portfolio can have diversification benefits due to reported low correlation to equity markets and lower overall portfolio volatility. However, due to the staleness of pricing, the difficulty of marking to market of illiquid assets, and market volatility, the effect will vary and in some cases not be as great. In addition, PE can have a much longer time horizon of investment and therefore have lower volatility than the broad equity markets. The uncertainties surrounding the pricing of companies in private markets can offer very attractive upside potential. This is particularly true in the climate change universe, where the complexities surrounding clean energy regulation, market access, technology, finance, commodity risk management, and taxation require a sophisticated understanding in order to properly manage investment risk.

PE can offer a number of advantages, including a potential for high risk-adjusted returns and diversification relative to other asset classes. Climate change PE adds diversification to an investment portfolio due to potential improvements to the risk-return characteristics of a portfolio. It also offers access to a rapidly growing segment within the economy.

Infrastructure

At a global level, changing demographics and economic development are driving demand for improved infrastructure worldwide. Climate change only enhances this growing demand and therefore the risk-return profile of any investment. Demand for energy will be increasing, even without the added pressure of a carbon limit, and will therefore require value-added strategies not only for developers but for their investors as well. Due to historical underspending on public infrastructure in energy, water, and transportation, climate change regulations will make the supply-and-demand imbalance more acute. Governments are finding that they cannot fund infrastructure demand through traditional sources and must use private capital in a responsible way.

Climate change portends new constraints and opportunities for infrastructure developers and therefore investors. For example, electric utilities are now faced with renewable portfolio standards, and new efficiency standards are leading to smart grid installations. Parking garages and storage facilities are now being outfitted with solar cells in order to send energy back into the grid. Constraints on water resources as a result of climate change will challenge water infrastructure developers. Successful infrastructure fund managers will have a unique understanding of potential regulatory arbitrage across jurisdictions, as well as a keen understanding of the global interplay between traditional energy generating sources, renewable energy sources, and the impact of a future price for carbon.

Most infrastructure funds raising capital today are focusing primarily on "mature" infrastructure investment opportunities, broadly defined as all developed infrastructure and not "greenfield" development opportunities. Mature infrastructure funds generally have somewhat different strategies and targeted returns, but most share certain unique features, including: (1) a focus on investments generating stable cash yield, (2) moderate but steady asset appreciation, (3) portfolio diversification and risk mitigation, and (4) returns hedged against inflation. In general, infrastructure investors will seek to acquire investments that are not exposed to a large degree of technology risk.

Clean energy, in particular, offers investment opportunities that will fit well with infrastructure funds' risk-reward investment profiles. Clean energy developments can offer investors a fixed income stream, as they typically sell their generated energy through attractive power purchase agreements with established creditable counterparties. Another area in which climate change investors are interested is transmission and distribution (T&D). T&D assets provide many of the investment characteristics desired by infrastructure investors. The opportunities for climate change investors are widespread, all-encompassing electricity grids and power generation, energy storage, and water infrastructure.

For instance, as climate change continues to impact meteorological patterns, additional water infrastructure investments will be required, especially in drought-ridden areas. A large portion of infrastructure funds that target industrialized countries focus on improving existing assets. By contrast, infrastructure funds targeting emerging companies may have a greater focus on greenfield assets. In many of the emerging economies, before government authorities give concessions to private operators/developers to build new infrastructure, they are increasingly requiring "sustainable" or "green" types of developments to be used. The credit crisis has created headwinds for debt-heavy infrastructure for a while. However, the American Recovery and Reinvestment Act of 2009 has focused on infrastructure, including those related to climate change, and infrastructure projects are in the process of receiving funds. We expect some sectors to begin rebounding faster than others as funds are disbursed from the stimulus plans globally. Additionally, markets are exploring ways to get investors more directly involved in project finance opportunities.

Sustainable Timberland and Forestry Investing: Reforestation

Forests offer the climate change investor the opportunity to sequester carbon and even potentially derive valuable and tradable carbon credits. The key to this is using a sustainable approach to managing the forest and ensuring that the end use of the timber reduces carbon emissions (e.g., second-generation biofuels, housing, furniture). Reforestation of degraded lands would be particularly positive for carbon sequestration. Therefore, from a climate change perspective, forestry and timberlands offer a tremendous opportunity for investing.

Timberland investing offers uncorrelated returns with financial assets historically and also has served as an inflation hedge. Timberland investment is a subsector of the real estate investment class. Like real estate, timberland investors are able to invest in timberland-focused funds, pure-play timber companies, and the actual timberland itself. Timberland is generally differentiated from basic real estate investments, insofar as it is focused on the production of timber, a saleable asset. Unlike farmland, owners of timberland can choose to delay harvesting the wood on their land. Depending on the price of lumber and the state of the larger timber markets, this characteristic offers investors a possible mechanism to hedge against downturns in current timber prices. The long-term nature of timberland investing often matches the investment goals of the long-term pension liabilities it serves. Moreover, the biological growth of the forest of 5 to 15 percent per year and the harvest decision as a valuable option are also advantages of this investment. Studies prior to the recent market crash have shown that the major components of timber returns include land prices of 5 percent, timber prices of 33 percent, and biological growth of 61 percent. Over the long run, both inflation and timberland returns have been positively correlated, and the class is often cited as an inflation hedge, especially against unexpected levels of inflation. Timberland can also be a better risk-adjusted investment than equities.

BENEFITS OF INCLUDING CLIMATE CHANGE IN ASSET ALLOCATION

Investors seek investment managers that can provide returns within a certain tolerance of risk. While we believe that many investment managers have deep knowledge of market risk, the knowledge and sophistication required of the climate change universe offers a unique potential for information advantage. Recall that it is the understanding of governmental regulations, break-even analysis of technologies, and the various market trends that are the key factors that can lead to a manager's alpha, the value added above beta exposure that reflects a manager's skill. In addition to the selection of managers, studies have shown that strategic asset allocation explains greater than 90 percent of the variability of an investment plan's returns over time and that there is 35 to 40 percent variability of returns across investors. The unexplained portion of return is from a variety of sources, including style within asset classes.

Efficient Frontier: Balancing Risk and Return

The efficient frontier provides a method of comparing the relative risk-return profiles of different portfolios. With increasing risk, a portfolio is typically expected to return at a higher rate. A traditional portfolio consisting of a mix of cash, bonds, and public equities can be expected to generate a moderate risk-adjusted return. This return is tightly correlated to the amount of risk that the portfolio holds. When alternatives are introduced to the portfolio,

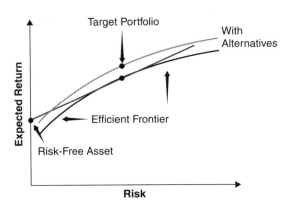

FIGURE 8.7 Overall investment strategy that seeks to construct an optimal portfolio by considering the relationship between risk and return.
Source: DB Climate Change Advisors analysis, 2009. For illustrative purposes only.

the potential return may increase. More importantly, however, a portfolio of alternatives can provide a potential increase in returns with a lower amount of corresponding risk. Additionally, the plot of the alternative risk-return profile is not as linear as the traditional portfolio. The return increases rapidly with commensurately less risk increase, until a point on the curve when the increase of return begins to slow and the increase of risk begins to speed up. Successful portfolio allocation can move the efficient frontier further and further above the line plotted by the traditional portfolio (see Figure 8.7).

CASE STUDY: Historical Performance of Introducing Listed Climate Change Companies in a Portfolio (Historical Data)

In order to understand the role that climate change strategies can play in a well-diversified portfolio, we conducted research using an efficient frontier analysis and a portfolio optimizer tool, Portfolio Choice™. We looked at the effect of adding climate change sectors into a portfolio based on the return, volatility, and correlation over the period January 2006 to September 2008, December 2008, and March 2009. We then constructed our efficient frontier using three scenarios of 1 percent, 3 percent, and 5 percent allocation to each strategy, with a total of three strategies for a total climate change allocation of 3 percent, 9 percent, and 15 percent. Due to their unusually low returns, we used our strategic return views and historical volatilities for the MSCI world and the Citi WorldBIG indices, and the historical returns and volatilities for the climate change strategies (Table 8.3). We used the following indices to represent our three climate change strategies: water (PIIWI Index Palisades Global

TABLE 8.3 Inputs for Efficient Frontier Analysis

Asset Class	Number of Periods	Return View	Historical Ann. Return	Historical Ann. Volatility	Projected Ann. Return	Projected Ann. Volatility
MSCI World	33	6.63%	0.60%	13.05%	7.06%	14.66%
Citigroup World Big (Hedged)	33	3.70%	3.82%	2.57%	3.62%	2.82%
Clean Energy	33	–	12.94%	26.98%	–	–
Agriculture	33	–	22.27%	26.35%	–	–
Water	33	–	4.82%	21.55%	–	–

Source: DB Climate Change Advisors analysis, 2009. For illustrative purposes only.

Water Index), agriculture (DXAG Index DAXGlbl Agribusiness), and clean energy (NEX, The WilderHill New Energy Global Innovation Index).

From a historical perspective, all three portfolios with climate change indices did shift to the right on the volatility measure; the increase in total portfolio volatility versus the traditional portfolio was relatively small compared to the increased returns. The addition of climate change assets improved the efficient frontier through September 2008, yet collapsed in our analysis through December 2008 and March 2009. We attribute any incremental increases in return to be associated with each increase in allocation with similar measures of risk. This indicates to us the positive impact of climate change sectors on portfolio performance through September 2008 (Figure 8.8).

However, from September 2008 through the present, the economic crisis has brought nearly all sectors and correlations into a tight correlation of negative returns. While our original analysis, conducted through the third quarter 2008, showed that adding climate change sectors added significant value to a traditional portfolio, our analysis through December 2008 shows that due to the market downturn, the climate change investment sectors were also strongly negatively impacted. As seen from the correlation study as well as our performance study, climate change sectors overall dropped with the rest of the market. We did expect that the climate change investment theme would reestablish itself quickly postcrisis due to the strong regulatory frameworks in place as well as the economic stimulus packages being enacted across the globe. This is, indeed, what occurred post crash. (See Figures 8.2 and 8.3.)

We therefore constructed a scenario that is forward looking, with an inflationary hedge. Our assumptions of returns are equities at 8 percent, bonds at 3 percent, and our climate change sectors at 15 percent. These assumptions, of course, represent a return in a positive market environment. We estimated the average additional returns of the climate change sectors over the MSCI world based on the moving 12-month average of the HSBC Climate

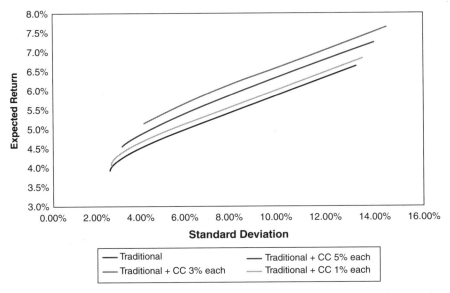

FIGURE 8.8 Efficient frontier (Pre-Crash): Adding climate change sectors could potentially add benefits to portfolios.
Source: DB Climate Change Advisors analysis, 2009. For illustrative purposes only.

Change Index since 2006. When looking at the average additional returns up to the end of October 2008, which represents a positive market time frame (see Figure 8.3), we found that the Climate Change Index outperformed by about 15 percent. The resulting efficient frontier in Figure 8.9 shows that by adding climate change sectors to a portfolio in the 3 percent, 9 percent, and 15 percent weights, our efficient frontier reflects improved risk-adjusted returns.

MARKET DEMAND AND SUPPLY

In the absence of major regulatory changes, the International Energy Agency (IEA) forecasts a 70 percent increase in oil demand by 2050, and a 130 percent increase in CO_2 emissions, compared to 2005 levels. However, with ambitious new policies, emissions can be reduced significantly and oil demand can be reduced by 27 percent compared to the 2005 level. Renewable energy sources have the potential to comprise 46 percent of total electricity supply by 2050. To meet increasing renewable energy demand, along with a greenhouse gas emissions reduction target of 50 percent from 2005 levels by 2050, the IEA finds that $45 trillion of investment will be needed from present day through 2050. Capital would need to be heavily deployed into

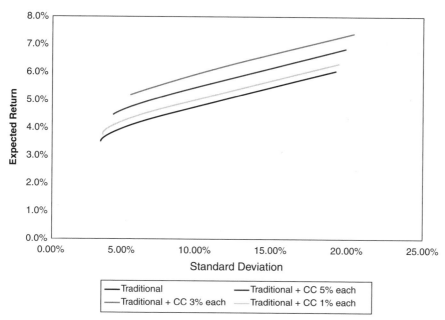

FIGURE 8.9   Efficient frontier with expected returns of 15% for climate change sectors: Adding climate change sectors can potentially add benefits to portfolios
Source: DB Climate Change Advisors analysis, 2009. For illustrative purposes only.

the development of next-generation technologies to create energy efficiency and low carbon options. The world population is also expected to grow to over 9 billion in 2050, causing significant effects on food and energy resources. Driven by higher inevitable demand, water, agriculture, and other resource depletion will lead to carbon emissions and climate change as key consequences. Therefore, the need for further investment into climate change sectors is critical.

In 2007, the clean technology sector saw approximately $148 billion of new worldwide investment, a 60 percent increase from 2006 investment levels, according to New Energy Finance (2008). Investment capital was allocated across a number of markets: research and development, VC/PE, project/asset financing, and public markets. The rise in clean technology investment over the past year depicts a greater interest in the advancement of next-generation technologies, as well as an increase in renewable energy capacity in areas outside developed nations. Importantly, in 2008, many markets continued to show growth in the face of the credit crisis resulting in 5 percent growth up to $152 billion (Figure 8.10). While this is a good start for deploying capital into climate change markets, it falls short of the

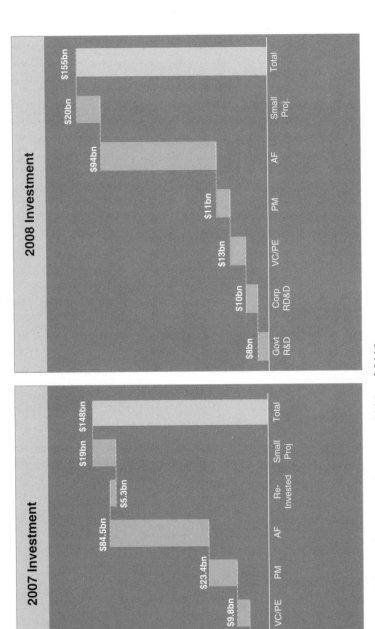

FIGURE 8.10 Total global new clean energy investment 2007 and 2008.
Source: New Energy Finance, 2008 and 2009.

requisite funds needed to avert catastrophic climate change. As investors, we will continue to deploy capital into these sectors and therefore expect significant growth to continue. We are also compelled by government regulations being developed across the globe to stimulate these markets in the face of the combined economic and climate crises.

CONCLUSION

As we have discussed, environmental investing can be suitable for all asset classes. Different asset classes provide distinctly different risk-return profiles, and investors should pay close attention to their individual portfolio asset allocation needs and goals. While the economic crisis has impacted the entire market, the climate change crisis continues to take hold of our economies and continues to receive attention from regulators. Therefore, the fundamental attributes of environmental investing will persist through our economic recovery. We did show in a case study the environmental investing can add significant value to a portfolio, even though in the market collapse, the listed equities in climate change sectors followed the rest of the market.

The key takeaway for institutional investors is that environmental investing is a mega trend, will persist, and may lead the economic recovery due to support of governmental regulations and fiscal stimulus. Second, while many opportunities for investments are necessary, financing of projects may prove challenging until lending returns to the marketplace. Additionally, there are stand-alone opportunities for investments, and institutional investors may simply add climate change–based investments to their portfolio to enhance the diversification of their investments.

DISCLAIMER

DB Climate Change Advisors is the brand name for the institutional climate change investment division of Deutsche Asset Management, the asset management arm of Deutsche Bank AG. The material does not constitute investment advice and should not be relied upon as the primary basis for any investment decision. All opinions and estimates herein, including any forecast returns, reflect Mark Fulton's and Bruce Kahn's judgment on the date of this report and are subject to change without notice. Such opinions and estimates, including forecast returns, involve a number of assumptions that may not prove valid. Opinions expressed herein may differ from the opinions expressed by departments or other divisions or affiliates of Deutsche Bank. This document may not be reproduced or circulated without our written authority. The manner of circulation and distribution of this presentation may be restricted by law or regulation in certain countries. Persons into whose possession this material may come are required to inform themselves of, and to observe such restrictions. Deutsche Asset Management is the marketing name for asset management activities of certain affiliates of Deutsche Bank AG.

I-012187-1.1

REFERENCES

Chia, C. P., L. R. Goldberg, D. T. Owyong, P. Shepard, and T. Stoyanov. 2008. *Is there a green factor?* Journal of Portfolio Management.

Deutsche Asset Management. 2007. *Investing in climate change: An asset management perspective.* Deutsche Asset Management, October.

Cleantech Venture Network. 2005. *Cleantech venture investment: Patterns and performance.* Cleantech Venture Network, March.

New Energy Finance. 2008. *Towards low carbon energy—the vision, the numbers.* New Energy Finance Database, February.

BIBLIOGRAPHY

DB Advisors. 2008. *Investing in climate change 2009: Necessity and opportunity in turbulent times.* Deutsche Bank Group, October. http://dbadvisors.com/climatechange.

CHAPTER 9

Carbon as an Investment Opportunity

Jurgen Weiss, PhD, and Véronique Bugnion, PhD

The term *carbon* is increasingly associated with any perceived direct or indirect financial impact of greenhouse gas (GHG) legislation or, more broadly, climate change. This chapter explores existing international and regional approaches to limiting the emissions of GHGs, and the business opportunities or "carbon markets" that these schemes are creating. The discussion in this chapter will in general be limited to these markets and will not extend to more general impacts of climate change on business (already covered in Chapters 4 and 5).

THE KYOTO PROTOCOL AND EMISSION TARGETS

The Kyoto Protocol is the landmark treaty by which its signatories agreed to reduce the emissions of six GHGs in industrialized countries by 5.2 percent relative to 1990 levels by the end of the 2008–2012 compliance period. The Kyoto Protocol has now been ratified by 184 countries. Commitments by individual industrialized countries range from an 8 percent reduction relative to 1990 levels for the European Union to an 8 percent allowed increase for Australia. The United States agreed in Kyoto to a 7 percent reduction in emissions, but has since failed to ratify the Protocol. The Kyoto Protocol

Jurgen Weiss is managing director of Watermark Economics. Véronique Bugnion is managing director of Point Carbon.

put in place three so-called "flexibility mechanisms" to help countries meet these targets:

1. Trading of allowances assigned to each target bearing signatory country (assigned amount units [AAUs]). Thus far, very few trades in AAUs have actually taken place—fewer than 20 million tons in 2008. Would-be buyer countries that are emitting more than their target cap will likely wait until close to the end of the Kyoto compliance period, 2012, to assess their compliance position before engaging in substantial AAU transactions.
2. The sale of project-based reduction credits for emissions-reducing projects located in developing countries.
3. The sale of project-based reductions originated in countries with an emissions reduction target.

In parallel to these flexibility mechanisms, the Kyoto Protocol also introduced the concept of supplementarity, the concept that purchases of carbon credits from abroad should only supplement domestic policies, but should not form the basis for complying with the Kyoto Protocol.

Member parties to the Kyoto Protocol meet annually. Current discussions are focused on defining a successor framework to be implemented after the end of the first compliance period of the Kyoto Protocol ending in 2012.

THE CLEAN DEVELOPMENT MECHANISM AND JOINT IMPLEMENTATION

The Clean Development Mechanism (CDM) and Joint Implementation (JI) are project-based flexibility mechanisms instituted by the Kyoto Protocol. They differ in their details, but both rely on individual projects to achieve emission reductions. Under the CDM, a project developed in a country with no binding emission reduction targets can sell its "output" or certified emission reduction credits (CERs) to a party, private or government, seeking to achieve GHG reduction targets. JI is similar but relates to projects in countries that have taken on mandatory reduction targets.

These mechanisms are peculiar in that the project output—emission reduction credits—is initially tied to individual projects that carry with them all of the project's risks. This includes risk that the project will be rejected by the United Nations board set up to approve these projects, risks associated with the project developer, and risk that the project will underdeliver the reduction volume relative to the plan submitted to the UN. We will refer to this market, where each project and its output is unique, as the "primary" market. However, once CERs are issued and approved by the UN, these credits become fully fungible and tradable with each other and hence form a commodity, which we will refer to as "secondary" market.

THE EUROPEAN UNION EMISSIONS TRADING SCHEME

The European Union Emissions Trading Scheme (EU ETS) was set up to achieve compliance with the Kyoto Protocol. The scheme covers only large stationary sources of emissions, or slightly less than half of the EU countries' emissions: power plants, chemical plants, most metal sectors, the pulp and paper industry, and refining emissions. The European Commission directive setting up the scheme was passed in October 2003 with two initial phases. The initial three-year trial period from 2005 to 2007 turned out to be overallocated: emissions in the covered sectors were approximately 5 percent less than the cap. The trial period was, however, successful in establishing a clear emissions baseline for the second compliance period from 2008 to 2012 and set to coincide with the Kyoto Protocol's compliance period. The scheme covers 28 countries in the second compliance period. Europe has already outlined targets and rules for the third compliance period, which will extend from 2013 to 2020 and will increase the stringency of the cap by aiming to reduce emission by 20 percent below 1990 levels by 2020.

EMERGING CARBON SCHEMES

A number of cap-and-trade schemes are being put into place around the world:

- The Regional Greenhouse Gas Initiative (RGGI) started on January 1, 2009. It caps the emissions of power plants of 10 states in the northeastern United States.
- The Western Climate Initiative (WCI) is a joint effort by nine western states and Canadian provinces to set up a multisector cap-and-trade scheme to start in 2012 with the objective of reducing emissions to 15 percent below 2005 levels by 2020. California aims to use the WCI in part to achieve the target of returning emissions to 1990 levels by 2020, passed into state law in 2006.
- The Midwestern Greenhouse Gas Reduction Accord is a collaborative effort by seven Midwestern States to reduce emissions by 15 to 25 percent below 2005 levels by 2020.

- Australia's cap-and-trade scheme is set to start on July 1, 2010, and will seek to reduce emissions by a minimum of 5 percent below 2000 levels but possibly as much as 15 percent below 2000 levels if other industrialized countries commit to similar goals in an international agreement.

FLAVORS OF CARBON

The carbon space can be broken down into a set of discrete markets and investment opportunities:

- Trading in allowances of carbon dioxide equivalent as a commodity. This commodity is created by governments setting a cap on or other mechanisms to constrain emissions. A number of examples of such commodities markets already exist, including the EU ETS and the RGGI. Trading in emissions allowances represents perhaps the most direct and conventional form of investment in the carbon space.
- Flexibility provisions built into cap-and-trade schemes such as the CDM allow individual emissions reducing projects in sectors or countries not covered by the cap on emissions to monetize these reductions in the capped market; this leads to two forms of investment opportunities:
  - Direct project finance investment in the emissions reduction projects.
  - Trading opportunities in emissions reduction credits in the so-called "secondary" reduction credit markets.
- Direct equity investments in the "carbon" space are limited, especially among listed companies. Nonetheless, a few companies have specialized in investing in emissions reducing projects. Beyond this handful of listed companies, a number of privately held investment funds also invest in these projects and offer their investors returns either in the form of delivery of carbon credits, or more conventionally as cash returns. A broader set of service, information, and technology companies offer support services to the carbon space on an exclusive basis and can thus still be characterized as carbon "pure plays," even though their financial performance might be driven by factors other than the actual traded price of carbon the commodity.
- A number of sectors are directly affected by carbon caps:
  - Entities regulated by cap-and-trade schemes such as power plants, industrial facilities or refineries. As noted later in this chapter, the fact that an entity is regulated does not mean that the impact on its bottom line will necessarily be negative.
  - Companies building new business lines to meet the demand of regulatory carbon programs. Examples include the project verification

business, financial exchanges listing carbon contracts alongside other commodities or news services providing carbon coverage.

■ Beyond those directly affected by carbon caps, climate change will over time affect the entire economy, from sea-level rise and storm impacts on real estate and insurance to changes in weather patterns' impacts on agriculture and energy usage.

The discussion in the following sections will focus primarily on the first three categories described in this list, with a focus on the impacts associated with compliance cap-and-trade programs. General and indirect impacts on the economy of climate change are expected to be far reaching but exceed the scope of this chapter.

Commodities

Commodities trading is generally divided among spot trading, which requires the immediate delivery of the product; carbon allowances or credits; and forward trading, which allows for the future delivery of these products based on agreed-upon terms.

Spot Trading

The immediate exchange of allowances or credits upon execution of a spot transaction requires registries to be in place to move allowances from one transaction party to the other. A number of registries now exist to allow spot transactions. Under the Kyoto framework, CER, emissions reduction unit (ERU), and AAU accounting take place within the International Transaction Log (ITL), which is maintained by the United Nations Framework Convention on Climate Change secretariat. Spot transactions of secondary CERs thus take place within the ITL. Spot transactions in EUAs take place within the Community International Transaction Log (CITL), which is maintained by the European Commission. The annual issuance of allowances under the EU ETS program takes place on February 28 of each year (i.e., year 2009 "vintage" allowances are issued on February 28, 2009). This restricts which allowance vintages can be spot traded to those already issued.

BlueNext, based in Paris, emerged in the second half of 2008 as the dominant spot market for European Union Allowances (EUAs).

Futures Trading

Futures or, more generally, derivatives instruments, on carbon are traded at a number of exchanges in Europe and the United States. The dominant players in EUA futures trading are the European Climate Exchange (ECX) based in The Hague, the European Energy Exchange (EEX) based in Leipzig, and Scandinavia's Nordpool exchange. In the emerging North American markets, the Chicago Climate Futures Exchange (CCFE) dominates the RGGI market. In 2008, futures and options trading in EUAs represented 37 percent of the market linked to the EU ETS.

Investing in Carbon Commodities: Carbon ETFs

Spot transactions and many futures contracts in EUAs or Regional Greenhouse Gas Allowances (RGAs) require each transaction party to have a registry account, therefore limiting the accessibility of the carbon commodities markets to sophisticated institutional investors. To address that limitation, a number of institutions have created carbon-linked indices that track these markets. One example is the Barclays Global Carbon Commodities Index Total Return™ (BGCITR) is designed to measure the performance of the carbon market and is designed to be an industry benchmark for carbon investors. The index, launched in 2007, currently tracks the most liquid and tradable carbon-related credit markets based on the EU ETS Phase II and the Kyoto Protocol's CDM. As other regions develop cap-and-trade schemes to address climate change, Barclays plans to add additional liquid instruments to the BGCITR. It is comprised of futures or forward contracts on a carbon emissions credit from each mechanism that are listed on the ECX.

Important to investors, in June 2008, Barclay's issued an exchange-traded note, the iPath Global Carbon ETN (GRN; www.ipathetn.com/GRN-overview.jsp?c=JAC03), to track the BGCITR. The GRN gives investors a convenient and simple way to access the carbon market.

Project Finance

Most carbon project finance investment vehicles are structured as funds, private or government managed. The World Bank pioneered offset project investment. The first fund dedicated to investing in projects to reduce the emissions of greenhouse gases around the world was aptly named the "Prototype Carbon Fund"; it still exists and has $125 million under management. In aggregate, the World Bank manages over $1.4 billion in eight carbon investment funds. Among the private funds, the Climate Change Capital Carbon Funds, Sindicatum Capital's Carbon Funds, and Natsource's GG-Cap Fund are among the largest, each with $400 million to $1 billion under management. Whether public or private, these funds compete with each other to find the best projects and build a diversified portfolio. The CDM pipeline of projects now includes over 6,500 projects in over 100 countries. Most of these projects are under various stages of development; less than

10 percent of these projects have actually achieved the stage of delivering carbon credits.

A few companies listed on the alternative investment market in London focus heavily on carbon project investment and are typically cited as carbon "pure plays." These firms include Trading Emissions plc, ecoSecurities, and Camco. All three have suffered from delays in project implementation and, for some project types, project underperformance.

Equities

A number of companies are involved in the carbon space by providing support services to these emerging markets. Two prominent categories of service providers stand out: verifiers and exchanges.

Verifiers are companies that have made a business of validating that forecasted emissions reductions are actually taking place. With each one of the 6,500 projects in the CDM pipeline requiring services, both assess that their project reduction plan is sound and that the ensuing reductions are real; this has become significant business for the verification business. The dominant market players, DNV, SGS, and Tüev-Süd, are all large companies for which the carbon verification business represents a small though rapidly growing fraction of their business.

Only one company, Carbon Exchanges plc, is truly a carbon exchange. The other exchanges cited earlier, BlueNext, Nordpool, and EEX, all list a diversity of primarily energy-related contracts. Carbon Exchanges plc is structured as a holding company with a number of daughter companies: the ECX dominates the EU carbon futures market, while the CCFE dominates the small but growing RGGI marketplace. Both the ECX and the CCFE are part of Climate Exchanges Group plc, listed on London's alternative investment market and itself an interesting carbon investment opportunity.

Risk Assessment

Searching for "alpha" in the investment categories presented above represents very different propositions. Many investors are seeking to ride the carbon "wave." However, care must be taken not to confuse the anticipated growth in volume of the carbon commodities with automatic profits. Like other commodities, carbon is driven by supply and demand, with the additional role played by regulatory drivers. EUA prices in 2008 provide interesting examples on both counts. During the first half of the year, EUA prices rose steadily alongside the rest of the energy complex. With crude oil and, by extension, natural gas prices rising in Europe, the market was seen as increasingly underallocated. This bullish mood was compounded by an announcement in January 2008 by the European Commission proposing a climate and energy package that would extend the EU ETS market through 2020 with increasingly stringent targets. The EUA contract peaked at close to €30/ton in July 2008. The first cracks were felt in the summer as the oil price bubble popped, driving European natural gas prices down with it. This made natural gas power generation more competitive with higher-emitting coal generation. A wave of selling from industrial market participants in the fall as the recession set in put further pressure on EUAs, which closed out the year at little over €13/ton.

Carbon equities have, in general, underperformed and disappointed the expectations of early investors who thought that share prices would rise along with the growth of the market. The fate of one of the early pioneers in offset project development, agCert, formerly listed in London, illustrates this point. The company overestimated the volume of emission reduction credits that would be generated by its portfolio, while at the same time aggressively forward selling the output of those same projects. The combination eventually led the company into bankruptcy, and it was absorbed by U.S. energy giant AES.

These two examples illustrate how investing in the carbon space offers no guarantee of returns, even as the policy progresses ever so slowly towards more stringent targets.

WHY INVEST IN THE CARBON SPACE *TODAY*?

In a carbon-constrained world, the traditional view of asset risk may have to change. Different classes of assets will likely be affected by carbon constraints asymmetrically or even inversely. To understand the changing relative risks of multiple asset classes will require a focus not only on individual companies' carbon footprints (i.e., direct or indirect carbon emissions within the boundaries of any company), but also on the variation in revenues and costs stemming from carbon constraints or climate change at any point along the supply chain.

It is useful to distinguish between asset classes that will be affected by regulatory or voluntarily imposed carbon constraints and those that that will be affected by climate change itself. Electric power generation is an example of the former, whereas the real estate and building sectors are examples of the latter.

Among industries most likely affected by carbon prices, those where a large amount of total value added depends on carbon emissions–intensive activities seem to be the most important. Industries already regulated under the EU ETS and those being considered for regulation provide a good starting point: electric power generation, pulp and paper manufacturing, aluminum and iron smelting, as well as oil and gas refineries are among the industries affected because of their direct emissions.

As carbon constraints on the economy increase, one might assume that the electric power generation sector as a whole will suffer more than the broader market. However, such a view would be too simplistic. For one, subsectors of the electric power generation sector may benefit tremendously: for example, low carbon power generation technologies (and all companies involved in a material way in the value creation of the corresponding products) will benefit from stricter carbon limits. High-carbon-content power generation technologies, however, are more likely to suffer from tighter carbon limits, but even in this seemingly obvious case, the financial impact on, say, coal-fired power generators (or their manufacturers) is not as straightforward as it may appear. Rather, as we have seen in the EU ETS, decisions about allocation versus auctioning of CO_2 certificates impact the power sector's bottom line. In the case of the first phase of the EU ETS, where allowances were allocated for free to the power sector, the power sector in general and coal-fired generation in particular benefited from windfall profits—costs did not increase while revenues did as a result of carbon being priced into electricity prices.

The same principles can be applied to other regulated industries. While a carbon-regulated sector's exposure to carbon will make the sector overall more risky, it is likely that a differentiated view is needed to assess the relative strengths and weaknesses within the sector, with low-carbon players having a chance to gain market share and become more profitable as a result.

It is, however, equally important to assess the impact of carbon constraints on the large majority of businesses likely not directly subjected to carbon limits. Assessing the exposure of any such industry to carbon risk requires a detailed analysis of the carbon profile of a firm or sector's entire supply chain. The recent demise of the traditional U.S. car manufacturers may serve well as a case study: while the carbon footprint of making cars may be relatively small when compared to that of a coal-fired power plant, U.S. car manufacturers—and car manufacturers in general—may still be at risk because the use of their products leads to GHG emissions. Should those emissions be capped or taxed, the cost of driving would increase as the tax or cap would translate into higher gasoline prices and hence there would be less demand for driving or, at the very least, a shift of consumer demand away from inefficient and toward efficient models. The spike in oil prices in the summer of 2008 provided a preview of the result on the demand for cars from various manufacturers, and it turned out that GM, Ford, and Chrysler, with relatively inefficient fleets, were hit hardest and saw revenues and profits decline sharply.

Even industries seemingly unaffected by carbon prices may face increased risk due to climate change directly. Portions of the real estate industry are likely a bit in the middle. On the one hand, with tighter carbon constraints the demand for energy-efficient buildings is likely to increase and owners of inefficient real estate portfolios may see the value of their assets at risk. On the other hand, the value of real estate assets may shift due to climate change directly: as average temperatures shift, storm activity increases in parts of the world, and, ultimately, sea level rise endangers shoreline properties, the value of real estate assets is likely to change as well. The insurance industry is an example of an industry where carbon emissions are likely to be low along the entire supply chain, but climate change itself may still have a substantial impact on various companies in the sector. Depending on the profile of an insurance company's portfolio of policies, an insurance company may be exposed to rapidly increasing claims as a consequence of weather-related events. Also, differences in the portfolio of assets may influence any specific insurance company's exposure to carbon risk for the reasons described so far—does the insurance company have a large portion of its assets in, say, the U.S. automobile industry or in the wind industry?

Finally, as may have become obvious in the context of discussing the potential benefits accruing to individual companies in seemingly risky segments, there are likely entire new industries with essentially "negative carbon betas," that is, companies whose returns will be inversely correlated with carbon prices. Obvious examples include the current and likely future substitutes for carbon-intensive products, be they wind turbines, electric hybrid vehicles, or some new insulating material.

But there are also assets that have emerged as a result of some of the regulatory aspects of carbon markets and who are likely to exhibit similar risk characteristics. Chief among those are projects that fall under the acronym CDM (Clean Development Mechanism). The CDM is one of the means by which companies in countries with binding targets under the Kyoto Protocol can meet their carbon limits. CDM projects reduce or avoid emissions in countries that do not have binding targets. Examples of CDM projects are windmills installed in China, or some kind of methane capture in India. It is clear that the tighter carbon limits will be in industrialized countries, the more pressure will exist to allow for CDM-style activities to be used to comply. As a result, the demand (and correspondingly the price) for such projects will increase and with it the returns to the project owner.

CARBON WILL BE A HUGE MARKET

The current consensus—to the extent it exists—in the scientific community seems to center around the idea that to avoid the risk of catastrophic climate change, global average temperatures should be kept from rising by more than 2 degrees Celsius. There is an evolving discussion about the target concentration of CO_2 in the atmosphere that will accomplish this, but generally levels between 450 parts per million (ppm) and 550 ppm are being discussed. Taking into account the emissions growth expected from developing countries over the next 30 to 40 years, the resulting policy prescription for the industrialized world is generally that GHG emissions need to be lowered by between 70 percent and 80 percent relative to the beginning of the millennium by midcentury to give the world a chance at stabilizing CO_2 concentrations at manageable levels.

Lowering carbon emissions by 70 to 80 percent over the next 40 or more years means that the major industrialized economies will essentially have to decarbonize their economic activity. To do so, vast investments will certainly be necessary. For example, to decarbonize the electric power sector in the United States would require replacing the roughly 750 gigawatts (GW) of fossil-fired power generation with renewable (or nuclear) generation, even before accounting for growth in electricity demand. Also, to the extent fossil-fired power plants get replaced with renewable power, it is likely that more capacity will be needed, since much renewable power has a lower capacity factor than traditional power generation, due to the intermittency of the renewable resource used—the wind does not blow all the time, and the sun does not shine all the time. Just to get orders of magnitude right, let's assume that we will need to build 1,000 GW of renewable and nuclear power plants to replace most of the existing stock of fossil-fired power plants and accommodate load growth over the next 40 years. Let's assume further that the average capital cost for renewable or nuclear power plants will be $10,000 per kilowatt, or $10 million per megawatt of capacity. This means that the capital cost alone of transforming the U.S. power sector would be on the order of $10 trillion, or one year of U.S. gross domestic product. Similar investments will be needed in other parts of the industrialized world, not to speak of the need to shift large developing countries like China and India onto a lower carbon power generation path.

In addition to large investments needed in the power and transportation sector, carbon markets have already led to the creation of totally new activities such as CDM development, monitoring, verification and certification, climate modeling, carbon price forecasting, offset development support including methodology development, and so on. Some of these activities are likely to be enormous in scope. For example, whether or not the CDM will continue to exist at the current level, the activities covered by it (primarily carbon-lowering activities in developing countries) will certainly be required and become increasingly important as low-cost carbon abatement options in industrialized countries are exhausted. In other words, it is entirely plausible that the market size for these types of activities will ultimately exceed the size of the markets for low-carbon technologies in the major industrialized countries.

ALPHA, INFORMATION, AND RESOURCES

In a world with carbon constraints, there will thus be a need to dedicate a sizable portion of resources to addressing the climate problem. In all likelihood, this process has already led to significant "alpha opportunities," that is, opportunities to earn superior returns for equal risk.

The simplest models of competitive economics suggest that in the long run equilibrium firms earn just their economic cost of capital. The only means of earning excess returns is for those markets to be not quite as efficient as the theoretical model assumes. In markets affected by carbon constraints, two sources of such potential excess returns are likely be particularly important: information advantages and control of scarce resources. In addition, capital market imperfections, taking the form of limited access to capital, and various barriers to entry due to high up-front fixed costs of investments or regulatory hurdles may also play a role.

Information is a key source of alpha in all markets: those with superior information are more likely to identify and invest in businesses and projects with the highest profit opportunities. In general, information advantages are harder to explore the larger and more mature a given market; making money through information advantages in soybean markets is much harder than to make money through information advantages in, say, gene therapy for some obscure disease. The reason for this general tendency is that the larger the market, the more people there are who are trying to identify opportunities—and the resulting profit opportunities tend to get arbitraged away quickly. Similarly, in mature markets, the uncertainties that drive performance are generally well understood, and therefore there are simply fewer pieces of critical information not yet well understood.

As mentioned above, carbon markets are likely to be very large. Yet, because they are likely to become large relatively quickly, there will be significant alpha opportunities, especially early on. They are also just about the opposite of a mature market when it comes to the role of information. Carbon markets are and likely will be incredibly complex: not only are the regulatory rules required to design and maintain a market system based on an international agreement to curb GHG emissions complicated, the carbon price formation process itself, and with it the economics of just about any project related to carbon markets, is about as complex as it gets. This is because the natural and economic interactions involved in determining GHG emissions themselves are complex. Existing models designed to forecast carbon prices require thousands of inputs, the modeling of complex interactions between those inputs, tricky intertemporal feedback loops that result not only from the physical world but also depend on the shape and form of climate regulation, and many assumptions about the physical and economic world not well understood (by anybody).

A simple example illustrates some of the complexities and hence some of the opportunities that come with having more information, broadly defined: Assume a potential investor is trying to assess the value of an investment opportunity in a reduced emissions from deforestation and degradation (REDD) project in Brazil. (REDD projects are discussed in detail in Chapter 10.) Physically, forests provide a carbon sink. Therefore, deforestation results in CO_2 emissions—trees that absorbed CO_2 before no longer do so after they are cut. It is assumed that the avoided emissions can be monetized by selling a carbon offset into a carbon market. The value of the project depends on the value of the offsets, which in turn depends both on the price of carbon and on the quantity of such offsets generated by the project, minus the cost of the project, where, economically, the opportunity cost (i.e., the highest valued alternative use of the project land) should be taken into account. Some of the costs of the project are not yet well understood since there are not very many REDD projects with a long operating history, but costs such as the costs of maintaining the forest, monitoring emissions (or, in this case, carbon sequestration) and verifying those can be reasonably estimated. However, estimating the opportunity cost of such a project is much harder. It is likely that at least a potential best alternative use of the forest land is for conversion to biofuel, producing crops such as sugar cane, for example. The value of such a transformation again depends on many factors, but one of them is also the price of carbon, since biofuels will be more attractive when compared to conventional transportation fuels the higher the price of carbon. So figuring out the opportunity cost will be very complex and in essence would require a model of carbon price formation that takes into account exactly the types of choices faced by our hypothetical investor. And this is only the physical portion of the problem. Many regulatory/market uncertainties that make the investment decision harder: what kind of limits exist of may exist in the future with respect to earning credits for REDD projects? What kind of complementary support mechanisms will exist to avoid deforestation? How about mandates (or limits) for biofuel production, which in turn could have a serious impact on biofuel prices and therefore the value of the alternative use of the forest? How about additional biofuel subsidies?

We hope this simple example illustrates that the information required to make an optimal investment decision in a carbon market–related asset can be very substantial. At present, many of the data and relationships between various data inputs are poorly understood. This suggests that players with superior data/analysis will find significant alpha investment opportunities.

Because of the tremendous information and analysis requirements in immature carbon markets, specialized players have emerged to help potential investors navigate the intellectual maze related to carbon. Point Carbon is one leading example of such a company, but others exist, and many more will likely emerge. While those companies themselves may represent alpha opportunities for potential investors, their services will likely play an important role in helping investors understand carbon markets and thus create the information advantages needed to exploit alpha opportunities.

Access to scarce resources is the other significant potential source of alpha in existing and emerging carbon markets. There exists a large literature on the role of resources as a source of sustainable competitive advantage: patents may be the best example of a scarce resource (intellectual property that nobody can copy without paying the owner), but nonpatented trade secrets, managerial talent, even company culture are often cited as resource-based sources of competitive advantage (which ideally translates into superior performance).

The most basic source of superior returns is, of course, to have access to a scarce natural resource. The owners of the coal mines closer to population centers may be able to benefit from their location advantage in a way that cannot be replicated—there is only a limited amount of land, and some land is worth more than other land. Carbon markets are likely characterized by many such resource advantages. Carbon abatement is driven by at least two important factors: the existing potential for GHG abatement on the one hand, and the rate and direction of technological progress on the other hand. In our fight to curb global GHG emissions, we will need to lower carbon emissions across a broad spectrum of activities. In the absence of technological change, we would likely start with the cheapest available options and, as we attempt to further lower emissions, be forced to incur increasingly higher costs for further emissions reductions. We would also expect that the price of carbon would rise over time with the cost of abatement. In such a world, the investors in early carbon reduction projects may well reap the benefits of future higher cost abatement (through higher carbon prices). Early investors in hexafluorocarbon (HFC, one of the most potent GHGs) were able to achieve reductions in HFC emissions—and since one ton of HFC is the equivalent of 21,000 tons of carbon dioxide, very significant carbon credit levels—at relatively low cost. Given that the total demand for offsets quickly exceeded the supply from HFC reduction projects, which is limited by the number of HFC emitting sources and other constraints, the price of offsets is no longer driven by the cost of reducing HFCs, but rather by some other, more expensive set of carbon offsetting projects. As a result, the investors in the early HFC projects benefit from "inframarginal" rents; that is, the value of the carbon offsets produced by their projects far exceeds the cost of producing the offset. It is likely that many similar opportunities exist: much of GHG abatement will depend on lowering emissions or offsetting carbon emissions through location-dependent technologies: wind, solar, geothermal, tidal, and other renewable energy sources are likely much more economical in certain locations than in others, forests more productive carbon sinks in some parts of the world than others, and so on. Whoever identifies those "best locations"—in turn likely requiring superior information (see above)—should be able to create a sustainable resource-based competitive advantage or, put differently, achieve superior alpha performance.

The story is somewhat more complex, however, due in particular to technological change, but potentially also due to regulatory change. The quest to lock up the best sites would suggest that moving early is important. However, technological and potentially regulatory change provides an argument for waiting and potentially reduces or, in some cases, eliminates, resource-based competitive advantages of the kind described so far. For example, for a given wind turbine technology, a certain place in the desert may be the best site for a wind farm. Even though long transmission lines will be needed to connect the wind farm to the grid, the constant and high wind speed of the desert site may appear very attractive. However, it is possible that future generations of windmills achieve significantly higher efficiencies. If the efficiency improvements are large enough, they may make it more attractive and cheaper to build a wind farm closer to the grid, thus making the original site no longer the most economically attractive. Understanding the relationship between technological progress and the quality of projects, based on their resource attributes, at any given time will be an important requirement for potential investors when assessing various projects.

A final potential alpha game changer relates, again, to the evolution of the regulatory framework governing carbon markets. As we discussed above, carbon markets are both created by regulatory fiat and are characterized by a large amount of regulation. It is, therefore, not surprising that regulatory changes can and do have a substantial impact on value. Because future regulation is, by definition, uncertain investments today suffer from more regulatory risk than investments postponed. Just like technological change, regulatory changes can thus also reduce or eliminate alpha related to unique and scarce resources. One example of such a change would be an investment in a carbon offset project under strict additionality guidelines, whose value subsequently declines as either the range of acceptable technologies is widened or the threshold for additionality is lowered through regulatory action. The uncertainty of future regulatory action suggests both an additional source of information-based competitive advantage and the need for options-based approaches to assessing the benefits of early action versus waiting.

In summary, assuming that the community of nations will ultimately tackle the problem of climate change, the resulting challenge and market will also be very large. This will undoubtedly attract many sophisticated market players, who will ultimately drive the markets closer to perfectly competitive conditions and in the process eliminate many opportunities to earn extraordinary returns. However, carbon markets are already and will likely be among the most complex when it comes to understanding the relationships of multiple variables that affect both the supply and the demand side and hence ultimately determine the price path of carbon. Not only do regulatory and political decisions also affect both supply and demand directly, they are also the source of additional and important risk and uncertainty. As a consequence, there will be tremendous opportunities to leverage information and resource advantages as sources of extraordinary returns on investment. It is likely that players investing early and sufficiently into systems and information needed to understand the complex linkages (or at least understand them better than others) will be the primary beneficiaries of the existing information asymmetries. They will be well equipped to identify those project opportunities that are based on scarce resources and they will be better equipped to estimate the value of the myriad of carbon related project opportunities offered in the market.

CONCLUSIONS AND OUTLOOK

Carbon markets around the world are still in the earliest stages of development. Given the enormity and complexity of the underlying problems these markets are designed to address, these markets will be large by any scale and in existence for decades to come.

Carbon markets likely represent enormous alpha opportunities because of the complexity of the underlying problem and the continuously evolving regulatory framework creating carbon markets. We have identified information and resource-related advantages as the two most likely sources of competitive advantage and hence superior financial returns, but other market imperfections also likely present profit opportunities. Because of the transgenerational nature of the climate change challenge and the complexity of the political process that is creating and adapting carbon markets, carbon-related investment opportunities will be continuously created. As of the writing of this book, the evolution of U.S. federal carbon regulation and the ability of the international community to agree to a far-reaching post-Kyoto agreement are two of the most important drivers of regulatory uncertainty. Another important long-term source of uncertainty is extent to which the regulatory framework of setting up carbon markets is vulnerable to dilution if the cost of mitigating catastrophic climate change risk becomes large enough to threaten shorter-term political willingness to address the problem. Whether or not this risk exists depends largely on uncertain scientific progress in carbon emissions–lowering technologies as well as the political backbone of key participants in carbon markets, notably the European Union, the United States, and major developing countries such as China, India, and Brazil. It appears that at least as of the time of writing, there is some reason for cautious optimism as the European Union has unilaterally imposed on further significant carbon reductions in one of the most difficult post–World War II economic climates.

CHAPTER 10

Market-Based Solutions to Reduce Emissions from Deforestation and Degradation (REDD)

Stefanie Engel, PhD, Charles Palmer, PhD, and Martin Berg

Thereis growing awareness of the need to put a value on the services nature provides and, more specifically, affix a price to the human activities that use these resources. These activities are typically classified as land use, land-use change, and forestry (LULUCF) and, in aggregate account for about 30 percent of global greenhouse gas (GHG) emissions. Of these activities, forestry receives the most attention because "forestry, as defined by the Intergovernmental Panel on Climate Change (IPCC), produces around 17 percent of global emissions, making it the third largest source of GHG emissions—larger than the entire global transport sector. Annual forest emissions are comparable to the total annual CO_2 emissions of the United States or China. If we do not tackle deforestation, it is highly unlikely that we could achieve a CO_2e stabilization target that avoids the worst effects of climate change" (Eliasch 2008).

More so than the other four major categories of environmental investing, LULUCF is a work-in-progress, primarily because current government policies place little or no value on protecting forests and other habitats. But policy makers are coming to understand the necessity of including LULUCF activities in post-Kyoto agreements. With this clarity and certainty

Stefanie Engel and Charles Palmer are members of the Institute of Environmental Decisions. Martin Berg is vice president of Merrill Lynch Commodities (Europe) Limited.

in policy and improvement in reporting and monitoring techniques comes the development of new financing mechanisms that link forest abatement to the carbon markets. The most common of these is reducing emissions from deforestation and degradation (REDD). The aim of REDD and other such solutions is to mitigate GHG emissions and provide other ecosystem services that are critical to many rural and urban economies. With the proper policies and governance structures in place—and a well-designed carbon market—these REDD solutions could reduce deforestation rates by up to 75 percent in 2030 (Eliasch 2008). However, these solutions require financing mechanisms that, when properly structured, could also prove to be attractive sources of environmental alpha for institutional investors.

This chapter is divided into two sections. The first, written Dr. Stefanie Engel and Dr. Charles Palmer, discusses the critical scientific and policy issues underpinning the development of REDD market-based solutions. Martin Berg then builds upon this foundation and delineates the investment opportunities and challenges with these REDD solutions.

SECTION ONE: SCIENCE AND POLICY

Evidence for anthropogenic warming of the climate system as a consequence of GHG emissions, including CO_2 (carbon dioxide), into the earth's atmosphere is unequivocal (IPPC 2007a). Annual CO_2 emissions from deforestation in tropical and subtropical countries accounts for up to a fifth of global emissions, the second largest source of all GHG emissions (Baumert, Herzog, and Pershing 2005). It also makes up more than a third of developing countries' emissions. Conserving carbon stored in biomass could be a cost-effective strategy to mitigate future climate change impacts (see Chomitz, Buys, De Luca, Thomas, and Wertz-Kanounnikoff 2006; and Stern, 2007). Reducing emissions from deforestation was, however, excluded from the climate change regime that resulted from the Kyoto Protocol negotiations held during the 1990s. The first commitment period of Kyoto is due to end in 2012. At the Bali Conference of the Parties (COP) in December 2007,

---

Much of the material in this chapter is taken from "Painting the Forest REDD," a working paper written for the Institute of Environmental Decisions (Engel and Palmer 2008a) and *Avoided Deforestation: Prospects for Mitigating Climate Change*, which assesses the potential of REDD mechanisms from the perspective of economics and policy making (Palmer and Engel 2009). Published by Routledge in April 2009, this volume highlights the importance of avoided deforestation as part of a global strategy to mitigate the buildup of anthropogenic GHG emissions in to the earth's atmosphere.

countries agreed to create a mechanism for "reducing emissions from defor-estation and degradation" (REDD)[1] as a potential component of a post-2012 climate change regime (United Nations Framework Convention on Climate Change [UNFCCC] 2007).

More precise rules and modalities are to be developed by COP-15, which is due to take place in Copenhagen, in December 2009. Many open questions remain on how reducing deforestation could be incorporated credibly into a climate regime. There is therefore a need to take stock and consider the merits of such a mitigation strategy and how it might be implemented on the ground.

This chapter aims to introduce the role of forests in mitigating climate change and summarize some of the key issues and research covered in Palmer and Engel (2009). This section presents the climate change problem, the role of forests, and the policy debate so far. It also focuses on research to as-sess the cost-effectiveness of avoided deforestation as a strategy to mitigate climate change, and examines the barriers to the adoption of such a strat-egy, primarily those related to policy and institutions. We look at policy design, with a focus on overcoming additionality and leakage constraints and maximizing the efficiency and effectiveness of potential avoided defor-estation schemes. The main findings are analyzed alongside some further issues for discussion. It should be noted up front, however, that provid-ing a full review of related literature is beyond the scope of this chapter.[2] Excellent studies are constantly adding to the existing body of knowl-edge relevant for REDD policy design. This section concludes by address-ing some of the key, remaining open questions along with suggestions for future research.

Forests, Climate Change, and Avoided Deforestation

According to widely cited data published by the World Resources Institute (see Baumert et al. 2005), global anthropogenic GHG emissions, dominated by CO_2, are mainly given off via the burning of fossil fuels, and from agriculture and land-use changes. Emissions from deforestation and forest degradation occur as carbon stock is depleted and released to the atmosphere through changes in forest and other woody biomass stock, forest and grass land conversion, the abandonment of managed land, and forest fires. A 20 percent decrease in forest area since 1850 has contributed to 90 per-cent of emissions from land-use changes (IPPC 2001). Throughout the 1990s around 1.5 billion tons of carbon (GtC) was released annually through deforestation (Gullison et al. 2007). Two countries, Indonesia and Brazil, dominate CO_2 emissions released through deforestation and as a result are, respectively, the third and fourth largest GHG emitters in the world, behind the United States and China (cited in Baumert et al. 2005; Houghton 2005).

Impacts of Climate Change Anthropogenic interference in the climate system is a real and growing threat to people, economies, and the environment (Chomitz et al. 2006). On current trends, the average global temperature could rise by two to three degrees Celsius within the next 50 years. This rise is likely to rapidly change the Earth's climate, for example, leading to rising sea levels and a higher frequency of heat waves and heavy precipitation (IPPC 2007a). Business-as-usual or "baseline" climate change implies increasingly severe economic impacts if action is not taken to mitigate the worse effects.

Climate change can perhaps be characterized as the world's largest "market failure" (Stern 2007). The earth's atmosphere, into which anthropogenic GHGs are emitted, is a global public good; that is, it is nonrival and nonexcludable. These emissions are an externality in that those who produce them impose social costs on the world and future generations but do not face the full consequences of their actions. The actual source of emissions, whether producer or consumer, rich or poor, is irrelevant to the overall growth in global GHG stocks and the corresponding future changes in the climate. Nevertheless, the worst impacts of climate change are expected to fall disproportionately on people living in some of the poorest regions of the world. People living in these regions are the most vulnerable to adverse changes in, for example, food production and water resources.

Climate Change Policy The global causes and consequences of climate change imply the need for international collective action for an efficient, effective, and equitable policy response. The first global attempt to put a price on the social costs of emissions by stabilizing the amount of GHG in the atmosphere was seen in the formation of the United Nations Framework Convention on Climate Change (UNFCCC). Ratified by 182 Parties as of May 2008 (UNFCCC 2008), the Kyoto Protocol of the UNFCCC originally entered into force in 2005. It committed Annex I, mainly industrialized countries, to reducing their collective GHG emissions by about 5 percent below their 1990 levels by 2008–2012. In fulfilling these commitments, countries are able to achieve reductions in their emissions through several mechanisms, including the Clean Development Mechanism (CDM). The CDM allows entities in non–Annex I countries to develop "offset" projects leading to verified reductions in GHG emissions emitted from Annex I countries. So-called certified emissions reductions (CERs) are then transferred to Annex I countries at a price set by the carbon markets.

Reducing GHG emissions in order to stabilize the climate requires the deployment of a portfolio of GHG emissions-reducing technologies along with the application of appropriate and effective incentives (IPCC 2007b). These include adaptation and mitigation measures such as carbon storage and capture and reducing deforestation, all with varying, generally uncertain costs. None of these measures on their own, for example, the halting of all deforestation, would achieve the UNFCCC's goal (Pacala and Socolow 2004). But conserving forest carbon could likely be an important part of the climate change solution, particularly if it proves to be cost-effective compared to other mitigation options (see Chomitz et al. 2006; and Stern 2007). Negotiations on the types of admissible projects in Kyoto included a range of options for increasing forest stock and removing carbon from the atmosphere. Reducing emissions from deforestation was discussed, but was finally excluded from the CDM (discussed later in this section).

Forests as Carbon Sources and Sinks   The Forest Resources Assessment (Food and Agriculture Organization of the United Nations [FAO] 2006) estimated that a third of the Earth's land surface, up to four billion hectares, is covered by forest. Of this, around half is located in the tropics and sub-tropics. The largest intact tropical forests are found in the Amazon Basin (Brazil), the Congo Basin (Democratic Republic of the Congo) and in the Indo-Malayan region (Indonesia, Malaysia, and Papua New Guinea). These forests provide important traded and nontraded environmental goods and services, including carbon. Tropical forests have particularly high carbon stocks, perhaps holding as much as 50 percent more carbon per hectare than forests in other regions (Houghton 2005). In terms of economic value, even relatively low traded carbon values have been found to comfortably dominate the nonmarket values of other tropical forest environmental services (see Pearce, Putz, and Vanclay 2002). These include direct use values, although perhaps excluding the returns from unsustainable timber extraction.

Over the past century, tropical deforestation and forest degradation have increased dramatically. The former occurred at an average rate of 13 million hectares per year, between 1990 and 2005 (FAO 2001, 2006). Brazil and Indonesia accounted for, on average, around 40 percent of annual deforestation by area over this period. The causes of the continuing loss and degradation of tropical forests are many, varied, and complex (Chomitz et al. 2006; Geist and Lambin 2002; Kaimowitz and Angelsen 1998). However, understanding these is important for the design and implementation of policy to reverse their effects, whether related to policy to reduce CO_2 emissions or not. This requires identifying the underlying market and policy failures and understanding how these relate to activities both inside and outside the forest sector. The latter include those related to agriculture, migration, and infrastructural development. Recent government and nongovernment efforts to slow down or reverse overall deforestation and degradation trends, either through forest policy or policy made in other sectors, have been relatively unsuccessful for various reasons (see Bulte and Engel, 2006). Given the many interlinked pressures on forests, the challenge now for climate policy is to design a strategy for capturing the carbon value of natural forest stock that is not only effective but also efficient and equitable.

Avoided Deforestation and Climate Change Policy   Without effective policies to slow deforestation, business-as-usual tropical deforestation could release up to 130 GtC by 2100 (Houghton 2005). "Avoided deforestation" is a concept wherein countries are compensated for preventing deforestation that would otherwise occur (Chomitz et al. 2006). Reducing emissions by slowing deforestation could be a substantial and important component of climate mitigation policy, and has been discussed as such by researchers and policy makers for a number of years (see, e.g., Brown, Sathaye, Cannell, and Kauppi 1996; Schneider 1998). The available evidence shows that potential carbon savings from slowing tropical deforestation could contribute substantially to overall emissions reductions. Moreover, forests protected from deforestation could persist in the coming decades despite "unavoidable" climate change (Gullison et al. 2007). Possible side benefits from the realization of natural forest carbon values include other forest environmental values such as biodiversity.

Avoided deforestation projects were excluded from the 2008–2012 first commitment period of the Kyoto Protocol's CDM due to a number of concerns revolving around sovereignty and methodological issues (Fearnside 2001; Laurance 2007). The former arose as a consequence of forests *per se* not being considered as a global public good despite the public good nature of some forest services. Since exclusion, discussions have been ongoing to try to resolve these concerns through, for example, the UNFCCC's recent two-year initiative (Subsidiary Body on Scientific and Technical Advice [SBSTA]). This has acted as a useful forum for assessing new policy approaches and incentives for avoided deforestation in developing countries. Note that the SBSTA process included discussions about emissions from forest degradation, thus expanding the scope of potential policy mechanisms from RED (reducing emissions from deforestation, or avoided deforestation) to REDD. Meanwhile tropical forest nations such as Papua New Guinea, Costa Rica, and Brazil have been floating various initiatives to protect forests through utilizing their value as carbon sinks. (For example, at the COP-11 in Montreal in 2005, a coalition of 15 rainforest nations led by Papua New Guinea and Costa Rica floated a proposal to allow CDM-type credits, bought by industrialized nations, in exchange for reducing deforestation. See www.rainforestcoalition.org/eng/.) Forest carbon finance has also been endorsed by the United Nations, the World Bank, and the majority of nation states, with the World Bank's Forest Carbon Partnership Facility (FCPF) aiming to attract US$300 million in donor funding for pilot REDD schemes (World Bank Carbon Finance Unit, 2008).

The Bali COP is part of an ongoing process that will carry on through 2008 and 2009. It is hoped that a post-2012 international climate regime will be agreed by the end of 2009 at COP-15 in Copenhagen. Whether avoided deforestation or REDD will be included in a final framework agreement and what this arrangement might look like is beyond the scope of this chapter. Instead, and for the most part, it looks at what might be gained from including REDD as a feasible option in a post-Kyoto agreement and at how some of the challenges of such inclusion could be tackled from the perspective of economics and policy making. The following three sections, respectively, investigate the cost-effectiveness of avoiding deforestation or REDD, the policy and institutional barriers to implementing REDD, and some insights for effective and efficient REDD policy.

The Cost-Effectiveness of Avoiding Deforestation

Different approaches have been employed by researchers to estimate the costs of avoiding deforestation. In general, because avoiding deforestation involves a change in land use, opportunity costs (i.e., the costs of foregone net benefits from the next best alternative activity) tend to constitute the most important source of costs. Building on background research carried out for the *Stern Review* (Stern 2007), Grieg-Gran (2009) examines the cost-effectiveness of avoided deforestation as a mitigation option using empirical data for eight tropical countries. Her estimates of average opportunity costs per tonne of CO_2 avoided range from US$1.20 to $6.70, depending on the scenario under consideration. Grieg-Gran also highlights the spatial variation in cost estimates across countries. While perhaps "cheap" in, say, parts of Africa, avoided deforestation may turn out to be less cost-effective in Indonesia, depending on land use. Additionally, Grieg-Gran incorporates administration costs into her estimates. These particular transactions costs range from US$0.10 to US$0.20 per tonne of CO_2. Average cost estimates overall compare favorably with most other mitigation options, although higher transaction costs could potentially account for a substantial proportion of the total cost per tonne of CO_2 avoided through reducing deforestation.[3]

Combating deforestation will, of course, still require substantial funds. Rametsteiner, Obersteiner, Kindermann, and Sohngen (2009) develop a global land-use model, which indicates that a 50 percent reduction of carbon emissions from deforestation over the next 20 years would require financial resources of some US$33 billion per year. This is a figure that easily exceeds all current annual Overseas Development Assistance (ODA) spending on forestry. Given that ODA alone would fall short of funding requirements, a combination of funding sources and policy mechanisms would be required for reducing emissions from avoiding deforestation. Carbon trading, now active in various forms all over the world, could be one potential component of funding. (See Chapter 8 for a discussion of carbon trading.) Using a global timber market model, Sohngen (2009) confirms the conclusion that carbon credits for reductions in deforestation may be cheap in comparison to other options both in forestry (e.g., afforestation) and in the energy market. In a related study, Tavoni, Sohngen, and Bosetti (2007), found that reductions in deforestation can achieve levels of annual sequestration similar to those achieved with carbon capture and storage, but earlier and at lower prices. These studies highlight the need to include avoided deforestation or REDD as a mitigation option in global climate change models. Currently, these models often ignore this option.

Policy and Institutional Barriers

There are a number of important policy, institutional, and methodological barriers that prevented the inclusion of REDD strategies in the Kyoto Protocol, although progress has been made since to overcome these (Johns and Schlamadinger 2009). It seems that the most contentious issue relates to the financing of REDD activities. The challenge is to provide adequate, consistent, long-term funding of REDD activities, while also providing real and additional climate benefit. Developing countries have been arguing strongly that financial assistance for REDD should not be drawn from existing development-funding streams.

Current proposals mostly fall into one of three categories: (1) trading REDD credits in the carbon market (similar to CDM afforestation/reforestation credits); (2) a voluntary, fund-based approach not linked to the carbon market; and (3) indirect market approaches, drawing proceeds from the market but without a direct link to market credits. Hybrid approaches appear promising in that they could capture the larger financing potential of the market while benefiting from advantages of a fund, such as allowing for equity and biodiversity considerations in targeting. Alternative or complementary funding could be obtained through a tax on Kyoto mechanisms and/or on emissions from international air and maritime transport or through an obligatory contribution by Annex I countries to a revolving compliance fund (Dutschke and Wolf 2007). Another issue is whether a REDD mechanism should follow a CDM-type project-based approach, a national approach, or a "nested" approach. A strong argument for a national approach relates to the issue of "leakage," that is, the possibility that REDD in one area may be at least partially offset by increases in deforestation and degradation elsewhere. While international leakage can still occur under a national approach, it could be addressed in different and separate ways compared to the leakage that might occur at the subnational project scale (see "Insights for Effective and Efficient REDD Policy" later in this chapter).

Significant progress has been made in identifying and analyzing the large range of drivers of deforestation at varying spatial scales (Chomitz et al. 2006). It is clear that any feasible REDD mechanism needs to be built on an understanding of these drivers and be flexible enough to support solutions tailored to specific local and regional conditions. A limited capability to monitor deforestation and estimate forest-based emissions has long been another barrier to the inclusion of REDD in an international climate regime. Recent advances in the field of remote sensing in combination with appropriate ground truthing provide a solid basis, as described both in Johns and Schlamadinger (2009) and by Moutinho, Cenamo, and Moreira (2009). The remaining uncertainties can be treated by taking lower-bound estimates of the quantity of emissions reduced. Yet improvement in access to data and training in new technologies is still required. The degree to which forest degradation and forest regrowth are to be considered as activities included in a REDD mechanism is another crucial issue still to be resolved. While there are good reasons to include both activities, the difficulty and increased cost involved in emission monitoring and estimation poses a real challenge. Baseline setting—that is, the estimation of change in, say, deforestation rates or emissions levels in the absence of policies to change these—is another highly political issue, which is discussed in further detail later in this chapter.

Institutional barriers, while they have not been the center of negotiations thus far, have now increasingly become a focus of attention and discussion, as countries, nongovernmental organizations, and financing institutions grapple with the so-called readiness process. This is supposed to ascertain what it will take for a country to be ready to participate in an international REDD mechanism (Dutschke, 2008). Institutional barriers include, for example, appropriate land tenure and forest protection law; adequate capacity for monitoring and enforcement; and effective engagement of civil society, including forest-dependent and indigenous communities.

Moutinho et al. (2009), along with Palmer and Obidzinski (2009), discuss some of the preceding issues as they relate to the two main contributors of CO_2 emissions from forestry and land-use changes in the world: Brazil and Indonesia, respectively. In addition to illustrating some of the country-specific policy and institutional challenges in more detail, these researchers highlight the fact that REDD requires a coordination among different levels of governance of the implementing country (from local to national) as well as across different sectors of the economy. Policy reforms and initiatives are ongoing in both countries that would go hand in hand with an international mechanism.

Moreover, the important point is made that pressures on forests, whether in the absence or presence of an effective REDD mechanism, are likely to increase. Increases in agricultural product prices such as biofuels, exchange rates effects, and road construction may all enhance pressure on forests, yet are often ignored in baseline estimation. In Indonesia, the situation is particularly dramatic as growing demand for timber products and biofuels has driven Indonesian government plans for a massive expansion of these sectors. This expansion is supported through substantial government subsidies, including the use of timber stands as collateral in plantation development. These have the effect of reducing deforesters' costs with implications for the estimation of deforestation baselines. Palmer and Obidzinski (2009) draw attention to the potential that the prospect of an international REDD mechanism may induce perverse incentives, by slowing the reforms necessary to correct government failures, in order to achieve higher baselines. This may, however, be more of an issue with the adoption of a business-as-usual baseline projection rather than one drawn from a historical reference period. In Brazil, land speculation and "land grabs" by prospective landowners, driven by agricultural commodities' and livestock prices, have long dominated deforestation behavior at the Amazon frontier. Property rights claims underlying such behavior raises similar baseline issues as in Indonesia but may also present opportunities in the context of a REDD payment regime, as discussed further in "Issues Related to REDD" later in this chapter.

Michaelowa and Dutschke (2009) pick up on one of the major concerns related to including REDD credits in the carbon market: the fear that the potentially large supply of carbon credits from reducing emissions through avoiding deforestation could upset the balance of the market. The authors estimate and project expected credit supply and market demand scenarios for carbon until 2020. Their supply analysis incorporates governance problems plaguing most countries with high deforestation rates, while demand is determined by the relative stringency of the climate policy regime. They find that, due to governance problems in many tropical countries, the credit supply from REDD would pick up slowly during the initial years of program implementation. Nevertheless, any integration of credits from reducing emissions through avoiding deforestation into the carbon market should be accompanied by long-term target setting.

Insights for Effective and Efficient REDD Policy

This section addresses the issues of leakage estimation, baseline setting, dealing with trade-offs in objectives, and increasing the efficiency of current mechanisms compensating for avoided deforestation.

Murray (2009) focuses on the importance of recognizing, estimating, and, where possible, ameliorating the risks of leakage from compensation policies that are likely to be applied to a subset of countries with deforestation potential. He illustrates some of the different approaches that can be used to estimate leakage empirically and of the relatively small number of studies that have attempted to do so to date. The results of these studies suggest that international leakage from avoided deforestation policies could be substantial if not addressed in policy design. One way to reduce leakage is to expand the scope of policy coverage as widely as feasible. Scope expansion could involve covering more countries or more activities.

However, an important point made by Murray is that expanding the number of countries involved in a voluntary system involves the balancing act of enhancing incentives for their participation through, among other things, generous baselines against the need to maintain the environmental integrity of the system by not crediting "hot air." Expanding the scope beyond deforestation may both help lure countries with low baseline deforestation rates into the system and ensure that deforestation emissions are not reduced at the expense of carbon losses elsewhere in the forest sector. Covering all forest carbon in an international compensation system, however, raises some concerns about spurring land-use changes that could potentially undermine other environmental objectives such as biodiversity and water provision unless addressed via agreed-upon protocols, for example, discouraging the conversion of native ecosystems to plantations.

As mentioned earlier, one crucial issue for any REDD mechanism is the setting of the hypothetical baseline (or business-as-usual projection) against which REDD progress is measured. A baseline for forest conservation has two main components: the projected land-use change and the corresponding carbon stocks in applicable pools in vegetation and soil. For the latter, there are now standard values recommended by the IPCC for different vegetation types that can be used (Dutschke and Wolf 2007). The most commonly discussed method for baseline estimation is the use of some sort of national historical reference period of emissions from deforestation. Alternative, more sophisticated approaches have been proposed, as described both by Johns and Schlamadinger (2009) and Murray (2009). These include more or less sophisticated projections of past trends into the future or a normative baseline.

Harris, Petrova, and Brown (2009) emphasize the need for a standardized, scalable baseline approach that is accurate, transparent, credible, and conservative. They briefly review three specific models for baseline estimation applied elsewhere and evaluate how they differ in terms of transparency, accuracy, and precision; applicability at various scales; compatibility with international requirements; and cost-effectiveness in terms of data, time, and expertise needed for application. They then describe in detail a spatial modeling approach that ranked highest in their evaluation. The so-called GEOMOD approach is interesting because it can be used to estimate a deforestation baseline at the project, regional, or national scale and to predict the spatial location of deforestation. A weakness of the approach is that estimated overall rates of deforestation are based purely on historic rates, while driving factors are used to predict location only. Such weaknesses, however, need to be considered in light of the overriding need for setting the most objective, transparent, and comparable REDD standards possible at the international level.

Pfaff and Robalino (2009) further demonstrate the importance of correct baseline estimations. They explain how impact evaluation and policy planning are complicated by several factors: the inability to observe how land choices would have differed without a policy; the fact that policy location may be affected by private and public choices; and the spatial and temporal interactions among land use choices. Using empirical examples from Costa Rican policies, specifically the widely cited payments for environmental services (PES) policy and Costa Rican parks and protected areas, the authors convey how impact analysis could address these hurdles. Pfaff and Robalino's results also show that forest conservation policies that appeared to have been very successful at first sight may have added much less in the way of conservation benefits once the appropriate baselines are considered. In other words, once policy impacts were evaluated in a thorough manner, conservation policy resulted in much lower levels of "additionality" than was originally expected. These results have two important implications for REDD. First, they cast doubt on very simplistic baseline approaches, for example, those based on simple historic deforestation data. Second, they highlight the need for more efficiency in conservation spending.

Finding ways to increase the efficiency of forest conservation spending, whether at the international, national, or subnational level, is important for several reasons. First, the actual cost of REDD will depend on how efficiently available funding is used, which in turn depends on the design of a future REDD mechanism. If REDD is to be achieved through the establishment of some type of international fund like the Amazon Fund (www.amazonfund.org/), the fact that financial sources are limited requires a procedure for deciding which countries, regions, or projects are selected for REDD funding. Moreover, increasing the efficiency of current forest conservation spending can be seen as an important complement to a strategy of raising additional funds for reducing carbon emissions through avoided deforestation. As argued by Engel, Wünscher, and Wunder (2009) in work on funds' targeting in Costa Rica's PES scheme, demonstrating efficiency can be important in attracting new funding sources, particularly from the private sector. By increasing the efficiency of existing programs, funds can be freed up for additional programs or for inclusion of additional sites in a given program ("achieving more bang for the buck").

Both increasing cost-effectiveness of funding and dealing with institutional or policy-related barriers on the ground requires a careful consideration of policy choice and policy design. There is a variety of policies that could be applied to avoided deforestation from so-called "command and control" instruments such as state-protected areas to ones based on a market mechanism (see Gupta et al., 2007). Policy choice may depend on a number of factors, including the source(s) of market failure and, in the particular case of deforestation, the identification and level of understanding of the drivers and agents of deforestation (Engel, Pagiola, and Wunder 2008).

PES, the focus of both Engel et al. (2009) and Alix-Garcia, de Janvry, and Sadoulet (2009), is an increasingly used instrument for both financing and implementing forest conservation and thus has potential in application to payments for REDD. The relevance of PES to the REDD debate is demonstrated by cost studies such as Grieg-Gran (2009), which tend to assume that some type of PES will be put in place to compensate landowners (or land users) for the profits forgone by avoiding deforestation. The defining characteristic of PES lies in its conditionality (Wunder 2005): Payments are made by an environmental service buyer conditional on the environmental services provided by an environmental service seller. Such "beneficiary pay's" positive incentives have not only been shown to be relatively more cost-effective compared to more indirect conservation approaches (see Ferraro and Kiss, 2002) but may also be politically more acceptable than instruments such as taxes on forest products or land clearance.

The Costa Rican national PES scheme is often considered a pioneer and leading model of PES. Payments there are made largely for avoided deforestation, although as shown by Pfaff and Robalino (2009), these have not had the impact on deforestation rates claimed by the scheme's proponents. In establishing a national or regional PES scheme, questions arise on how land parcels are selected for program inclusion, and about the size and allocation of conservation payments. Voluntary PES projects within a country could lead to the leakage of emissions to nonenrolled parcels. While national accounting may capture this, there will still be efficiency issues, which could be at least partially dealt with through improved payments targeting. Engel et al. (2009) show that the amount of environmental services achieved with a given budget for a region in Costa Rica could be nearly doubled through improved targeting in-site selection. In particular, they develop a tool for selecting among applicant sites on the basis of three criteria: the amount of environmental services provided by the site; the probability that these services would be lost in the absence of PES (additionality); and the cost to landowners of providing the services. Alix-Garcia et al. (2009) find similar efficiency gains when targeting is considered for the national PES scheme in Mexico.

Engel et al. (2009) along with Alix-Garcia et al. (2009) both deal with issues related to national-level schemes.[4] The lessons drawn from these chapters can also, to some extent, be applied to international-level mechanisms. The CDM is an example of an international PES scheme. One of the main declared objectives of the World Bank's prototype FCPF is to test a system of performance-based incentive payments for REDD services. There is considerable spatial variation in the carbon content of forests. Moreover, a fund-type REDD mechanism may also want to consider additional environmental services like biodiversity conservation or equity arguments, all of which can in principle be integrated as targeting criteria for PES, as demonstrated by Engel et al. Threat levels and opportunity costs may vary even more in space. Some of the methods presented elsewhere in Palmer and Engel (2009) could be applied here, for example, the estimation of location-specific deforestation baselines à la Harris et al. (2009). Engel et al. and Alix-Garcia et al. also discuss scientific, administrative, and political challenges of targeting and how these may be overcome. Such challenges may be even greater at the international level. Efficiency gains need to be weighed against political feasibility and increases in implementation costs.

DRIVING CAPITAL TO THE RAINFOREST CANOPY: A Case Study in Iwokrama, Guyana

By Hylton Murray-Philipson and Andrew Mitchell

In a deal announced in March 2008, investors bought into a private-equity company called Canopy Capital, which had paid for the rights to develop value for the ecosystem services (ES) produced by the 371,000-hectare Iwokrama rainforest reserve. Services included in the deal

Hylton Murray-Philipson is managing director and Andrew Mitchell is executive director of Canopy Capital Ltd, London, U.K. Canopy Capital was established in 2007 to drive capital to the rainforest canopy. Twenty percent of the company is held by the Global Canopy Programme, a U.K. charity dedicated to the research and preservation of tropical forests. The remaining 80 percent is funded by a dozen international investors, including the Waterloo Foundation. The Global Canopy Programme provides advice and technical expertise to Canopy Capital. All of Canopy Capital's investments will benefit local communities and conservation efforts in tropical rainforests as well as financial investors.

are rainfall generation, climate regulation, carbon storage and sequestration, biodiversity maintenance, and water storage. Gifted by Act of Parliament to the Commonwealth for research into sustainable forest management, the reserve lies at the heart of the Guiana Shield, the most intact rainforest system left in the world.

Funds received from Canopy Capital are used to continue the management of the Iwokrama forest in accordance with its philosophy of conservation through sustainable best practice, providing livelihoods and business partnerships for the 7,000 Makushi people living in the forest and the surrounding area. Income from the deal will help to make Iwokrama financially independent of institutional donors by 2010 in accordance with the reserve's business and research plans. In the longer term, up to 90 percent of any investment upside will go to Iwokrama and the people of Guyana.

Canopy Capital is exploring various approaches to securing substantial investment in ES. In particular, it is looking at marketing them through an Ecosystem Service Certificate attached to a 10-year tradable bond, the interest from which will pay for the maintenance of the Iwokrama forest.

The Canopy Capital/Iwokrama deal opens the way for financial markets to price the "eco-utility value" of rainforests. Simply put, why would the world pay an oil company US$100 per tonne to perform industrial carbon capture and storage (CCS) while failing to pay for the natural CCS function of the standing forest? This makes even less sense when bearing in mind the co-benefits associated with living carbon in the canopy. In order for such markets to work at scale, however, governments must create the framework to drive demand. It is still open to debate whether the role of tropical forests, and furthermore that of standing forests, in the battle against climate change will fall within the UNFCCC, within the Convention on Biological Diversity (CBD), or within some other mechanism.

The key scientific issues to overcome are to demonstrate the value of tropical forest ES such as rainfall generation at local to global scales and to create methods to monitor, report, and verify them. Politically, nations must be convinced of the need to create a mechanism for proactive investment in the ES delivered by standing forests that will ultimately be cheaper than dealing with the costs if they are lost.

Issues Related to REDD

In this section, some further thoughts are presented relating to the role of REDD in climate change mitigation, permanence in REDD carbon benefits, the importance of incentives and avoiding "hot air" credits, and governance.

Role of REDD in Global Climate Change Mitigation   There are several reasons why the inclusion of REDD, or at least RED (avoided deforestation), in an international climate regime should be considered as part of a portfolio of mitigation options alongside an agreement containing stringent curbs in global GHG emissions.

First, as demonstrated by Michaelowa and Dutschke (2009), modest emissions reduction targets imply that in the midterm a glut of REDD credits may lead to low carbon prices that remain low. Cheap prices for combating climate change, while intuitively a good thing for climate policy (see Chomitz et al. 2006), may dampen incentives for more long-term investments in other mitigation options such as improving energy efficiency (Kremen et al., 2000; Schneider, 1998). This may have serious implications for long-run climate policy objectives, although technological change needs to be complemented with public investment in research and development as well as price incentives. For the worst predicted effects of climate change to be overcome, sharp curbs of perhaps up to 70 to 80 percent of current global emissions may be required (IPPC 2007b).[5] In 2008, cuts of 50 percent were agreed, in principle, by the Group of Eight (G8) at its annual meeting, which are expected to be achieved by 2050 (G8 2008). Since deforestation accounts for around a fifth of current emissions, it is obvious that incentives will also be necessary to ensure emissions reductions in other sectors, particularly energy production, transport, and industry, in order to realize ambitious global targets (IPPC 2007b). A glut of REDD credits could potentially send out price signals that would not be sufficient to push producers and consumers toward a low-carbon economy over the coming decades. However, today's cheap REDD potential may decrease with every year that it is not taken advantage of while the reduction of energy-related emissions could become more accessible over time with technological advancement. It is possible, however, that uncertainty about the supply of REDD credits, particularly in the early years of scheme implementation, might make it difficult to tune the supply-and-demand balance in a CDM or allowance-type approach.

Second, in the event of considerable global warming occurring, there is a risk that forests, even if conserved by society, may be severely damaged by climate change, which could trigger a chain reaction of forest die-off and carbon release that would be difficult to stop (see Nepstad, Stickler, Soares-Filho, and Merry 2008). Thus, an important point made by Chomitz et al. (2006) and emphasised by Michaelowa and Dutschke (2009) is that REDD credits can contribute most in a climate change mitigation scenario of high stringency.

Third, the inclusion of REDD as a low-cost mitigation option may be needed in order to provide incentives to bring more emitters into a collective climate agreement post-2012.

Given the relative cost-effectiveness of REDD as compared to other mitigation options, including REDD (or at least RED) in a global strategy to combat climate change could increase the likelihood of both getting industrialized countries to agree to stricter targets (if REDD credits can be used to meet these targets) and getting developing countries on board as well. In this sense, the large potential magnitude of REDD credits may be seen as a hope rather than a concern (Chomitz et al., 2006). For example, developing countries inspired by the Brazilian government's initiative (see Moutinho et al. 2009) are keen to see industrialized countries reduce rather than simply offset their emissions elsewhere. While some major emitters such as those in the EU might agree with this position, others such as the United States and Japan, may like to see a larger role for emission offsets than is presently allowed under Kyoto. A compromise between the extreme "domestic reductions" and "offsetting" positions may be the best hope of getting not just as many countries as possible to agree on a single climate regime but also one that commits the Parties to stringent emissions reductions.

Permanence in REDD   An issue that was only marginally addressed in Palmer and Engel (2009) and one that is closely related to the previous point is that of permanence, that is, whether emissions from deforestation and degradation are reduced for good and not simply shifted to another period (Murray 2009). A lack of permanence can, in principle, be viewed as a form of "temporal leakage," similar to spatial leakage discussed earlier. As forest systems interact with climate and hydrological systems, unforeseen changes may occur including feedback effects and forest "die-back." Moreover, local deforestation may vary with market conditions, leading to unexpected outcomes. For example, biofuel policies may increase the demand for arable soils, thereby increasing emissions from deforestation. The approach of temporary crediting applied under the CDM could also be an option for REDD. However, the flipside is that the market value of a temporary emission allowance can be very low, as it depends on price expectations for the subsequent commitment period. Countries might thus prefer to take over liability for longer periods, while insurance could help reduce the risk of nonpermanence in emissions reductions. The former point implies the need for institutions that would be able to hold countries to their long-term commitments.

Most current proposals include a carryover of commitments to the subsequent commitment period in case deforestation has increased beyond the agreed-upon level, combined with some obligatory banking of some share of the credits (Dutschke and Wolf 2007). Averaging emission reductions over longer commitment periods (e.g., of 10 years) would also help to deal with difficulties in predictability (Dutschke and Wolf). Within countries, incentive mechanisms like PES could be more directly linked to market prices. Even if a significant portion of REDD turns out to be nonpermanent in the longer run, REDD may still serve an important role in bridging the time to a less CO_2-intensive global economy (Lecocq and Chomitz 2001).[6] The idea of "carbon rental" may also get around the problem of locking in certain land uses in perpetuity, which has been perceived by some countries as an infringement of sovereignty over their natural resources (Laurance, 2007). Moreover, the idea of perpetuity is simply not feasible in many developing countries given unstable political and economic conditions, all of which implies that at most REDD can create temporary carbon credits in these countries.

Avoiding "Hot Air"   The issues of additionality, permanence, and leakage have been cornerstone concerns for project-based GHG mitigation policy (Murray 2009). The fear is that REDD could become a feel-good market, achieving insignificant real emissions reduction (CIFOR 2008). Indeed, recent empirical evidence on the lack of additionality of existing forest conservation policies, some of which has been summarized above, reinforces the need for solid baseline assessment. As Harris et al. (2009) put it, the development of an accepted, standardized baseline approach for avoided deforestation activities is therefore a key step toward the adoption of any future REDD mechanism. Such an approach also needs to balance the gains from more accurate baseline estimates against the associated costs. A major challenge in this regard is to agree on a method that could effectively incentivize emissions reductions in high-deforestation countries, while still supporting the maintenance of forests that may be under more threat of deforestation or degradation in the future (Johns and Schlamadinger 2009). Potential REDD mechanisms that can minimize hot air at the international level should be given serious consideration (e.g., see Strassburg, Turner, Fisher, Schaeffer, and Lovett 2008).

Another important point made by Murray (2009) is the potential trade-off between increasing additionality and decreasing leakage: the less stringent baselines are set, the greater the incentive for a large number of countries to participate in a REDD mechanism, which will help control international leakage. There is also some urgency in agreeing on baselines. Palmer and Obidzinski (2009) clearly demonstrates the potential for perverse behavior, with the adoption of a business-as-usual baseline approach leading to increased deforestation and reduced incentives for policy reform. A related issue, and one that may partly help to address the above challenges, is the optimal length of baseline projection. For example, Harris et al. (2009) propose a project length of 20 to 60 years, but with baselines "locked in" for 10 years only (see also Sohngen 2009).

Changing Results Requires Changing Incentives   The importance of incentives is stressed throughout Palmer and Engel (2009). It is important to acknowledge that at the local level, deforestation is usually a profitable activity. To some degree this also holds at the country level, as halting deforestation can imply forgone economic development. Yet policy failures are also widespread at that level, resulting in above-optimal deforestation rates even from the national perspective. A PES-type mechanism appears promising both at the international and the within-country level. Particularly, making incentives conditional on actual REDD performance is an essential part of avoiding hot air. Paying nations contingently on their REDD performance also opens up options to leverage policy reforms (CIFOR 2008). At the local level, bundling REDD with other environmental services like biodiversity conservation or hydrological services may help raise additional funds for avoiding deforestation. Again, whether a national PES scheme is the best approach for individual countries in achieving compliance with national REDD commitments will depend on the underlying sources of deforestation and on the governance system in place. For example, where deforestation is driven by credit market imperfections or perverse incentives in other sectors, it would be preferable to address these issues directly (Engel et al. 2008). PES is likely to work best in a situation of secure property rights to forest lands and requires some basic quality of governance. This also holds, however, for other types of conservation policies.

Governance and Readiness Participating in an international REDD mechanism and setting up an effective and efficient local incentive system (whether through PES or other measures) is possible only if basic institutional prerequisites are satisfied. These include, for example, a system of secure and well-defined property rights over forest lands, the capacity to quantify forest inventories and assess future land-use trends and related carbon flows, a functioning legal system, the capacity to monitor and enforce existing rules and regulations, and the political will to establish new institutions for forest conservation.

The FCPF explicitly aims to help countries build up the necessary capacity for participating in a REDD mechanism. About a third of the FCPF funding is earmarked to a so-called "readiness fund," which would: (1) help interested developing countries to arrive at a credible estimate of their national forest carbon stocks and sources of forest emissions; (2) assist in defining their reference scenario based on past emission rates for future emissions estimates; (3) offer technical assistance in calculating opportunity costs of possible REDD interventions; and (4) designing an adapted REDD strategy that takes into account country priorities and constraints (World Bank Carbon Finance Unit 2008). Such an approach appears promising to facilitate participation of least developed countries in a future REDD mechanism. Another approach is the establishment of bilateral forest partnerships between an Annex I Party and a developing country (Dutschke and Wolf 2007).

While capacity building is necessary, it should be acknowledged that improvements in governance take time. In the meantime, prospective landowners, whether local communities, government agencies, or firms, are likely to continue to claim *de facto* (and sometimes *de jure*) property rights in remote and poorly governed forest areas. Could such speculative behavior, typically made in anticipation of earning future rents from the land, also occur in the context of a local REDD payments mechanism? And if so, would it matter? We might expect similar rent-seeking behavior, although with a system of conditional payments, the "new" forest owners would have incentives not to convert forest. They may even be expected to proactively protect it. REDD payments could then potentially have a positive impact on the environment, particularly where there are weak, endogenous property rights (Engel and Palmer 2008b). Adding carbon values to landowners' value of the standing forest may increase their ability to protect their *de facto* property rights against intrusion. Although this would to some extent deal with the open access problem of forests, there could be distributional problems if richer actors colonize forest areas at the expense of poorer ones. A nationally administered, carefully targeted payments scheme could be one way around this problem (see Hall 2008; Moutinho et al. 2009).

Policy Conclusions

There appears to be a strong case for including REDD in a global climate change mitigation strategy post-Kyoto. Significant progress has been made in addressing previous concerns to such an inclusion. It is now time to synthesize approaches and develop an integrated REDD mechanism. The success of REDD will depend on the ability to show that it can be done. The establishment of the FCPF as a prototype for REDD measures as well as other current pilot activities have an important role to play in this regard. In doing so, it will be crucial that these initiatives incorporate the lessons from recent studies highlighting the complexities and weaknesses of existing forest conservation policies. Upscaling policies such as PES without improving on scheme additionality and cost-effectiveness could undermine the success of a performance-based incentive payment system for REDD services and raise costs of REDD beyond expectations.

In focusing on REDD as a potential strategy for mitigating climate change, this analysis has neglected some key elements of the climate change policy debate. First, adaptation strategies have been ignored. It is clear that neither adaptation nor mitigation alone will avoid climate change impacts (IPPC 2007b), and that forests play a crucial role in adaptation as well. Second, while potential mechanisms for including REDD in an international climate framework have been considered, the practical and legal arguments, for example, of whether to include REDD in an extension to the Kyoto Protocol or to create an entirely new Protocol altogether were not (see Forner, Blaser, Jotzo, and Robledo 2006; Gupta et al. 2007). Related to this, the international political economy of REDD inclusion was only briefly touched upon when discussing the possible preferences of different nation states and the political trade-offs being made at the international level in forums such as the UN and G8. For example, recent U.S. legislation considers a potentially important role for REDD and other international forest carbon offsets in domestic climate policy.[7]

REDD should, however, be viewed through a prism of scarcity and trade-offs between competing uses for the world's resources. With the world's population forecast to reach 7 to 11 billion by 2050 (United Nations 2004), the global demand for energy and food will continue to rise in the coming decades. Ultimately, decisions over the allocation of natural resources will probably be political ones. At the very least, allocations based on economic criteria will be substantially affected by political forces. In this context, the introduction of REDD, or at the minimum, RED, should be seen as an opportunity to reverse deforestation trends and capture forest carbon values, but only as one, perhaps particularly cost-effective way of mitigating climate change. For it to work, it must remain competitive with other land uses. Other mitigation options also require further opportunities for development and implementation; avoided deforestation should not be allowed to stunt investment in other mitigation technologies and economic sectors.

SECTION TWO: MARKET-BASED SOLUTIONS

Emissions from LULUCF are back on the agenda of international climate policy. As noted earlier, emissions from deforestation and degradation are estimated to account for 17 percent of annual global GHG emissions or about eight gigatons of CO_2 equivalent ($GtCO_2e$) per year (Stern 2007). It is generally agreed that this segment must be included in any post-Kyoto agreement if the envisaged emission reduction targets are to be achieved. Reducing emissions for deforestation and degradation (REDD) is a key mechanism in achieving these emission reductions.

This section builds on Engel and Palmer's assessment that the policy, institutional, and methodological barriers that prevented REDD from being included in the Kyoto Protocol appear to more manageable today, and that the most contentious issue is how to finance REDD activities. It takes the view that public-sector finance alone will not be able to achieve the level of reductions required in the timescales necessary and will therefore focus on how private-sector investors could engage into REDD-related investments.

The views expressed by Mr. Berg are personal and do not represent the views of Merrill Lynch Commodities (Europe) Limited.

This section will not examine the potential of LULUCF, even though it is estimated that these activities can contribute 2.7 $GtCO_2e$ emission reductions per year by 2030 (McKinsey & Company 2009). The main reason for the exclusion is that LULUCF activities are generally embedded in timber investments, with any carbon revenues representing an ancillary cash flow to such investments, and do not necessarily represent a stand-alone investment opportunity. Current carbon accounting allows for only temporary credits to forestry projects due to concerns over nonpermanence of the emission reductions, presenting an additional hurdle for such investments.

Overview of the REDD Mechanism

The attractiveness of REDD from an investor's perspective is that it can be compared to a mitigation project and therefore overcomes one major challenge of LULUCF activities. Unlike other forestry projects, REDD decreases a source of emissions by conserving forest rather than creating a sink (Ebeling & Yasué 2008). The REDD mechanism is therefore not very different from, for example, emission reductions created by the replacement of a fossil fuel power source with a renewable energy project. A REDD project will have to establish a baseline for the business-as-usual scenario and account for the difference between the verified emissions in the project area and the baseline.

The mechanism accounts for the emission reductions associated with delaying deforestation activities that would occur under a business-as-usual scenario; that is, even a one-year delay in deforestation would lead to a real emission reduction. Engel and Palmer (2008a) describe the methodological and operational challenges, such as the selection of reliable baselines, addressing leakage, ensuring verification with remote sensing devices and dealing with catastrophic events, but highlight that many of those challenges could be addressed through policy design.

Market Potential

The market potential of REDD credits is significant, which does not come as a surprise when looking at the overall emission from deforestation. McKinsey (2009) estimates that REDD could contribute to 5.1 $GtCO_2e$ of emissions reductions per year of which 3.6 $GtCO_2e$ are low cost.

Some authors fear that cheap REDD credits could flood the market and push prices down, removing the incentive to invest in renewable energy in developing countries through the CDM of the Kyoto Protocol (Livengood & Dixon 2009). This fear appears to be exaggerated, as any negotiations of a REDD mechanism should go hand-in-hand with discussing the need for countries to ramp up demand for emission reductions by taking on tougher emission reduction targets [see Engel and Palmer (2008a) for more details]. This risk can also be managed with appropriate policies on how forestry credits can be used for compliance (Parpia 2009); that is, they may not directly compete with the CDM.

Furthermore, it is not expected that all avoided deforestation projects could be funded through a private-sector mechanism as a result of certain institutional barriers such as unclear legal title of the rainforest (*The Economist* 2009). However, even a small fraction of the estimated potential could provide significant investment opportunities compared to the CDM. The mechanism has issued 272 million CERs since 2005 and is estimated to generate 1.4 billion CERs during the Kyoto Protocol's first commitment period from 2008 to 2012 (UNEP Risoe 2009). Consequently, the total projected CDM pipeline would not even achieve 30 percent of the estimated annual REDD potential by 2030.

Investment Opportunities in REDD

Most REDD investment opportunities arise from reducing or avoiding logging for subsistence and intensive agriculture, cattle ranching, and unsustainable timber extraction. Any REDD project has to provide an alternative income for the local population and provide co-benefits for biodiversity and human development (Ebeling & Yasué 2008). A typical REDD project will focus on sustainable forest management. This could include sustainable agriculture and timber production. Moreover, the investor would not only finance carbon emission reductions but also provide payments for environmental services (PES), which has been proven to be an effective approach with positive implications for the poor (Ortega-Pacheco, Lupi, and Kaplowitz 2009). The breadth of the projects' impact beyond pure carbon may also help attract a wider investor base, as some purchasing companies may increase their focus by focusing on their "environmental" rather than just their "carbon" footprints.

Project-Based Approach From an investor's perspective, PES are difficult to monetize and consequently carbon credits present the main, if not the only, source of investment return of REDD projects. Carbon credits generated by REDD projects appear attractive as their value could appreciate significantly in a carbon-constraint world. However, it also presents investors with the risk that these credits could decrease in value, thus requiring investors to manage the associated price risks.

The main features of a good REDD project will be clear legal title of the forest, sound data on past and projected deforestation rates, local and national political support, rigorous third-party audits, and a sound project management team that can handle the methodological and technical challenges, such as the monitoring of the reduced deforestation rates. In many instances, the forest is owned or inhabited by indigenous people, so the involvement of a nongovernmental organization can ensure that the project protects their rights and that the payments for the carbon credits reaches the local community.

Unlike with energy projects, the main costs associated with REDD projects are the ongoing costs to protect the rainforest rather than large up-front payments. In project finance terms, REDD projects have low capital expenditures but significant operational expenditures. The main challenge for the project-based approach is that an investor takes both delivery and price risk. These risks could be mitigated by not investing directly in the project but instead paying for the carbon credits on delivery. However, this approach may jeopardize the overall financial viability of the project. An investor of the project-based approach will have to deal with this trade-off and generally aim for higher returns on capital to be compensated for the risks associated with this type of investment. (See the carbon financing case study for an example of a REDD project.)

CARBON FINANCING CASE STUDY: The Ulu Masen Ecosystem Project

By Abyd Karmali and Matthew Hale

The Ulu Masen Ecosystem Project is an example of carbon finance in action and highlights the potential for broadening investment flows beyond carbon toward ecosystem services.

With an annual contribution of around 17 percent of total global GHG emissions, tropical deforestation is the most significant climate change challenge not currently included within the Kyoto Protocol. Developing a market for carbon credits that reward efforts to reduce emissions from deforestation and degradation to facilitate the flow of capital and finance toward more sustainable land use and agriculture will form a core part of the successor agreement to the Kyoto Protocol.

The Ulu Masen Ecosystem Project is groundbreaking in providing carbon financing and structuring for the world's first independently validated avoided deforestation project that is compliant with the rigorous Community, Climate, and Biodiversity Alliance (CCBA) standard. (See www.climate-standards.org/ for more information.) The project is the product of collaboration between various stakeholders:

- The government of Aceh, Indonesia, is responsible for the overall direction, management, and supervision of the Ulu Masen project, including special planning and

Abyd Karmali is managing director and global head of carbon markets at Bank of America Merrill Lynch. Matthew Hale is head of environmental sustainability for the EMEA region of Bank of America Merrill Lynch.

provincial law, implementation of illegal logging controls, and facilitation with the Indonesian government.

- Fauna and Flora International (the world's oldest conservation organization) and Oxfam International, manage the community-based conservation and special planning in select districts and areas, capacity building, geographic information systems, and biodiversity expertise.
- Carbon Conservation, an Australian company specializing in avoided forestation projects, is responsible for the project design, carbon stock and flux estimates, assistance with legal structures and partner relations, and engagement of the private sector.
- Bank of America–Merrill Lynch is the anchor risk taker and carbon off taker and responsible for selling and distributing carbon credits to its investment banking, commodities, and wealth management clients.

Aceh, located on the northern tip of Sumatra, has 3.5 million hectares of tropical forests, which are essential to the ecosystem of Indonesia. Since 1998, Sumatra has lost over 6.5 million hectares of forest, a 30 percent reduction, and since the tsunami of December 2004, demand for timber in Aceh has further increased, fueling further logging. The Ulu Masen Ecosystem Project is specifically seeking to protect 750,000 hectares of forest by providing financing that will enable land planning and reclassification, increased monitoring and law enforcement, reforestation, and restoration and sustainable community logging to significantly reduce the levels of deforestation and thus avoid CO_2 emissions.

The project seeks to reduce deforestation in the region by up to 85 percent, reduce carbon emissions by over 3 million metric tons per year, finance the development of a sustainable timber harvesting industry in the area, and also provide the necessary bridge financing to seed a sustainable soft commodity sector in Aceh. The project also is being undertaken in conjunction with, and will partly fund, the government's broader "Green Aceh" program, designed to establish environmentally, economically, and socially sustainable land uses that build sustainable local communities, help mitigate leakage concerns, and help avoid the negative impacts of the project on certain community constituencies (e.g., furniture markers reliant on locally sourced timber).

The project is ambitious, with three key design goals. Firstly, reduce the amount of legal logging, through a community-based zoning process. This will be done by revising provincial and district special plans, reducing the forest area classified as conversion forest, and increasing the area under a range of formal permanent forest estate categories. Secondly, reduce illegal logging in the area by 85 percent, through increased monitoring and enforcement. Third, introduce a range of support activities, including sustainable logging, reforestation, agro-forestry and other livelihood initiatives that reduce carbon emissions and reduce the risk of future emissions, capacity building, and direct community payments.

However, ultimately, it is the project's financing that has the potential to be replicated in countries across the world. The UN Framework Convention on Climate Change in 2007 concluded that by 2030 approximately 86 percent of financial flows required to address climate change will need to come from the private sector. The Ulu Masen Ecosystem carbon financing deal demonstrates the positive impact that can be achieved from attracting investment capital to carbon and ecosystem service-driven projects. The carbon credits that will be sold to investors will provide audited evidence not only of the CO_2 savings from avoided deforestation but also include the broader sustainability benefits to the region's economy, population, and wildlife. In addition to the CCBA standard, the credits hope to achieve the Voluntary Carbon Standard (VCS), which has strong stakeholder buy-in and has recently published guidelines for avoided deforestation projects.

The projects expects to attract a broad range of investors—from companies and individuals who wish to use the carbon credits to offset their corporate or personal carbon or footprints to traders who foresee this type of credit's becoming liquid and potentially rising in value in the event that a mechanism for REDD is included in the post-Kyoto Protocol in December 2009 or in the emerging domestic U.S. carbon market, which will almost certainly include the opportunity for compliance entities to meet part of their compliance obligation through importing forestry carbon credits from avoided deforestation projects.

While at present the only commodity specified in the project's emission reduction purchase agreement is carbon, opportunities to develop and tap markets for other ecosystem services may also emerge. These may include endemic species, carbon stocks, watershed protections, and ecotourism value. It is hoped that the project will not only act as a catalyst for the development of markets for REDD but also help pave the way for monetizing a broader range of biodiversity-related investment themes.

Structured Transactions   Investors are able to mitigate some risks, depending on how the investment is structured. For example, price risk can be eliminated by including a sales agreement with a financial institution that will off-take any carbon credits that are generated by the project at a preagreed price. The investor could invest into either a prepayment structure or a bond that is linked to the performance of the project.

In a prepayment structure, the investor would prepay the expected cash flows from the fixed sales agreement with the purchaser of the carbon credits. The main risk is the underperformance of the project, which could be at least partially mitigated through insurance products. The prepayment structure could be extended by obtaining funding from a senior and a mezzanine tranche that would allow investors to take different risk levels and obtain different returns on their investment.

The bond structure would allow an investor to invest in a bond backed by the proceeds of a portfolio of REDD projects. In order to diminish the risks for institutional investors, the bond can be structured to include a principal protected feature. This approach might facilitate the use of public funds to help underwrite the risk, which should, in turn, attract greater private-sector investment flows. The bond structure may be an ideal product to combine ethical investment with investing in an alternative asset class if a governmental or international financial institution issues the bond and provides principal protection on a long-dated tenure.

CONCLUSION

There is a growing agreement that action must be taken to reduce GHG emissions for land use, land-use change, and forestry activities, and this sector must be a central part of any future deal on climate change. The proper supportive policies will allow for the creation of cost-effective financing solutions that "could not only reduce carbon emissions significantly, but also benefit developing countries, support poverty reduction, and help preserve biodiversity and other forest services" (Eliasch 2008).

A REDD mechanism has the potential to achieve significant and cost-effective emission reductions. The current debate focuses on how REDD could best be financed. The sheer scale of the challenge dictates that public funds will not be sufficient to achieve the required emission reductions. REDD-related investments represent an important conduit for private sector investment flows, with the similarities to emissions mitigation projects and the opportunity for ethical investment combining with attractive potential returns. In particular, a REDD bond should be appealing to institutional investors and could possibly serve as a model for policy makers on how public funds can be combined with private-sector funds to attract the necessary financing for REDD projects.

NOTES

1. "Avoided deforestation" is otherwise known as reducing emissions from deforestation (RED). Inclusion of forest degradation extends this definition to REDD. Both RED/avoided deforestation and REDD are covered in Palmer and Engel (2009), depending on the topic under discussion.
2. For example, a recent overview of the issues can be seen in Dutschke and Wolf (2007). A discussion of avoided deforestation in the context of climate change and deforestation can be seen in the relevant chapters of Stern (2007) and Chomitz et al. (2006), respectively. Policy-related issues alongside the legal and technical considerations of avoided deforestation are covered by Moutinho and Schwartzman (2005). See also Murdiyarso and Herawati (2005) for chapters relating to livelihood issues, in addition to those focusing on the CDM and carbon sequestration from a bioscience perspective. International policy issues and how REDD might fit into climate policy from a practical perspective are covered by Streck, O'Sullivan, Janson-Smith, and Tarasofsky (2008).
3. Note, however, that the price of avoided deforestation as a mitigation option will be determined by the marginal cost and not the average cost. We expect there to be differences between the two, depending on the shape of the marginal cost curve.
4. Alternatively, and instead of making direct payments to potential deforesters (assuming that they can be identified in the first place), any transfers received

by a particular country, say, from an international fund like the FCPF could simply be invested in systems that discourage deforestation behavior (e.g., for monitoring and enforcement).
5. In order to stabilize the CO_2 concentration in the atmosphere at around 400 to 450 ppm (IPPC, 2007b).
6. A time delay in emissions reduced by abatement measures could result in permanent climate benefits if the cumulative atmospheric concentrations of GHGs are lower at any future point in time (Ebeling and Yasué 2008).
7. The Lieberman-Warner Climate Security Act of 2008. See www.epa.gov/climatechange/downloads/s2191_EPA_Analysis.pdf.

REFERENCES

Alix-Garcia, J., A. de Janvry, and E. Sadoulet. 2009. The role of risk in targeting payments for environmental services. In C. Palmer and S. Engel (eds.), *Avoided deforestation: Prospects for mitigating climate change*. Oxford, UK: Routledge.

Baumert, K., T. Herzog, and J. Pershing. 2005. *Navigating the numbers: Greenhouse gas data and international climate policy*. Washington, DC: World Resources Institute (WRI).

Brown, S., Sathaye, J., Cannell, M. and Kauppi, P. 1996. Management of forests for mitigation of greenhouse gas emissions. In R. T. Watson, M. C. Zinyowera, and R. H. Moss (eds.), *Climate change 1995. Impacts, adaptations and mitigation of climate change: Scientific-technical analyses. Contribution of Working Group II to the Second Assessment Report of the Intergovernmental Panel on Climate Change*. Cambridge: Cambridge University Press.

Bulte, E., and S. Engel. 2006. Conservation of tropical forests: Addressing market failure. In R. López and M. Toman (eds.), *Economic development and environmental sustainability*. Oxford, UK: Oxford University Press.

Center for International Forestry Research (CIFOR). 2008. *CIFOR's Sven Wunder discusses how REDD can learn from PES* (online). Available at www.cifor.cgiar.org/Highlights/redd_interview.htm (accessed July 2008).

Chomitz, K. M., P. Buys, G. De Luca, T. S. Thomas, and S. Wertz-Kanounnikoff. 2006. *At loggerheads? Agricultural expansion, poverty reduction and environment in the tropical forests*. World Bank Policy Research Report, Development Research Group. Washington, DC: World Bank.

Dutschke, M. 2008. *Simply REDD? Konzeptionen, modelle, vorschläge zur emissionsverringerung aus entwaldung und walddegradierung*. Paper presented at the Workshop *Wald und Klima*, GTZ Eschborn, Germany, February 12.

Dutschke, M., and Wolf, R. 2007. *Reducing emissions from deforestation in developing countries: The way forward*, Climate Protection Programme, Federal Ministry for Cooperation and Development (GTZ), Eschborn, Germany: GTZ.

Ebeling, J., and M. Yasué. 2008. Generating carbon finance through avoided deforestation and its potential to create climatic, conservation and human development benefits. *Philosophical Transactions of the Royal Society B* 363:1917–1924.

The Economist. 2009. The Brazilian Amazon: Preventing pillage in the rainforest. *The Economist,* print edition, February 26.

Eliasch, J. 2008. *Climate change: Financing global forests: The Eliasch review.* London: Earthscan.

Engel, S., T. Wünscher, and S. Wunder. 2009. Increasing the efficiency of forest conservation: The case of payments for environmental services in Costa Rica. In C. Palmer and S. Engel (eds.), *Avoided deforestation: Prospects for mitigating climate change.* Oxford, UK: Routledge.

Engel, S., and C. Palmer. 2008a. Painting the forest REDD? Prospects for mitigating climate change through reducing emissions from deforestation and degradation. Institute for Environmental Decisions. Swiss Institute of Technology Zurich.

Engel, S., and C. Palmer. 2008b. Payments for environmental services as an alternative to logging under weak property rights: The case of Indonesia. *Ecological Economics* 65(4):799–809.

Engel, S., S. Pagiola, and S. Wunder. 2008. Designing payments for environmental services in theory and practice—an overview of the issues. *Ecological Economics* 65(4):663–674.

Fearnside, P. 2001. Saving tropical forests as a global warming countermeasure: An issue that divides the environment movement. *Ecological Economics* 39(2):167–184.

Ferraro, P. J., and A. Kiss. 2002. Direct payments to conserve biodiversity. *Science* 298(5599):1718–1719.

Food and Agriculture Organization of the United Nations (FAO). 2001. *Global forest resources assessment 2000.* Rome: Author.

Food and Agriculture Organization of the United Nations (FAO). 2006. *Global forest resources assessment 2005.* Rome: Author.

Forner, C., J. Blaser, F. Jotzo, and C. Robledo. 2006. Keeping the forest for the climate's sake: Avoiding deforestation in developing countries under the UN-FCCC. *Climate Policy* 6(3):275–294.

Geist, H., and E. Lambin. 2002. Proximate causes and underlying driving forces of tropical deforestation. *BioScience* 52(2):143–150.

Grieg-Gran, M. 2009. The costs of avoided deforestation as a climate change mitigation option. In C. Palmer and S. Engel (eds.), *Avoided deforestation: Prospects for mitigating climate change.* Oxford, UK: Routledge.

Group of Eight (G8). 2008. *Summary of the Hokkaido Toyako summit* (online). Ministry of Foreign Affairs of Japan. http://www.mofa.go.jp/policy/economy/summit/2008/news/summary.html (accessed August 3, 2008).

Gullison, R. E., P. C. Frumhoff, J. G. Canadell, C. B. Field, D. C. Nepstad, K. Hayhoe, *et al.* 2007. Tropical forests and climate policy. *Science* 316(5827):985–986.

Gupta, S., D. A. Tirpak, N. Burger, J. Gupta, N. Hoehne, A. I. Boncheva, et al. 2007. Policies, instruments and co-operative arrangements. In B. Metz, O. R. Davidson, P. R. Bosch, R. Dave, and L. A. Meyer (eds.), *Climate change 2007: Mitigation. Contribution of Working Group III to the Fourth Assessment Report of the Intergovernmental Panel on Climate Change.* Cambridge and New York: Cambridge University Press.

Hall, A. 2008. Better RED than dead: Paying the people for environmental services in Amazonia. *Philosophical Transactions of the Royal Society B* 363:1925–1932.

Harris, N. L., S. Petrova, and S. Brown. 2009. A scalable approach for setting avoided deforestation baselines. In C. Palmer and S. Engel (eds.), *Avoided deforestation: Prospects for mitigating climate change*. Oxford, UK: Routledge.

Houghton, R. A. 2005. Tropical deforestation as a source of greenhouse gas emissions. In P. Moutinho and S. Schwartzman (eds.), *Tropical deforestation and climate change*. Belém, Brazil: Amazon Institute for Environmental Research (IPAM), and Washington: Environmental Defense.

Intergovernmental Panel on Climate Change (IPCC). 2001. *Climate change 2001: Synthesis report. A contribution of Working Groups I, II, and III to the Third Assessment Report of the Integovernmental Panel on Climate Change.* Cambridge and New York: Cambridge University Press.

Intergovernmental Panel on Climate Change (IPCC). 2007a. *Climate change 2007: Synthesis report. Contribution of Working Groups I, II and III to the Fourth Assessment Report of the Intergovernmental Panel on Climate Change.* Geneva: Author. IPCC

Intergovernmental Panel on Climate Change (IPCC). 2007b. *Climate change 2007: Synthesis report summary for policymakers.* Intergovernmental Panel on Climate Change. Fourth Assessment Report. Cambridge and New York: Cambridge University Press.

Johns, T., and B. Schlamadinger. 2009. International policy and institutional barriers to reducing emissions from deforestation and degradation in developing countries. In C. Palmer and S. Engel (eds.), *Avoided deforestation: Prospects for mitigating climate change*. Oxford, UK: Routledge.

Kaimowitz, D., and A. Angelsen. 1998. *Economic models of tropical deforestation: A review*. Bogor: Center for International Forestry Research (CIFOR).

Kremen, C., J. Niles, M. Dalton, G. Daily, P. Ehrlich, J. Fay, *et al.* 2000. Economic incentives for rain forest conservation across scales. *Science* 288(5472):1828–1832.

Laurance, W. F. 2007. A new initiative to use carbon trading for tropical forest conservation. *Biotropica* 39(1):20–24.

Lecocq, F., and K. Chomitz. 2001. *Optimal use of carbon sequestration in a global climate change strategy: Is there a wooden bridge to a clean energy future?* Washington, DC: World Bank.

Livengood, E., and A. Dixon. 2009. REDD and the effort to limit global warming to 2°C: Implications for including REDD credits in the international carbon market. *KEA* 3 (New Zealand), March 30.

McKinsey & Company. 2009. *Pathways to a low carbon economy. Version 2 of the Global Greenhouse Gas Abatement Curve.* McKinsey & Company.

Michaelowa, A., and M. Dutschke. 2009. Will credits from avoided deforestation in developing countries jeopardize the balance of the carbon market? In C. Palmer and S. Engel (eds.), *Avoided deforestation: Prospects for mitigating climate change*. Oxford, UK: Routledge.

Moutinho, P., M. Cenamo, and P. Moreira. 2009. Reducing carbon emissions by slowing deforestation: REDD initiatives in Brazil. In C. Palmer and S. Engel (eds.), *Avoided deforestation: Prospects for mitigating climate change*. Oxford, UK: Routledge.

Moutinho, P., and S. Schwartzman (eds.). 2005. *Tropical deforestation and climate change*. Belém, Brazil: Amazon Institute for Environmental Research (IPAM) and Washington: Environmental Defense.

Murdiyarso, D., and H. Herawati (eds.). 2005. *Carbon forestry: Who will benefit? Proceedings of Workshop on Carbon Sequestration and Sustainable Livelihoods*. Bogor, Indonesia: Center for International Forestry Research (CIFOR).

Murray, B. 2009. Leakage from an avoided deforestation compensation policy: Concepts, empirical evidence, and corrective policy options. In C. Palmer and S. Engel (eds.), *Avoided deforestation: Prospects for mitigating climate change*. Oxford, UK: Routledge.

Nepstad, D., C. Stickler, B. Soares-Filho, and F. Merry. 2008. Interactions among Amazon land use, forests and climate: Prospects for a near-term forest tipping point. *Philosophical Transactions of the Royal Society B* 363:1937–1946.

Ortega-Pacheco, D., F. Lupi, and M. Kaplowitz. 2009. Payment for services: Estimating demand within a tropical watershed. *Journal of Natural Resources Policy Research* 1(2) (April):189–202.

Pacala, S., and R. Socolow. 2004. Stabilization wedges: Solving the climate problem for the next 50 years with current technologies. *Science* 305(5686):968–972.

Palmer, C., and S. Engel (eds.). 2009. *Avoided deforestation: Prospects for mitigating climate change*. Oxford, UK: Routledge.

Palmer, C., and K. Obidzinski. 2009. Choosing avoided deforestation baselines in the context of government failure: The case of Indonesia's plantations policy. In C. Palmer and S. Engel (eds.), *Avoided deforestation: Prospects for mitigating climate change*. Oxford, UK: Routledge.

Parpia, A. 2009. *The impact of forestry on the global carbon market*. New Carbon Finance, Global Kyoto–Research Note, February 25.

Pearce, D. W., F. Putz, and J. K. Vanclay. 2002. Is sustainable forestry economically possible? In D. W. Pearce, C. Pearce, and C. Palmer (eds.), *Valuing the environment in developing countries: Case studies*. Cheltenham: Edward Elgar.

Pfaff, A., and J. Robalino. 2009. Human choices and policies' impacts on ecosystem services: Improving evaluations of payment and park effects on conservation and carbon. in C. Palmer and S. Engel (eds.), *Avoided deforestation: Prospects for mitigating climate change*. Oxford, UK: Routledge.

Rametsteiner, E., M. Obersteiner, G. Kindermann, and B. Sohngen. 2009. Economics of avoiding deforestation. In C. Palmer and S. Engel (eds.), *Avoided deforestation: Prospects for mitigating climate change*. Oxford, UK: Routledge.

Schneider, S. 1998. Kyoto Protocol: The unfinished agenda. An editorial essay. *Climatic Change* 39(1):1–21.

Sohngen, B. 2009. Assessing the economic potential for reducing deforestation in developing countries. In C. Palmer and S. Engel (eds.), *Avoided deforestation: Prospects for mitigating climate change*. Oxford, UK: Routledge.

Stern, N. 2007. *The economics of climate change: The Stern Review*. Cambridge, UK: Cambridge University Press.

Strassburg, B., K. Turner, B. Fisher, R. Schaeffer, and A. Lovett. 2008. *An empirically-derived mechanism of combined incentives to reduce emissions from deforestation.* CSERGE Working Paper ECM 08-01, Centre for Social and Economic Research on the Global Environment (CSERGE), University of East Anglia.

Streck, C., R. O'Sullivan, T. Janson-Smith, and R. Tarasofsky (eds.). 2008. *Climate change and forests emerging policy and market opportunities.* Washington, DC: Brookings Institution Press.

Tavoni, M., B. Sohngen, and V. Bosetti. 2007. Forestry and the carbon market response to stabilize climate. *Energy Policy* 35(11):5346–5353.

UNEP Risoe. 2009. *Overview of the CDM pipeline.* March 1, UNEP Risoe.

United Nations Framework Convention on Climate Change (UNFCCC). 2007. *Decision-/CP.13 Bali Action Plan* (online). Available at http://unfccc.int/files/meetings/cop_13/application/pdf/cp_bali_action.pdf (accessed June 15, 2008).

United Nations Framework Convention on Climate Change (UNFCC). 2008. *Kyoto Protocol status of ratification* (online). Available at http://unfccc.int/files/kyoto_protocol/status_of_ratification/application/pdf/kp_ratification.pdf [accessed August 1, 2008].

United Nations. 2004. World population in 2300. *Proceedings of the United Nations Expert Meeting on World Population in 2300.* New York: Authors.

World Bank Carbon Finance Unit. 2008. *About forest carbon partnership facility (FCPF).* Washington: World Bank (online). Available at http://carbonfinance.org/Router.cfm?Page=FCPF&ft=About (accessed July 20, 2008).

Wunder, S. 2005. *Payments for environmental services: Some nuts and bolts.* CIFOR Occasional Paper 42, Bogor, Indonesia: Center for International Forestry Research (CIFOR).

BIBLIOGRAPHY

The Economist. 2008. Paying for the forest. *The Economist,* August 9.

Nicholas Institute for Environmental Policy Solutions. 2009. *Forest and climate: The crucial role of forest carbon in combating climate change.* Duke University, March.

Richards, K., and C. Stokes. 2004. A review of forest carbon sequestration cost studies: A dozen years of research. *Climatic Change* 63(1–2):1–48.

Wunder, S., S. Engel, and S. Pagiola. 2008. Taking stock: A comparative analysis of payments for environmental services programs in developed and developing countries. *Ecological Economics* 65(4):834–852.

CHAPTER 11

Effective Clean Tech Investing

Russell Read, PhD, CFA and John Preston

"**C**lean" technologies have been compelling among a rarified set of investors, policy makers, environmentalists, and scientists since the commercialization efforts for wind power among American farmers early in the twentieth century and photovoltaic energy in the late 1970s. However, for the most part, clean technologies have evolved slowly over recent decades and have been viewed as cost-ineffective compared with "conventional" technologies involving fossil fuels, mineral extraction, building construction and operation, and the deployment of public infrastructure. Today, however, we are increasingly constrained in our ability to supply resources needed for economic growth. In addition, the assault on the environment by current practices threatens our well-being and the viability of the quality of life. Simultaneously, recent advances in science and technology enable new clean energy and materials to compete economically with prior dirtier practices. The combination of these conditions sets the stage for revolutionary opportunities for both technology innovation and investment.

As defined throughout this volume, environmental investing generally includes carbon trading, water projects, sustainable property, and projects related to reduced emissions for deforestation and degradation in addition to clean tech investments. Conceptually, *clean tech investments are those that introduce, develop, deploy, and scale the technologies that transform the way we produce, distribute, or consume natural resources in a way that protects or enhances the natural environment.* It is clean tech investing, therefore, that has the greatest potential for altering in an ecologically beneficial way those technologies that rely on our natural resources to produce energy and goods worldwide. Note that a narrow focus on specific environmental

Russell Read is chief executive officer and John Preston is managing partner at C Change Investments.

factors such as CO_2 concentration is not needed or even desired, as such an exclusive focus could crowd out other ecological imperatives such as species preservation and particulate matter emissions. Also, it is the technological transformation of the way in which people produce, distribute, and consume natural resources, including fuels, metals, food, concrete, timber, and water, that is at issue to the clean tech investor. Well-known and intuitive examples of clean tech investments include solar energy companies and wind farm projects. However, clean tech investments also involve those technologies that improve (sometimes radically) the efficiency in which natural resources are produced, distributed, or consumed, including those that utilize fossil fuels and the extraction of minerals.

Two relatively recent developments have changed the prospects for clean tech investing for many years to come. First, escalating demand for natural resources (particularly from the emerging markets of Asia) is straining production and distribution capacity worldwide, resulting in highly volatile—but generally higher—prices for commodities, including energy, metals, and agricultural products. Second, accelerating ecological strains (including rising CO_2 levels, particulate matter emissions, and rainforest deforestation) have arisen from this combination of escalating demand for energy and raw materials, coupled with the inefficiency and ecological by-products of today's energy- and materials-related technologies, many of which have not changed in over a century.

The combination of generally higher commodities prices and environmental damage also creates new opportunities for both technological innovation and strong investment returns. Indeed, the post–World War II era has generally been characterized by plentiful and cheap supplies of energy and raw materials on world markets. What is different now (and for the past several years compared to the rest of the post–World War II era) is the growth in demand for energy and raw materials in the "emerging" or "developing" markets. Sustained economic growth, particularly in China and India, has fueled a much higher demand for natural resources worldwide due to the high populations in those countries compared with many industrialized economies. In contrast to economic growth in industrialized nations, where the demand for natural resources is subject to relatively little change, economic growth in the emerging markets leads to people buying their first homes, cars, and major appliances such as refrigerators and washing machines. Figure 11.1 reveals how this relatively recent protracted growth in emerging market economies has led to an entirely different trajectory for the demand for energy worldwide. The bottom portion of the bar in Figure 11.1 (non-OECD Asia) has now reached the same energy consumption as the rest of the world's nations, and by 2020 will exceed today's total world energy. This change in energy demand is driven predominantly by growth in China

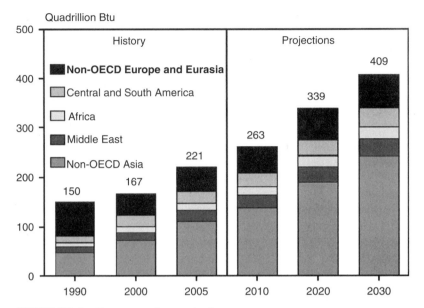

FIGURE 11.1 Energy market outlook.
Source: Energy Information Administration 2008.

and India. In addition, this accelerated trajectory in energy demand is mirrored by the demand for raw materials, including metals, food, concrete, and timber.

Inexpensive and plentiful natural resources have undoubtedly been helpful in fueling the overall growth in the world economy and stock market gains over the past many decades. However, this overall growth pattern in the world economy and stocks has also encountered some important speed bumps along the way. Supply shocks during 1973–1974, for example, led to much higher energy prices accompanied by a decline in U.S. stock prices of approximately 45 percent. Moreover, these price declines in U.S. stocks were mirrored by international stock markets with London-based stocks retreating by some 73 percent. Indeed, the 16-year period from 1964 to 1980 can be generally characterized as flat for equity investors overall. Importantly, when cheap and plentiful energy and raw materials cannot be supplied to world markets, both economic growth and stock market gains can suffer. That said, it is precisely during these periods of strain in producing and distributing natural resources when investment in new energy- and materials-related technologies has been and is expected to be especially effective.

The last time investment capital and innovation were focused on energy and materials were in the 1960s and 1970s. During that time, energy companies, for example, were considered "growth" companies with high price-to-earnings (P/E) ratios. Indeed, by 1980, energy companies represented nearly one-third the capitalization of U.S. and international equity markets and the majority of growth stocks. Also reflecting their overall importance, 6 of the 10 largest capitalization companies in the United States were energy companies.

Over the ensuing two decades, however, an entirely different story emerged. Companies involved in the production and distribution of energy and raw materials became "deep value" with low P/E ratios in the eyes of world stock markets, with little to no new capital invested in new technologies or even in the maintenance of existing infrastructure. By 2000, energy stocks dropped to a low of less than 7 percent of the capitalization of U.S. and international stock markets. Slack supply for crude oil (i.e., the difference between potential full-throttle worldwide production and current demand) dropped during this period from some 17 million barrels per day (mBPD) to 1 to 5 mBDP in recent years.

Expanding fuel and raw materials supplies to meet accelerating worldwide demand has also increased strains on the natural environment. Since the advent of the industrial revolution, atmospheric CO_2 has risen from approximately 280 parts per million (ppm) to some 373 ppm today. Most of this increase occurred in recent decades as the absorptive capacity for CO_2 from oceans and other sources reached saturation levels, leading to serious concerns regarding the prospects for sustained global warming stemming from the increase in this and other greenhouse gases (GHGs). Although not mirrored so far in the Antarctic region, the Arctic has experienced significant melting of the ice pack over the past decades, coupled with decreasing salinity of Arctic waters and changes in ocean conveyers in the northern hemisphere (Figure 11.2).

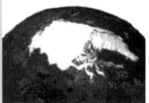

Side-by-side comparison of sea ice from 1979 and 2003

FIGURE 11.2 Impact of "global melting."
Source: NASA Goddard Space Flight Center,
www.nasa.gov/centers/goddard/news/topstory/2003/1023esuice.html.

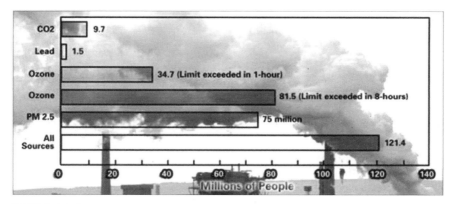

FIGURE 11.3 Resurgence in U.S. air pollution.

The accelerating and inefficient use of natural resources also has led to serious increases in conventional pollution in the United States and especially internationally in recent years (Figure 11.3 and Table 11.1). This increase in conventional pollution obviously exerts a negative influence on human health. However, it also represents a reversal from previous significant improvements in conventional air and water pollution in the United States from the 1950s and 1960s. Much of this increase in pollution, in fact, is attributable to emissions from the emerging markets, which are producing an accelerating share of the world's goods. That industrial production in these countries is not typically accompanied by the environmental standards in many industrialized economies has undoubtedly been a major contributor to pollution problems worldwide, with an increasing impact on the United States. Indeed, some 25 percent of particulate matter air pollution in Los Angeles is now attributable, by some estimates, to industrial emissions from Asia.

Air pollution is tied to numerous health problems, particularly for the young and old. Pollution is tied to increased asthma in children and emphysema, lung cancer, cardiovascular diseases, and respiratory diseases.

Looking forward, the general prospects for clean tech investing appear to be good precisely because of this combination of accelerating demand for natural resources; the inefficiency in their production, distribution, and use; and the alarming ecological consequences of current industrial, commercial, and residential practices involving fuel and raw materials. Yet, what is clean tech investing, how will it evolve in the future, and how can investors take advantage of emerging opportunities as part of an effective and diversified portfolio? These are the topics of the following sections of this chapter.

TABLE 11.1 More Americans Die from Inhaling Particulates than from Auto Accidents

Metropolitan Area	Estimated Annual Cardiopulmonary Deaths Attributable to Particulate Air Pollution
1. Los Angeles–Long Beach, CA	5,873
2. New York, NY–NJ	4,024
3. Chicago, IL	3,479
4. Philadelphia, PA–NJ	2,599
5. Detroit, MI	2,123
6. Riverside–San Bernardino, CA	1,905
7. San Francisco–Oakland, CA	1,270
8. Pittsburgh, PA	1,216
9. St. Louis, MO–IL	1,195
10. Cleveland, OH	1,161
11. Phoenix, AZ	1,110
12. Anaheim–Santa Ana, CA	1,053
13. San Diego, CA	999
14. Atlanta, GA	946
15. Houston, TX	939
16. Tampa–St. Petersburg, FL	938
17. Baltimore, MD	861
18. Newark, NJ	819
19. Boston–Lowell–Brockton, MA	792
20. Dallas–Fort Worth, TX	743
21. Miami, FL	645
22. Cincinnati–Hamilton, OH–KY–IN	617
23. Nassau–Suffolk, NY	605
24. Washington, DC–MD–VA	588
25. Milwaukee, WI	518
26. Kansas City, MO–KS	507
27. Seattle–Everett, WA	501
28. Minneapolis–St. Paul, MN–WI	495
29. Indianapolis, IN	494
30. Fresno, CA	488
30. Sacramento, CA	488

Source: National Resources Defense Council.

CLASSIFYING THE SCOPE OF CLEAN TECH INVESTMENTS

Although the case for clean tech investing is intuitively compelling, defining what it is exactly and how to accomplish it effectively requires a new perspective. For example, one important conceptual distinction can be drawn by differentiating (1) "clean" technologies, whose use is intended to have little or no ecological impact, from (2) "cleaner" technologies, whose adoption is intended to improve the utilization of natural resources compared with current practices. Examples of cleaner technologies include more efficient transmission of electricity throughout a national grid system, major efficiency improvements at power stations and refineries, smart building construction and operations, and the sequestering/remediation of CO_2 or other GHGs (including methane) during conventional power generation or fuel production processes. Clean technologies, in contrast, typically involve producing power or fuels with little or no environmental impact and can include solar, wind, and geothermal technologies as well as renewable fuels including bio-based fuels and hydroelectric power. In addition, recycled materials, in lieu of newly extracted or produced materials, also qualify as clean technologies. For example, fiber-based composites (comprised of recycled plastics and sawdust) used in place of steel-based seawalls offer the prospect of a clean alternative to an otherwise resource-intensive project. Of course, significant and important debate can and should ensue about how truly "clean" any technology solution can be in absolute terms. For example, wind power can alter avian migration patterns, hydroelectric projects can disrupt fisheries, biofuel production can compete with food production, and even solar technologies (viewed by many as "perfectly" clean) can also promote the production and disposal of toxic materials associated with energy storage and solar panel manufacturing.

That is not to say, however, that there is little difference between what can be characterized as "cleaner" versus "clean" technology. Generally, most investors, policy makers, environmentalists, and scientists would agree that a bright line distinction exists between these two classes of natural resources–based technologies. Indeed, such a distinction has led many to argue that only "clean" technologies should be considered to be "green." However, as we shall see in the next section, both cleaner and clean technologies will play critical (if sometimes transitional) roles in the development of the world's energy- and materials-dependent economy and natural environment for decades to come. Indeed, in terms of economic and ecological impact over the coming decade, cleaner technologies are far more likely to have a more beneficial impact than clean technologies and thus cannot be discounted from investment or policy perspectives.

A second important means of classifying clean tech investments is by function. One such taxonomy, leveraging off of major investment consultants and institutional investors, differentiates among the following clean tech areas:

1. Alternative energy (including solar, wind, geothermal, hydro, biofuels)
2. Energy efficiency (including high-efficiency lighting, improved electrical transmission, improved conversion from fossil fuels into electricity)
3. Energy storage (including batteries, heat/cold capture, fuel cells, hydrogen, super-capacitors)
4. Materials efficiency (including those that improve the utilization of metals, water, concrete, agricultural products, and wood)
5. Remediation (including carbon capture and storage, recycling, climate adaptation and mitigation technologies)

What is readily apparent from this taxonomy is that clean tech has moved from a fringe area in both technology and investing to a vast and central role for technology and natural resource innovation. What is also apparent from an investment standpoint, however, is that clean tech has evolved from a fringe role for energy- and materials-related investments to one of central importance rivaling or even exceeding those investments made for new petroleum exploration and production and the mining of metals and other materials. Moreover, the success of many industrial companies, including those in the automotive industry, will be critically dependent upon how effectively those companies can utilize clean tech advances.

Of special note is the major (but oftentimes underappreciated) area for clean tech investing related to the construction and operating of buildings. Buildings play a major role in the world's consumption of both energy and raw materials. For example, buildings account for roughly 40 percent of global energy use alone, and nearly a billion tons of earth are moved each year to create the needed building materials. Thus, improving the efficiency for energy and materials involved in building construction and maintenance (including commercial, industrial, and residential spaces) presents a major clean tech investment opportunity. Moreover, the construction industry is unique in that for virtually every country, construction has experienced little to no growth in worker productivity during the past two to three decades.

Although the energy and raw materials intensity involved with constructing and operating buildings presents a major opportunity, the escalating demand for new and better housing across the emerging markets of the world is also presenting some unprecedented challenges. For example, if we were to attempt to provide conventional wood "stick-frame" construction for the approximately 1 billion people in the world expected to move from subsistence-level to moderate-consumption-level housing over the coming two decades, we would create an ecological, economic, and raw materials disaster of epic proportions. Thus, the construction industry, having witnessed so few innovations over the past several decades in terms of the building process, materials used, and energy required, is poised to enjoy some of the best innovations in any industry in the coming years. This veritable revolution in green building approaches, materials, and energy technologies undoubtedly will afford investors unprecedented opportunities in this seemingly most predictable of industries. Already, advances in light-emitting diode (LED) lighting, "smart" systems that economize on energy use, and the use of composite materials to enhance and reduce the need for wood, concrete, and steel show the potential for improving dramatically the natural resources footprint associated with buildings worldwide.

A third important classification for clean tech investments is how they are made. Essentially, clean tech investments can be made in at the following four stages:

1. Identification and sourcing (research, venture capital, applying proven technologies to new problems)
2. Development (proving out the commercial potential, late-stage venture capital, commercialization)
3. Deploying (commercialization, late-stage venture capital, growth capital)
4. Scaling (infrastructure, widespread commercial adoption)

From an investment perspective, these investments can be made by industrial companies (to improve and develop clean tech for their own industrial uses), early-stage venture capital investors (focused on identifying the most promising clean tech innovations from research labs and small private companies), late-stage venture capital investors (focused on developing the commercial potential for clean tech innovations), and infrastructure investors (focused on deploying commercially proven clean technologies in scale). To be certain, the investment types, expertise, and capital requirements involved in successful clean tech investments differ substantially from what prevailed for information technology (IT) and biotechnology (biotech). One particular challenge is that many clean tech opportunities require substantially more investment capital in order to prove out their full commercial potential. For example, new cleaner coal and carbon sequestration projects may require billions of dollars simply to demonstrate their full potential. Once proven, however, the commensurate benefit for many of these clean tech opportunities is the scale at which they can be deployed and the potential revenues that they can generate. The eventual leaders in LED lighting, for example, can be expected to enjoy billions of dollars in annual revenues with proportionately compelling profits. Looking forward, what is particularly telling about the prospects for clean tech is that clean tech opportunities have become of central importance for technology research in general, for early- and late-stage venture capitalists, and for infrastructure investors after being off their radar screens entirely in recent decades.

The vast scope and scale of clean tech investments expected over the coming years can be appreciated through the classifications of (1) "clean" versus "cleaner" technologies, (2) the industrial roles/functions for clean tech, and (3) how clean tech investments are actually made. Clean tech innovations involved with the construction and operations of buildings alone hold the potential for hundreds of billions of dollars of compelling investments despite the slow development of new materials, technologies, and practices in the building industry in many years.

THE PACE OF CLEAN TECH INNOVATION AND FUTURE PROSPECTS FOR INVESTING

Of course, the most recent (and perhaps only other) period for major clean tech development and investment occurred during the late 1970s. Born out of a widespread belief that petroleum supplies were soon to be exhausted (a belief shared by the U.S. president) and that conventional fuel prices were poised to explode without respite, research and development for clean tech solutions, including solar photovoltaics, became a major national focus in the United States. By 1980, however, the actual slack supply in daily petroleum production had swelled to over 17 million barrels per day or nearly one-third of world consumption at that time. Essentially, the overproduction of infrastructure for producing energy and materials throughout the 1960s and 1970s led to a protracted period of plentiful and inexpensive natural resources during the 1980s and 1990s. As discussed earlier, by the year 2000, slack supply (the difference between what the world can produce at peak capacity and what the world is currently demanding) had virtually disappeared across the wide range of natural resources. To compound the production issue, protracted growth in the emerging markets beginning around 2000 led to an unprecedented escalation in the demand for natural resources worldwide. Thus, unlike the natural resources "bubble," which prevailed in the late 1970s, today's natural resources challenges are likely to persist for many years to come. In addition, the international response to escalating environmental challenges increasingly has been to place a cost on emissions (including GHGs such as CO_2) and other environmental damage, further improving the long-term prospects for clean tech investing. Clean tech opportunities can thus be expected to be far more numerous and scalable than at any time in the past.

Since clean tech investing has, until recently, been of fringe interest to investors, policy makers, scientists, and environmentalists, an important question arises as to how relevant technologies and investment opportunities will develop over the coming years. To be certain, generally higher (if also more volatile) prices for energy and raw materials coupled with escalating stress on the natural environment have created the essential conditions for a revolution in energy- and materials-related technologies. A critical issue then arises regarding the expected pace of innovation and what types of investment opportunities will ensue.

There are essentially two distinctive views regarding the expected pace of innovation, which divide investors, policy makers, environmentalists, technologists, and scientists. The first view is formed by the general lack of innovation in energy and materials technologies (as well as building materials, construction techniques, and operating technologies) over the past few decades. The fact that so little has changed in basic energy refining and power generation efficiencies since the 1960s and 1970s, for example, reinforces an intuition that any innovations in clean tech will inherently be slow and incremental in nature. This view on the slow pace of technology innovation in energy- and materials-related technologies would lead investors to prefer later-stage clean infrastructure investments that seek to drive down the costs of well-known technologies currently available, including solar, wind, and geothermal projects.

The opposite view is that literally decades' worth of pent-up innovation will be unleashed over the coming years, and that the consequent change in energy- and materials-related technologies will be substantial and revolutionary in character. Such a "revolutionary" view of innovation in green technologies foresees changes in energy- and materials-related technologies as proceeding similarly to the dramatic changes witnessed in IT and biotech during the 1990s. If the pace of innovation in energy and materials-related technologies is rapid, then the strongest investment returns will be enjoyed by those investors able to participate in the process of innovation. As with the IT and biotech revolutions at the outset of the 1990s, it was extremely difficult for market participants and even scientists to foresee which technologies would prevail a decade or more into the future. Thus, successful investors in IT and biotech were those who could participate in the sometimes disruptive pace of technological change. Identifying the long-term winners (such as Google or Oracle) early on in their development was certainly helpful for investors. Yet, participating in technologies with instrumental or transitional value was also profitable so long as this transitional value was appreciated for what it represented.

Corn-based ethanol in the United States certainly can be viewed as a technology with only transitional value. At its core, corn-based ethanol utilizes a dubious bio-crop in corn, using a 120-year-old fermentation technology to refine it into a product (ethanol) that is not compatible with the nation's fuel pipeline infrastructure. Worse yet, corn-based ethanol production consumes 80 percent or more of the end product's energy value in its production. Overall, corn for biofuels use has driven up prices and crowded out consumption for food and feed, caused material soil degradation, utilized only 30 percent of the plant's BTU (British thermal unit) value that can be fermented, and produced only a similar amount of liquid fuel to what was required in the planting, harvesting, production, and distribution process. Corn-based ethanol—in no terms—can be viewed as a good long-term solution and would likely be viable as a long-term investment only with the indefinite provision of government subsidies. However, corn-based ethanol undoubtedly has had significant instrumental value and short-term investment value by detailing the requirements and infrastructure of what more successful long-term biofuel solutions will need to look like. So long as investors do not view corn-based ethanol as a long-term panacea, they may enjoy appropriate and compelling returns on these early projects in combination with government subsidies.

Corn-based biofuels in the United States may thus be viewed as Biofuels 1.0. Although there were profits to be made in Biofuels 1.0, the next generation of Biofuels 2.0 is likely to improve significantly on the shortcomings of corn-based ethanol. It is also possible that future generations of biofuels (such as 3.0 and 4.0+) in the United States could use feedstocks and fuel refining/conversion technologies that bear little resemblance to corn-based ethanol, potentially even using algae in a non-fermentation-based chemical process to create a fuel such as natural gas, gasoline, or Fischer-Tropsch (FT) fuels that could be pipelined at minimal cost throughout the United States. The real lessons are that during a period of technological change, successful investment involves being part of the process of technological change and also that the technologies and companies with long-term staying power are often difficult to identify during the early stages of a technology revolution. To be certain, regarding the change in energy- and materials-related technologies, we are still in the formative stages for such a revolution.

Another important question is whether (as defined in the previous section) "cleaner" energy technologies or "clean" technologies will dominate in the future and which will have the more attractive investment returns. Cleaner technologies such as those that improve the efficiencies associated with electrical power generation and liquid/gas fuel production and distribution, or those that improve engine efficiencies, fundamentally differ in character from those "clean" technologies, such as wind, solar, and geothermal,

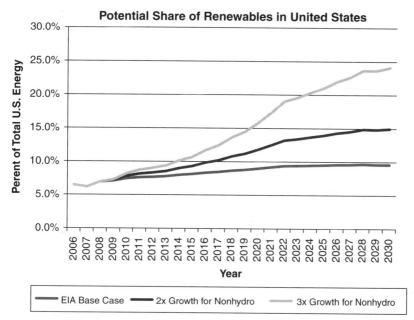

FIGURE 11.4 Potential share of renewables in the United States.
Source: Energy Information Administration 2009.

that attempt to provide investment with little environmental impact. The answer to the relative importance of cleaner versus clean energy technologies likely depends on the stage of investment and the time horizon for such investments.

A key insight is that given the enormous amount of energy produced from conventional sources, such as coal and petroleum, and the persistently low efficiencies associated with these fuels, cleaner technologies are likely to have a significant impact in the coming years. For example, the BTU efficiency of powering cars from petroleum or from generating electricity from coal has remained constant for over a century at generally less than 20 percent, taking into account all transportation, refining/conversion, transmission, and engine efficiencies. Figure 11.4 shows that fossil fuels (including coal, petroleum, and natural gas) will likely dwarf energy production from renewable sources, including biofuels, solar, geothermal, hydroelectric, and wind production, even in the most optimistic scenarios for at least the coming decade.

Thus, efficiency improvements associated with coal or petroleum of 50 to 100 percent or more is not only possible but likely, given its current level of inefficiency. Given the dominance of fossil fuels in at least the short term, such efficiency improvements among fossil fuels are likely to have a pivotal impact on the world's natural resources, natural environment, and financial system. Investments such as "cleaner" coal or petroleum (even if truly "clean" coal or petroleum may not be possible in the purest sense) are likely to have significant applications as they are developed in the near future. Such technologies are thus also likely to dominate in terms of providing infrastructure-type investment returns and wide-scale improvements to the natural environment compared with current levels of pollutants and other emissions.

This is not to say that clean technologies will be unimportant generally or that associated investment returns will be uncompelling—far from it. Rather, investing in "clean" technologies, including solar, wind, and geothermal, may in the long-run dominate investing in "cleaner" technologies like enhancing the efficient use of coal, but in the short term, both types of investments are needed and compelling. The reason we need both types of investments is that clean technologies are unlikely to dominate cleaner technologies until they become price competitive. Although the prospects are good that clean technologies will become increasingly price competitive in specific applications, particularly wind in consistently windy areas and solar in consistently sunny areas, it will likely take several more years before these technologies will generally prevail as more price effective than conventional energy sources on a widespread basis (Figure 11.5).

When specific clean technologies do consistently dominate conventional energy sources for providing electricity or liquid/gas fuels, they can be expected to enjoy meteoric growth rates for years to come. In the meantime, many competing technologies will likely attempt to gain a leadership position, particularly for solar, wind, and energy storage projects. Indeed, today some 200 different solar technologies are competing for a leadership position in the U.S. market alone. With only a handful of these technologies likely to have even transitional (let alone long-term) value, it can be perilous for investors to stake their success on a single winner at such an early stage. Rather, investments that can capture the transitional value of such technologies are likely to provide some of the most compelling returns.

In all, we are likely to be entering an era in which innovations in energy- and materials-related technologies will become a focal point both for technological innovation and for new capital formation. Although improving the current highly inefficient use of natural resources through "cleaner" technologies will likely have a dominant financial and ecological impact both now and for much (if not all) of the coming decade, true "clean" technologies will play an increasingly critical role for early-stage investors (including venture capitalists) for the foreseeable future and can be expected to dominate conventional ones, including fuel technologies in the long run.

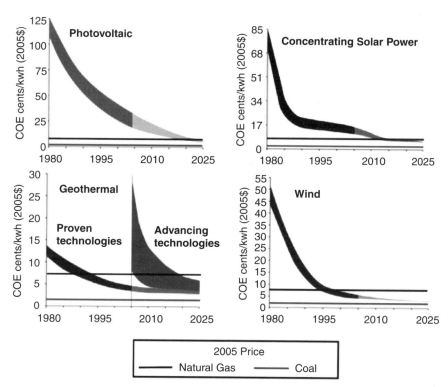

FIGURE 11.5 Renewable energy costs on path to be competitive with gas and coal. *Source:* National Renewable Energy Laboratory (www.nrel.gov), a division of the U.S. Department of Energy.

FITTING CLEAN TECH INVESTMENTS INTO AN EFFECTIVE AND DIVERSIFIED PORTFOLIO

In the past, both individual and institutional investors have typically viewed clean tech investments as opportunistic in nature and not fitting neatly into an asset allocation program. Indeed, they were viewed as comprising a portion of an environmental, social, and governance (ESG) portfolio that should occupy generally no more than 5 to 10 percent of a prudent and diversified portfolio. This relatively small allocation also reflected the relative paucity of attractive clean tech investments compared with conventional investments in stocks, bonds, private equity, real estate, and even infrastructure.

What is changing fundamentally, however, is that rather than being a fringe area of investment, clean tech investing is taking a central role *within* virtually every conventional asset class. For example, if energy stocks dropped from a high of almost one-third of the U.S. stock market in 1980 to a low of under 7 percent in 2000, energy (including a significant contribution from clean energy) can be expected to exceed some 20 percent of the value of U.S. and international stock markets in the coming years. Similarly, whereas neither energy- nor materials-related venture capital existed until recently, they now comprise some 50 percent of the venture capital pipeline for many programs. Furthermore, energy and raw materials costs were also considered to play a minor role in the construction and operation of real estate projects until quite recently. Today, the cost of energy and materials related to building and operating real estate has taken a central focus and can make the difference between a profitable versus an unprofitable project. Importantly, vast new capital requirements for infrastructure projects around the world include a significant portion associated with the production and distribution of energy and raw materials. Indeed, Goldman Sachs now estimates that some $17 trillion in new infrastructure investments alone will be needed to meet the world's escalating demand for gas and liquid fuels over the coming two decades.

Moreover, the ecological benefits of green investments over conventional ones have heretofore not generally been reflected in financial returns to investors. The landscape has begun to shift considerably, however, as environmental costs are increasingly being monetized in the form of carbon and pollution credits as well as taxes. Thus, for the first time on a widespread basis, the ecological benefits of green investments are beginning to be enjoyed by investors in financial terms, and these ecological benefits should play an increasingly important role for assessing and measuring investment success in the future.

CASE STUDY: Sovereign Wealth Funds and Clean Technology: The Masdar Initiative

By Massimiliano Castelli, PhD

Why should OPEC's third largest oil producer, which is pumping 2.4 million barrels of oil a day, develop a multibillion-dollar zero-carbon-emission city? This is what is happening in the capital of the United Arab Emirates (UAE), oil-rich Abu Dhabi, where one of the investment arms of the local government, a $10 billion sovereign wealth fund (SWF) named Mubadala Development Company, is investing $22 billion in the Masdar ("the source" in Arabic) Initiative. The Masdar Initiative is a multifaceted investment project consisting of several components, all centered around the flagship project, Masdar City, the world's first carbon-neutral, zero-waste city powered entirely by renewable energy. The other components of the initiative include establishing the region's first research institution in advanced energy and sustainability in cooperation with the Massachusetts Institute of Technology (MIT), developing alternative energy, and technology investments, and building

Massimiliano Castelli is executive director at UBS AG.

a portfolio of renewable energy assets for strategic investors. The long-term strategic goals of the Masdar Initiative is to position Abu Dhabi as a research-and-development hub for new transformative energy technologies, attract a cluster of innovative international firms operating in the sector, and diversify Abu Dhabi's oil-dependent economy.

The Masdar Initiative is a radical initiative as it departs from traditional SWFs' investment patterns and reflects broader changes that have occurred in the SWFs' asset allocation in the last few years. Until recently, in fact, with the exception of Norway's Government Pension Fund, which has a well-established track record in sustainable investments, SWFs—particularly those from oil exporting countries—have paid little attention to the growing trend among institutional investors to incorporate sustainability concerns into their global investment strategies. SWFs have traditionally invested their country's surpluses from commodity or manufactured good exports through global capital markets in industries such as real estate, aerospace, and financials, with a geographical bias toward developed countries, particularly the United States and Europe. This was seen as prudent because such investments typically meet SWFs' long investment horizon and desire for long-term return maximization.

In the last few years, however, SWFs have revised their investment strategy to incorporate the changes occurring in the global economy and in their domestic economies. In the global economy, the major change has been the rising attractiveness of emerging markets as a destination of funds. Asian, central and eastern European, and Middle Eastern countries have provided increasing direct and portfolio investment opportunities to SWFs searching for long-term return maximizations. The recent trend among Gulf-based and Asian SWFs toward investing a growing share of their wealth into emerging markets has also been reinforced by the negative political reaction that their high-profile investments in the U.S. and European financial sector have raised among Western policy makers.

With respect to their economies, oil-based SWFs have become increasingly involved in the implementation of the domestic economic policy agenda of their sponsor governments, whose priorities are domestic and regional growth, economic diversification away from oil, and job creation. The trend toward increasing asset allocation in their domestic economies has been sustained by the current economic and financial crisis and the sharp correction in commodity prices as SWFs are often required by their sponsor governments to replace private capital for strategic national investments or to provide emergency funding to state-controlled corporations and financial institutions.

Once domestic growth and economic diversification become strategic goals for SWFs, the adoption of a "sustainability-enhanced" investment orientation such as that adopted by Mubadala with the Masdar Initiative satisfies three long-term investment goals: (a) to reduce the domestic dependency on oil as the global economy embraces broader policies aimed at reducing carbon emissions and the oil reserves diminish; (b) to provide for a new domestic source of economic development and employment creation; and (c) to achieve return maximization. In fact, the Masdar Initiative fits very well into the diversification agenda of the oil-rich emirate that was recently outlined in "The Abu Dhabi Economic Vision 2030," a long-term road map elaborated by the local government to guide the evolution of its economy beyond oil that incorporates sustainable growth.

As oil corporations rebrand and raise their corporate profile by incorporating sustainability principles in their policies and invest into renewable energy sources to diversify their revenue stream, SWFs as the financial arm of their sponsor governments help to raise the profile of their countries in the international arena and provide an opportunity for economic diversification. Through the Masdar Initiative, Abu Dhabi is rebranding itself as an international center of excellence in renewable energy. The UAE was a founding member of the recently formed International Renewable Energy Agency—an international institution recently established and promoted by some major Western countries to support the development of alternative energy sources—and has put forward its candidature for hosting the agency's headquarters in the newly created Masdar City.

Abu Dhabi has been the first major hydrocarbon-producing economy to incorporate sustainability in its diversification agenda, but its pioneering project appears to be finding more followers across the region and in other oil-producing countries. Nigeria, another large oil producer, and Abu Dhabi have recently signed a cooperation agreement to develop projects to reduce carbon emission in the western African state and more cooperation initiatives across the Middle East region are in the pipeline. Thus, the Masdar Initiative has the potential to become a blueprint for commodity-exporting countries wishing to reduce their dependency on oil and adapt themselves to the evolving global energy market. Abu Dhabi and the UAE have an established successful track record in innovative projects, and should the Masdar Initiative be successful, similar initiatives could well be adopted by other commodity-exporting countries in the Gulf (e.g., Kuwait and Saudi Arabia) or in other regions (e.g., Russia). In the next decade, the huge wealth accumulated during the last oil price boom and currently managed by SWFs could well become the main engine of the "green revolution" in the oil-exporting countries. SWFs whose wealth is derived from noncommodity exports (e.g., China and Singapore) also might invest significant resources in this revolution because of the potential returns and because of their sponsoring governments' directives to reduce their growing dependency on oil and to reduce the environmental impact of their sustained growth.

The key lesson is that environmental investments will no longer be a fringe area in the capital markets, playing only a minor role in an individual or institutional investment portfolio. Instead, such investments will play an increasingly integral role in every major class of stock, bond, private equity, real estate, and infrastructure program for years to come. Thus, going forward, prudent investors should generally not consider "green" investments to constitute a separate investment class but rather should evaluate the attractiveness and risks of environmental investments for each investment class in which they fit.

CONCLUSIONS AND IMPLICATIONS

Clean tech investing has effectively come of age. Whereas clean tech has historically been considered a fringe investment to most individuals and institutions or, at most, part of a small ESG allocation over the past many years, clean tech investments are poised to play a key role in each major class of stock, bond, private equity, real estate, and infrastructure allocation for most prudent investors going forward. For the first time in decades, energy- and materials-related technologies will likely take center stage in catalyzing and promoting innovative technology research and development with a proportionately important role in venture capital, "commercialization-scale" investing, later-stage private equity, and infrastructure. Clean tech investments are also likely to play an increasingly pivotal role in global stock and bond markets. New scientific discoveries and innovative technologies that we see today will provide exciting investment opportunities for years to come, and technological breakthroughs in this space are accelerating.

What is inherently promising is that the world's accelerating thirst for natural resources does not have to continue even as it becomes more prosperous. Advances in clean tech represent the most promising mechanisms for controlling and economizing on the world's natural resources while protecting and enhancing our natural environment. To be certain, the world faces unprecedented economic, ecological, and natural resource challenges in the coming decades. Clean tech solutions, however, are poised to play a central role in meeting these challenges, and clean tech investors will play the central role in making these solutions possible and scaling them for meaningful impact.

BIBLIOGRAPHY

Climate Change Institute, n.d. "Human impact on the landscape," University of Maine www.climatechange.umaine.edu/Research/Contrib/html/22.html.
Dampier, M. 2003. Reading the stock market. *BBC News.*
Energy Information Administration. 2004. *Annual Energy Review, 2004.* Energy Information Administration. ww.eia.doe.gov.
Energy Information Administration. 2007. *Annual Energy Review, 2008.* Energy Information Administration. www.eia.doe.gov. (Retrieved on December 30, 2008.)
Energy Information Administration. 2009. *Annual Energy Outlook 2009 with Projections to 2030. Energy Information Administration.* http://www.eia.doe.gov/oiaf/aeo/electricity.html
Gore, A. 2006. *An inconvenient truth.* New York: Rodale Press.
Murray, I. 2008. *The really inconvenient truths.* Washington, DC: Regnery.
National Energy Renewable Laboratory. 2008. www.nrel.gov. (Retrieved on December 30, 2008.)
Simmons, M. R. 2008. The era of cheap oil is over. www.simmonsco-intl.com.
Woodard, D. *1973–1974 stock market crash.* About.com. (Retrieved on September 11, 2007.)

CHAPTER 12

Sustainable Commercial Property

Tim Dixon, PhD

Property is a very important aspect of environmental investing not only because of its value as an asset in portfolio diversification, but also because of its global impact on carbon emissions. The focus in this chapter is on "commercial property," which, as the largest and strongest performing global property investment sector, includes retail, offices, and industrial buildings. *Property* in this sense is taken as being synonymous with the term *real estate*.

The overall aim of the chapter is to define what is meant by *sustainable commercial property* and to explain why this asset class now forms a growing investment opportunity in the context of the wider growth of property investment and responsible property investment (RPI). It will also:

- Analyze the important interrelationship between commercial buildings and climate change.
- Identify the potential market for sustainable commercial property.
- Examine the key drivers and barriers in the sustainable commercial property market.
- Identify the rationale for investing in such property.
- Identify the range of real estate investment opportunities (both direct and indirect) currently available.
- Examine future trends and risks in the sector.

Tim Dixon is professor of real estate and director of the Oxford Institute for Sustainable Development (OISD), Oxford Brookes University, U.K.

The chapter focuses on the United Kingdom (and Europe) and the United States, but will also review developments in emerging markets and elsewhere, in a global context.

BACKGROUND AND CONTEXT

The surge in interest in sustainable property (or green buildings) has coincided with a growing focus by investing institutions on environmental, social, and governance issues (ESG), and in turn ESG is being increasingly reflected explicitly within company valuations. This has also been mirrored in the growth of the RPI agenda into which the sustainable property agenda also dovetails. But what is meant by the terms *sustainable property* and *responsible property investment?*

Sustainable Property

Although some literature has tended to treat both sustainable and green buildings synonymously (see Kats 2003), the majority of the literature on the subject can be divided into two separate, but also linked, strands of thought, which map (although not exclusively so) onto a difference in emphasis between *sustainable* thinking in the United Kingdom and Europe and *green* thinking in North America and Australasia, and this is now explored in detail.

The differences between the terms are partly related to whether the focus is on new build (green) or new build plus existing buildings, including refurbishment (sustainable) (Dixon, Ennis-Reynolds, Roberts, and Sims 2009; Sayce, Ellison, and Parnell 2007). In a U.K. context, one possible definition of a sustainable building is:

> *Any building that exhibit(s), at a minimum, better environmental performance than buildings built to building regulation standards in England, and that, in addition, may or may not have any features that address social and economic sustainability principles. (Williams and Lindsay 2007)*

An alternative definition for a green building in the United States comes from the Office of the Federal Environmental Executive (OFEE 2003), which defines such a building as one that:

> *... increases the efficiency with which buildings and their sites use energy, water, and materials, and reduces building impacts on human health and the environment through better siting, design, construction, operation, maintenance, and waste removal through the complete building life cycle.*

In fact, there are numerous definitions of sustainable and green buildings, but little agreement on the terms, and there are also different emphases on different stages of the building life cycle, often with a variation in focus on different stakeholders and different factors under the umbrella term of *sustainability*. Although definitions have frequently been concerned exclusively with environmental impacts, this is changing as broader social and economic impacts are increasingly recognized within the term *sustainable building* (Kremers 2006). According to Williams and Lindsay (2007), sustainable buildings are now also measured by their ability to provide a "healthy environment for occupants," which supports "sustainable travel patterns"; provides "flexible" spaces, which can adapt to changing occupier needs; and contributes to "sustainable patterns of urban and rural development" (see also Edwards and Marsh 2001).

This increasing focus on a "triple bottom line approach" has led to a broader definition of sustainable buildings being developed by the U.K. Green Building Council as (UKGBC 2008a):

Buildings which are (1) are resource efficient (physical resources, energy, water, etc); (2) have zero or very low emissions, (CO_2, other greenhouse gases, etc); (3) contribute positively to societal development and well being; and (4) contribute positively to the economic performance of their owners/beneficiaries and to national economic development more generally.

This definition should be contrasted with the more limited definition provided by Williams and Lindsay (2007). However, it is also probably fair to say that both the terms *green* and *sustainable* frequently focus on environmental issues (Table 12.1), with *beyond compliance* (in terms of building regulations) being a key characteristic.

To summarize, there is a clear difference in the evolution of thinking on green and sustainable buildings partly created by cultural differences between North America and Australasia, where the term *green* is

TABLE 12.1 Sustainable and Green Buildings

Sustainable	Green
United Kingdom and Europe	North America and Australasia
New and existing buildings	New buildings
Beyond compliance	Beyond compliance
Aspires to triple bottom line but strong environmental focus	Frequently an environmental focus

commonplace, and the United Kingdom and Europe, where the term *sustainable* tends to be used. This is not an exclusive distinction, and the terms have frequently been used interchangeably. However, the main differences between the terms can partly be related to whether the focus is on new build (green) or new build plus existing (sustainable). In using the term *sustainable property* in this chapter, it important to note that there are various shades of green or sustainability in commercial property investment. In a sense, some definitions (for example, UKGBC, 2008a) are aspirational, because ultimately the reality of property investing means that greater importance is more frequently attached to environmental impact than, for example, the social dimension. Despite this, there has also been an increasing focus on the RPI movement, which places great importance on the triple bottom line approach to property investment, where a sustainable property is seen as integrating the environmental, social, and economic dimensions of sustainable development (Elkington, 1997).

Responsible Property Investment (RPI)

Sustainable buildings form the key investment category within the aegis of RPI, which is defined by UNEPFI (2007a) as:

> ... *an approach to property investing that recognizes environmental and social considerations along with more conventional financial objectives. It goes beyond minimum legal requirements, to improving the environmental or social performance of property, through strategies such as urban revitalization, or the conservation of natural resources.*

The growth of RPI should be seen in the growth of a wider responsible investment (RI) agenda. For example, RI is defined by the World Economic Forum (2005, p. 7) as:

> ... *most commonly understood to mean investing in a manner that takes into account the impact of investments on wider society and the natural environment, both today and in the future.*

In theory, RPI can be implemented throughout the property life cycle; for example:

- Developing or acquiring properties designed with environmentally and socially positive attributes (e.g., low-income housing or green buildings)

- Refurbishing properties to improve their performance (e.g., energy efficiency or disability upgrades)
- Managing properties in beneficial ways (e.g., fair labor practices for service workers or using environmentally friendly cleaning products).
- Demolishing properties in a conscientious manner (e.g., reusing recovered materials on-site for new development).

This has also recently led to the development of Principles for Responsible Investment (PRI) (United Nations Environmental Programme Financial Initiative [UNEPFI] 2008). Pivo and McNamara (2005) define RPI as:

> ... *maximizing the positive effects and minimizing the negative effects of property ownership, management and development on society and the natural environment in a way that is consistent with investor goals and fiduciary responsibilities. It requires both an understanding of how cities and buildings relate to these larger issues and knowing how to address them in a financially prudent manner.*

This view of responsibility is predicated on the fact that the built environment is a major contributor to carbon emissions and pollutants, but also that the social and economic impacts of property investment strategies need to be considered (UNEPFI 2007b, Pivo 2008; see other papers and work by Pivo at www.u.arizona.edu/˜gpivo/). There is therefore a strong link between RPI and the concept of sustainable development, which incorporates a focus on the triple bottom line approach (Pivo and McNamara 2005, 2008; Rapson, Shiers, and Roberts 2007).

However, it is also clear that there is often an emphasis on environmental issues within the RPI movement. Research by Pivo (2008), for example, revealed that the top three ranked criteria for RPI by investors was energy efficiency, public transport, and daylight and natural ventilation.

Ten Principles of RPI (adapted from UNEPFI 2008)
1. *Energy conservation:* green power generation and purchasing, energy efficient design, conservation retrofitting
2. *Environmental protection:* water conservation, solid waste recycling, habitat protection
3. *Voluntary certifications:* green building certification, certified sustainable wood finishes
4. *Public transport–oriented developments:* transit-oriented development, walkable communities, mixed-use development
5. *Urban revitalization and adaptability:* infill development, flexible interiors, brownfield redevelopment

6. *Health and safety:* site security, avoidance of natural hazards, first aid readiness
7. *Worker well-being:* plazas, child care on premises, indoor environmental quality, barrier-free design
8. *Corporate citizenship:* regulatory compliance, sustainability disclosure and reporting, independent boards, adoption of voluntary codes of ethical conduct, stakeholder engagement
9. *Social equity and community development:* fair labor practices, affordable/social housing, community hiring and training
10. *Local citizenship:* quality design, minimum neighborhood impacts, considerate construction, community outreach, historic preservation, no undue influence on local governments

In other words, it may not always be possible or desirable for investors to place equal emphasis on the three main dimensions of RPI (environmental, economic, and social). Moreover, although examples of RPI are growing, there is still an apparent reluctance to apply RI approaches directly to commercial property investment portfolios, and this is often linked to investment managers' concerns over their fiduciary responsibilities (Rapson et al. 2007). Nonetheless there are substantial potential benefits associated with "good" (or sustainable) buildings (Figure 12.1), and it is frequently the benefits to occupiers and investors (particularly in environmental terms) that make the

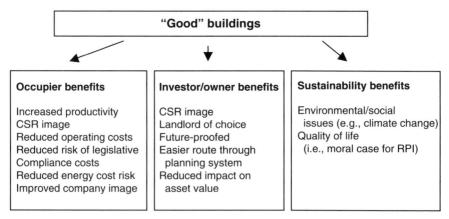

FIGURE 12.1 The benefits of sustainable buildings.
Adapted from D. Rapson, D. Shiers, and C. Roberts, Socially responsible property investment (SRPI): An analysis of the relationship between equities SRI and UK property investment activities. *Journal of Property Investment and Finance* 25(4): 342–358.

economic case for RPI a stronger one. For example, it is thought that as occupiers become aware of these benefits, their attitudes toward "bad" buildings are likely to change, leading to their avoidance. This could result in increased letting voids (or rental shortfalls) and reduced asset values for these properties, while those with better sustainability profiles enjoy higher demand and increased returns (see McNamara, 2008 and Roberts, Rapson, and Shiers 2007). Therefore, buildings that harm the environment less could theoretically be worth more, and this argument is even more salient when we understand the interrelationship between buildings and climate change globally.

In summary, we need to understand the importance of sustainable property investment within the context of RPI and the wider growth of RI, but also to understand that environmental investment strategies ultimately place a substantial emphasis on the environmental aspects of the property investment decision.

BUILDINGS AND CLIMATE CHANGE: A GLOBAL CHALLENGE

The building and construction sector is a vital sector globally employing 111 million people directly, and with 75 percent of those based in developing countries and 90 percent in micro firms. The sector also contributes to some 10 percent of global gross domestic product (GDP) with worldwide investment in the sector estimated at US$3000 billion (UNEP 2006), and so is an important part of the machinery of economic growth.

Over the past few years, the topic of sustainability has increasingly dominated discussion and debate in the building and construction sector. This has been largely fueled by a growing understanding that buildings are major contributors to increased carbon emissions, which most experts now acknowledge is a major cause of climate change, and that improving the energy efficiency of such buildings can help reduce emissions.

The facts are clear. For example, as UNEP (2006) reports, the built environment in its widest sense (including construction) is responsible overall for about 40 percent of CO_2 emissions, 30 percent of solid waste generation, and 20 percent of water effluents, as well as 40 percent of all energy used. Commercial property is also a major contributor within this overall context. In the United Kingdom, for example, energy use in nondomestic buildings accounts for about 18 percent of total carbon emissions in the United Kingdom (Carbon Trust 2008). Similarly, recent data (Commission for Environmental Co-operation 2008) suggests that buildings in North America use more than 2,200 megatons of CO_2, or about 35 percent of the continent's total, and data from the U.S. Green Building Council (USGBC 2008) suggests that buildings as a whole represent:

- 38.9 percent of U.S. primary energy use
- 72 percent of electricity consumption
- 13.6 percent of potable water supply

The developing nations, particularly China and India, also have a key role to play in helping the world achieve its global emissions targets. For example, official statistics suggest that buildings account for about 19 percent of China's total energy consumption, while others estimate the proportion at 23 percent, rising to 30 percent over the next few years (Fridley, Zheng, and Zhou 2008). This fact, combined with China's rapid urbanization program, will have profound consequences for global carbon emissions.

Against this backdrop, the publication of the HM Treasury's *Stern Review* (Stern 2007) therefore served to focus the minds of the property and construction sector not just in the United Kingdom, but globally, on the prospects for cutting emissions in buildings. The *Stern Review* was followed in 2007 by an Intergovernmental Panel on Climate Change (IPCC) report that updated its previous 2001 findings (IPCC 2007) and concluded that global climate change was "very likely" induced by man-made activity. Importantly, the IPCC report also identified real estate as having the largest "economic mitigation potential" of the major emissions sources, and far in excess of the potential of energy supply, forestry, and industry.

These two key reports have also paralleled two important global initiatives designed to highlight the importance of buildings and their impact on climate change. These are:

- *World Business Council for Sustainable Development (WBCSD) Energy Efficiency in Buildings Programme,* which is designed to develop a road map from a business perspective, outlining the critical steps needed to transform buildings' energy consumption. The program has brought together leading companies in the building industry to work across isolated "silos" and to help develop a cross-industry view of energy efficiency by identifying the approaches that can be used to transform energy performance (WBCSD 2008)
- *United Nations Environment Programme (UNEP) Sustainable Buildings and Construction Initiative (SBCI),* which is designed to promote a worldwide adoption of sustainable buildings and construction practices through the establishment of a common platform to tackle climate change by establishing baselines, developing tools and strategies and implementing these at a global level (UNEP 2006).

Given property's global importance as a major contributor to carbon emissions, the focus on investment risk caused by climate change and its impact on buildings has also been driven by the realization that climate change itself can also damage property asset values and pose significant risk to commercial property investors. This point was highlighted in a recent U.K. report funded by Hermes (Austin, Rydin, and Maslin 2008), which suggested, as a result of climate change, that occupiers of buildings would be more likely to suffer heat stress, causing disruption in shops, offices, and industrial premises; there would be an associated risk of flooding and flash flooding in urban areas; and that water shortages could be become more common, with ground movement also problematic. Cities such as Southampton, London, Bristol, Cardiff, and Cambridge were all seen in the report to be at risk.

Similar patterns of risk have also been identified globally: for example, the global costs of weather-related disasters have increased from an annual average US\$8.9 billion (1977–1986) to US\$45.1 billion (1997–2006), and in the coming decades, the number of people at risk from extremes will very likely grow, and extreme weather is likely to increase, posing severe risks for the world's major cities, particularly those located on seaboards (IPCC 2007).

SIZE OF THE MARKET FOR SUSTAINABLE PROPERTY

Against this backdrop of climate change, putting the size of the market for "sustainable property" into the context of the property investment market is important, not only because of the sheer difference in scale between the two sectors, but also because understanding property risk generically helps us understand the specific risks associated with sustainable property. This section begins with a brief overview of commercial property as a major asset class and the size of the global market, which serves to set the context for a discussion of sustainable property investment.

Commercial Property

Commercial property (i.e., primarily retail, offices, and industrial) offers important attractions as an investment class, which includes a secure and stable cash flow; relatively strong performance (over equities, gilts, and cash); low volatility; diversification benefits to spread risk; and the tangible aspects of "psychic" investment (Investment Property Forum 2007). Larger institutional investors (such as pension funds and insurance companies) own property directly[1] which provides them with rental income, but there are other indirect investment vehicles available for a range of investors (and which include property within them), such as, in the United Kingdom, limited partnerships and unit trusts. Real estate investment trusts (or REITs) also provide opportunities for investing in property, and they have spread globally since being introduced in the United States nearly 50 years ago, so that today there are more than 20 countries with some form of REIT structure resulting in a global securitized real estate market of $11,529 billion in 2006 (Charles Schwab 2007).

As to the overall market size, it is possible from recent estimates (see Hobbs and Chin 2007) to calculate that the total "invested" commercial property market was close to US$10 trillion at the end of 2006. This was property or real estate owned by investors, including pension funds, private investors, and financial institutions. When we include current owner-occupier space from large corporations (which might in time become leased as institutional space) the figure of "investible" stock is closer to $16 trillion. Geographically, within both categories, the U.S. dominates, followed by Europe and Asia, or put another way, the developed economies of the world have 85 percent of the global stock of investible real estate but only account for 20 percent of the world's population.

During the early part of this century global real estate markets performed well, driven partly by increased levels of cross-border investment activity and improvements in transparency and liquidity provisions in markets. Since 2001, for example, the amount of cross-border investment tripled, to reach US$116 billion in 2006, or 20 percent of the global total (Chin, Topinzi, and Hobbs 2007). However, more recently, with increased tightening of credit, and the onset of an economic recession, property markets themselves have also experienced the start of a cyclical downturn.

Sustainable Property

Within the growth of global commercial property assets, interest in sustainable property has also grown rapidly over the last five years, but the lack of an agreed definition for sustainable property and its relative novelty in the marketplace makes it difficult to estimate its size accurately.

Despite this, it is clear that there are now more than 4.2 billion square feet of commercial property, which now carries a Leadership in Energy and Environmental Design (LEED) certification (USGBC 2008). Moreover, data from McGraw-Hill Construction Analytics (2008) suggests that in the United States the commercial and institutional green building market size is expected to be about 20 to 25 percent of construction value by 2013.

Further evidence of growth potential comes from other sources. For example, of the 300 REITs in the United States, 41 percent are actively pursuing energy efficiency and green building upgrades, and another 27 percent plan to do so (Drummer 2007). Moreover, in the United States, there has been a move by a number of financial institutions toward green lending programs, and green loans themselves have also started to be incorporated within commercial mortgage-backed security pools, which could be expected to influence similar trends in U.S. underwriting, lending, and borrowing decisions (Tobias 2007a). Examples here include Citicorp, Bank of America, and Wells Fargo.

This trend has been mirrored in the United States by the emergence of pooled real estate funds that are devoted to green property investment, including Prologis, Liberty Property Trust, the Multi-Employer Property Trust, and the Urban Strategies America Fund (Tobias 2007a), and this view of growth is also supported by a recent survey by McGraw-Hill (2008a), which suggests that by 2013, 94 percent of responding construction firms will be building sustainable property (residential and commercial) globally on at least 16 percent of projects. This represents significant growth over a 10-year period (Figure 12.2).

In the property investment arena we have also seen international initiatives that have sought to raise awareness of the impact of climate change and related issues among investing institutions and other stakeholders. Examples include the Institutional Investor Group on Climate Change

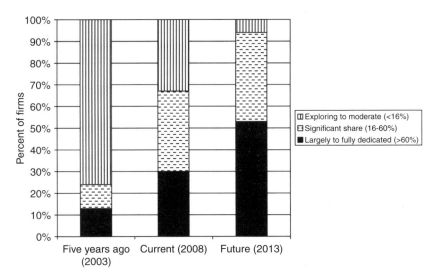

FIGURE 12.2 Global level of firm involvement over time in "green" building. Adapted from McGraw-Hill Construction Analytics *Global Green Building Trends* (2008).

(IIGCC), which is a forum for collaboration between pension funds and other institutional investors on issues related to climate change, and the UNEP Finance Initiative (Property) Programme, which is designed to encourage property investment and management practices that achieve the best possible environmental, social, and financial results. This has also been underpinned by a growth in the number of member countries of the World Green Building Council, which currently stands at 10, with a further 16 emerging national councils.

There are differences in market potential between the global regions, however. According to a recent and detailed analysis of green global trends (Nelson 2007), there is substantial variation between regions in the way in which green practices and regulatory environments have evolved, and hence how green property investment markets may move in the future. For example, green practices have grown most strongly in more advanced, slower-growth economies in contrast to more rapidly growing emerging economies. As Table 12.2 shows, the top markets overall for sustainable property construction and retrofits (and hence investment opportunities) are likely to be concentrated in North America and Europe, but with significant opportunities for growth in other economies such as China (see also McGraw-Hill Construction Analytics 2008).

To support this growth, certified property rating systems, which assess the sustainability characteristics of a particular property, are important for setting the standards against which investors may judge the extent to which an investment is truly "sustainable" (Dixon et al., 2007). Although only two countries (the United States and United Kingdom through their LEED and Building Research Establishment Environmental Assessment Method [BREEAM] schemes, respectively) have a substantial number of certified buildings, other countries have started to develop similar systems. For example, Germany is to introduce its Deutsches Gütesiegel Nachhaltiges Bauen (or German Sustainable Building Certificate) in 2009, and Holland already has Green Calc+, France its HQE system, and Switzerland has Minergie. Elsewhere, Australia uses Greenstar and Hong Kong the Building Environmental Assessment Method (BEAM) system.

In the United Kingdom, the BREEAM standard for commercial property is widely recognized as an industry standard and a simple way of identifying an office's sustainability credentials. According to BRE (2008), BREEAM is "the world's most widely used means of reviewing and improving the environmental performance of office buildings." In general terms, BREEAM currently assesses building performance in terms of:

- *Management*—overall policy, commissioning site management, and procedural issues.
- *Energy use*—operational energy and carbon dioxide emissions.

TABLE 12.2 Top Markets for Green Construction and Retrofits (Ranked in Order of Investible Opportunities)

	Projected Construction Volume	Green Share	Investible Green Opportunity
New Construction			
United States	Very Large	High	Very Large
China/Hong Kong	Very Large	Moderate	Large
United Kingdom	Large	Very High	Large
Germany	Large	High	Large
Japan	Large	High	Large
France	Moderate	High	Moderate
Canada	Moderate	High	Moderate
Australia	Moderate	Very High	Moderate
South Korea	Moderate	Moderate	Moderate
Spain	Moderate	Moderate	Moderate
Green Retrofits			
United States	Very Large	High	Very Large
Japan	Large	High	Large
United Kingdom	Large	Very high	Large
Germany	Large	High	Large
France	Moderate	High	Moderate
Italy	Moderate	Moderate	Moderate
Canada	Small	High	Small
Netherlands	Small	Moderate	Small
Spain	Small	Moderate	Small
Australia	Small	Very high	Small

Note: Size of investible real estate market adjusted for age of stock.
Source: Andrew Nelson, *Globalization and Global Trends in Green Real Estate Investment* (RREEF Research, No. 64, September 2008).

- *Health and well-being*—externally and internally.
- *Pollution*—including air and water.
- *Transport*—including carbon emissions and location-related factors.
- *Land use*—greenfield or brownfield.
- *Ecology*—including ecological value and enhancement of the site.
- *Materials*—including the environmental impacts.
- *Water*—including consumption and efficiency.

Credits are awarded for each area and combined and weighted, giving rise to a standard of either "Pass," "Good," "Very Good," or "Excellent."[2]

Data from BRE (Dixon et al., 2009) shows that from 1998 to the end of 2007, more than 1,000 offices, for example, had been assessed under the standard in the United Kingdom. This includes both new build and refurbished property, although the majority of BREEAM assessments apply to new build. However, to put the figures in some perspective, it is estimated that only about 7 percent of new offices in the United Kingdom are rated BREEAM "Good" or above annually, and new offices themselves comprise only 1 to 2 percent of existing stock. The number of offices reaching certification is currently lower still because of regulatory backlogs.

Similarly, some 1,700 projects have been certified under the LEED system since it began in 2000, and the system is also being used internationally (e.g., by U.S.-based multinationals) under the aegis of the World Green Building Council with more than 80 countries using LEED (Nelson 2008).[3]

In summary, therefore, the market for sustainable property is still relatively limited in extent but with an upward trajectory in growth. But what is driving the market and what might be holding the market back?

SUSTAINABLE PROPERTY DRIVERS AND BARRIERS

Drivers

The growth in interest and activity in the sustainable commercial property sector has been driven by a number of factors, including:

- Increasing importance of sustainability agenda
- Legislation and related guidance
- Stakeholder demand (occupiers and investors)

In recent years, there has been a key change in corporate attitudes toward environmental issues and sustainability. These issues have made the move into mainstream businesses, with many companies now recognizing the business benefits of sustainable business practices and processes. This change came as a response to government mandates aimed at reducing carbon footprints, public pressure, new business opportunities, and the shift toward greater corporate accountability, as companies start to recognize the benefits of "going green." Although energy still represents a relatively small proportion of overall business costs, more recently, rising energy prices has also focused attention in the property sector on how to improve the energy efficiency of buildings, and there has been a growing realization that buildings also impact substantially on water use, waste, and materials.

A further driver for sustainable offices and sustainable commercial property generally has been a changing legislative landscape bolstered by new and emerging guidance. In the United Kingdom and elsewhere, governments have been especially keen to focus the business community's attention on reducing carbon emissions to support the Kyoto Agreement.

For example, in the United Kingdom, the Climate Change Act (including a Carbon Reduction Commitment) was introduced in 2008, setting an 80 percent reduction target on 1990 carbon emission levels by 2050. This represented the first legally binding commitment by any country in the world, and the Act also introduced an interim target of a 26 to 32 percent reduction by 2020, as well as five-year, interim "carbon budgets," set in advance (Committee on Climate Change 2008). Any government failing to operate within the five-year budgets will have to report to parliament to explain its failure and could be subject to judicial review. This forms a triumvirate of new legislation (alongside the Energy Act 2008 and Planning Act 2008), which is also designed to strengthen the drive toward renewables in the United Kingdom.

In Europe, the Energy Performance of Buildings Directive (EPBD) is starting to have an impact. The principle underlying the Directive is to make energy efficiency transparent by the issue on sale, rent, or construction of an Energy Performance Certificate (EPC), which shows the energy rating, accompanied by recommendations on how to improve efficiency. In addition public buildings will have to acquire and display a Display Energy Certificate (DEC), and the Directive will be fully implemented by the end of 2008. EPCs and DECs are therefore expected to have an impact in the market on the price that investors will pay for property investments. Previous U.K.-based research (Investment Property Forum [IPF] 2007), for example, suggests that investors who are currently unprepared for the EPBD are likely to face difficulties as "the recommendations contained within the Energy Performance Certificate could be used for 'price chipping,' negatively impacting on the capital or rental value of the property."[4]

In the United States, the National Energy Policy (NEP), published in May 2001, was designed to improve the energy intensity of the U.S. economy by 20 percent between 2002 and 2012, and the Energy Policy Act of 2005 was the first omnibus energy legislation enacted in more than a decade, with its strong focus on improving energy efficiency and increasing the use of renewable energy sources (CA Chevreux 2008).

Key also to understanding the drivers for sustainable property is stakeholder demand, primarily from occupiers and investors, but also underpinned by legislation from government, and indeed its efforts to exemplify sustainable practice through public sector real estate strategies (e.g., the U.S. General Services Agency, which develops and manages most federal real estate, was an early champion of green technologies). Historically, the "vicious circle of blame" has been posited as a fundamental barrier to progress (Association for the Conservation of Energy [ACE] 2003; Cadman 2000; CORENET 2004), highlighting the more negative perceptions of investors, developers, occupiers, and constructors—for example, each stakeholder blaming the other for lack of progress. There are signs, however, that a "virtuous circle" could be emerging (CORENET 2004) (Figure 12.3).

As evidence of this, some recent surveys show changing requirements in factors influencing choice, at least in terms of the revealed preferences of occupiers. For example, GVA Grimley (2006) asked occupiers to rank, in order of importance, three key property-related environmental issues in relation to their accommodation strategy for the next five years. Some 64 percent rated "occupying energy-efficient buildings" as the most important, followed by "water-efficient buildings" and "proximity to transport networks (including public transport)." This view was supported by a more recent survey of U.K. property investors (GVA Grimley 2008), and other evidence from global markets points to a growing demand for sustainable

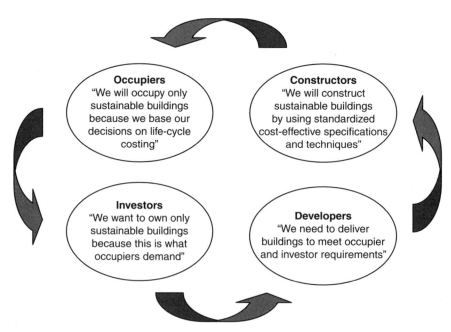

FIGURE 12.3   Virtuous circle of sustainability.
Adapted from CORENET, *Corporate Real Estate 2010: Sustainability & Corporate Social Responsibility* (2004).

property (McGraw-Hill 2008; McGraw-Hill Construction Analytics 2008). Despite this, however, recent work in the United Kingdom (Dixon et al. 2009) found that in the case of actual office moves over a two-year period, sustainability was still a less important factor for occupiers in choice of building than location, available stock and running costs (i.e., primarily rent and business rates). Nonetheless, it is clear that investor demand is an important driver, and an increasing number of examples of sustainable property investment practice are now emerging (Dixon et al.; UNEPFI 2008, 2009).

Barriers

Recent work by the All Party Urban Development Group (APUDG) (2008) suggested that a number of barriers were preventing the move toward more sustainable property in the United Kingdom. These included:

- Energy efficiency
- Poor knowledge/education
- Costs and physical barriers

The importance of energy efficiency should not be underestimated, and research by the Carbon Trust (HM Treasury 2005) suggests that there are a number of barriers to energy efficiency measures in both the public and business sectors. These fall into four main groups, which include:

- Investment costs of new technology set against energy savings.
- Hidden costs from adopting more efficient energy equipment.
- Market failures from "split incentives" (i.e., the landlord–tenant split, where tenants pay energy bills but landlords control the properties).
- Organizational inconsistencies, where there is a misalignment of return within an organization when differing parts of an organization may place different values on different rates of return. This may derive from managerial inertia or key decision makers lacking interest or motivation to improve energy efficiency.

Further evidence suggests that investors and lenders continue to associate a number of key risks with investment in sustainable property. These include (Dixon et al. 2009; Tobias 2007b):

- *Reliability of technology*—a tendency to consider "green" technologies as relatively untried and untested.
- *Uncertainty over costs and benefits*—lack of hard evidence on additional costs of building sustainably.

- *Uncertainty over green building performance over time*—lenders often do not have the information they need to make a risk assessment for loan.
- *Loan security and cash flow issues*—although initiatives such as the Clinton Climate Initiative (in alliance with the C40 Large Cities Climate Leadership Group) is helping tackle climate change in cities, other similar schemes are needed to structure finance deals to link capital provision explicitly with green improvement cash flows and energy savings.

In a more positive sense, this has resulted in the promotion of "green leases," which are designed to make the relationship between investor and occupier more explicit in terms of environmental responsibilities (Hinnells et al. 2008) and other codes or guidance designed to help occupiers benchmark environmental performance of their property, for example, the Investment Property Databank (IPD) Environment Code for occupiers (IPD 2008).

The lack of knowledge about the benefits of going green has also been highlighted as one of the core barriers preventing sustainable commercial development. This can operate at a number of levels. Nelson (2007), for example, found that few professionals have the specialized knowledge and experience to design and operate green buildings successfully (see also Dixon et al. 2007; World Business Council for Sustainable Development [WBCSD] 2007).

Finally, there are three ways that costs can act as a barrier to sustainable offices. The first relates directly to energy costs of companies. Energy usually accounts for between 1 and 6 percent of business operating costs and therefore can create a reduced incentive for change, although as energy prices increase, this effect is weakened. By contrast, staff costs can be as high as 85 percent, which means that the potentially biggest return on investment can arise when green buildings improve business productivity (Royal Institution of Chartered Surveyors [RICS] 2005). Second, key stakeholders in the development process have often perceived the costs of sustainable construction as being relatively high, and in some instances research has shown that the costs of building sustainable property is higher than conventional buildings, although much depends on the nature of the technology used. For example, a recent U.S. Green Building Council report (Galbraith 2008) shows that the additional costs of "going green" are 2.5 percent (based on 150 commercial, and some residential, buildings), although in theory the returns from investment are attractive.

The third way in which cost can act as a barrier to sustainability is through negative perceptions of payback periods. According to Nelson (2007), for example, increased payback periods may dissuade some investors from investing in sustainable property, because life-cycle costing, a central premise of the green building case, typically assumes a much longer

TABLE 12.3 Impact of Sustainability Agenda on Property Values

Factor	Investment Implications	Impact on "Nonsustainable" Buildings
Tenants prefer sustainable buildings to nonsustainable buildings.	Over time, rental differences emerge between the two types of building. Nonsustainable assets take longer to re-lease at lease end.	Rental growth lower, depreciation higher. Longer interruptions to cash flow, higher risk premium.
Nonsustainable buildings are more costly to run (energy costs).	Less money available for rent in nongreen buildings.	Rental growth slower.
Other investors prefer "conventional" buildings over "nonsustainable buildings."	Nongreen properties take longer to transact.	Greater illiquidity and opportunity costs, high risk premium required.

Adapted from P. McNamara, *The Fundamentals of Sustainable Real Estate: Towards Responsible Property Investment* (2008).

amortization period of 20 to 30 years more than the typical investor's holding period, and so longer payback periods could potentially create problems with shorter tenancies.

THE INVESTMENT THESIS FOR SUSTAINABLE PROPERTY

The benefits of sustainable commercial property are well documented (Green Building Council of Australia 2006; Newell 2008; RICS 2005). As Table 12.3 shows, these relate to potential factors that could improve rental performance and reduce overall risk.

Furthermore, according to a comprehensive 2008 survey of U.S. architects, engineers, and contractors by McGraw-Hill (2008), the value of the financial benefits from higher yields, higher liquidity, and capital gains over the lifetime of the buildings outweighs the initial additional costs. The survey highlighted the following financial benefits:

- An average expected decrease in operating costs of 13.6 percent
- An average increase in building value of 11 percent

- Occupancy rate expected to increase by 6.4 percent
- On average, rental yields expected to increase by 6.1 percent
- Average return on investment expected to improve by 10 percent

From the investment point of view, sustainable buildings might also be imputed to have longer economic lives (due to less depreciation) and lower volatility (because of lower environmental and marketing risks), which could lead to reduced risk premiums and higher valuations (Eichholtz, Kok, and Quigley 2008). Therefore, differentials are likely to occur in the market between sustainable and nonsustainable property.

In property valuation terms, this difference might be expressed as in Table 12.4, which represents a simplified example of a possible valuation, comparing a hypothetical sustainable office (e.g., BREEAM excellent office) with an office built to more conventional standards. This shows that the benefits of a more sustainable building feed through into a higher initial rent with better prospects of rental growth income and therefore a lower overall "all risks" yield. Over time, therefore, we might expect to see a clearer differentiation in value and return between a sustainable property and a "conventional" building.

In Table 12.4, it is assumed that the savings in energy costs are associated with a higher rent (which reflects the savings in energy costs), but, in practice, more sophisticated models of investment worth (based on discounted cash flow) would be needed to quantify the impact of functional performance on rental growth and property depreciation (Ellison, Sayce,

TABLE 12.4 Example of U.K. Market Valuation in Two Office Buildings

	BREEAM Excellent Building	Building Built to Minimum Regulations
Total occupation costs (£ per sq ft)	45	45
Energy costs	2.00	3.00
Repairs/service charge[5]	5.00	5.00
Business rates	10.00	10.00
Rent	28.00	27.00
All Risks Yield (%)	4.75	5.00
Capital Value (£ psf) (less acquisition costs)	555.50	508.90

Adapted from GVA Grimley, *Sustainability: Towards Sustainable Offices* (2007).

and Smith 2007) (Figure 12.4). Recent research by Compass (2007), for example, found that in Vancouver and Whistler in Canada LEED-certified buildings produced profitable investment with long-term internal rates of return at above 35 percent over 15 years. However, research by Pivo and Fisher (2008) suggests there is no substantial evidence of above-average returns for RPI-based investments based in the United States, although in property with a strong emphasis on energy efficiency, returns were found to be generated at lower risk.

Substantive and hard evidence for return and rental premia in sustainable commercial property remains elusive, however, partly because of the immaturity of many markets and the lack of market transaction data. However, there have been several studies recently that have used LEED and Energy Star data in the United States to assess whether a rental premium does exist, although the evidence is somewhat mixed. Studies by CoStar (2008); Eichholtz et al. (2008); Fuerst and McAllister (2008); and Miller, Spivey, and Florance (2008) (using CoStar data) suggest that certified buildings do have a rental premium. For example, according to the CoStar study, LEED buildings command a rental premium of $11.33 per square foot over their non-LEED peers and have 4.1 percent higher occupancy, although more recent work by Muldavin (2008) casts some doubts on findings of the CoStar study. Nonetheless, these studies suggest that rental premiums in LEED buildings are between 4.4 percent and 9.2 percent for LEED buildings and between 8.9 percent and 11.6 percent for Energy Star buildings.

Although such price differentials potentially reflect the relative shortage of green space relative to tenant demand, it should be remembered that valuers (or appraisers) are guided by the market (and by mandatory professional standards) and reflect market demand in their valuations only if the market itself is driving change. Recently, there has been a growing recognition that alongside other stakeholders, valuers therefore have a key role to play in helping create the environment for a "virtuous circle" rather than a "vicious circle." This was seen with the launch of the Vancouver Accord, a commitment by valuation standards organizations globally to begin the process to embed sustainability into valuation and appraisals, which was launched on March 2, 2007. (See www.worldgbc.org/default.asp?id=41&articleid=217 and www.vancouveraccord.org/.)

SUSTAINABLE PROPERTY INVESTMENT OPPORTUNITIES

As we have already seen, the market for sustainable commercial property is still relatively limited in extent. Given that existing stock continues to

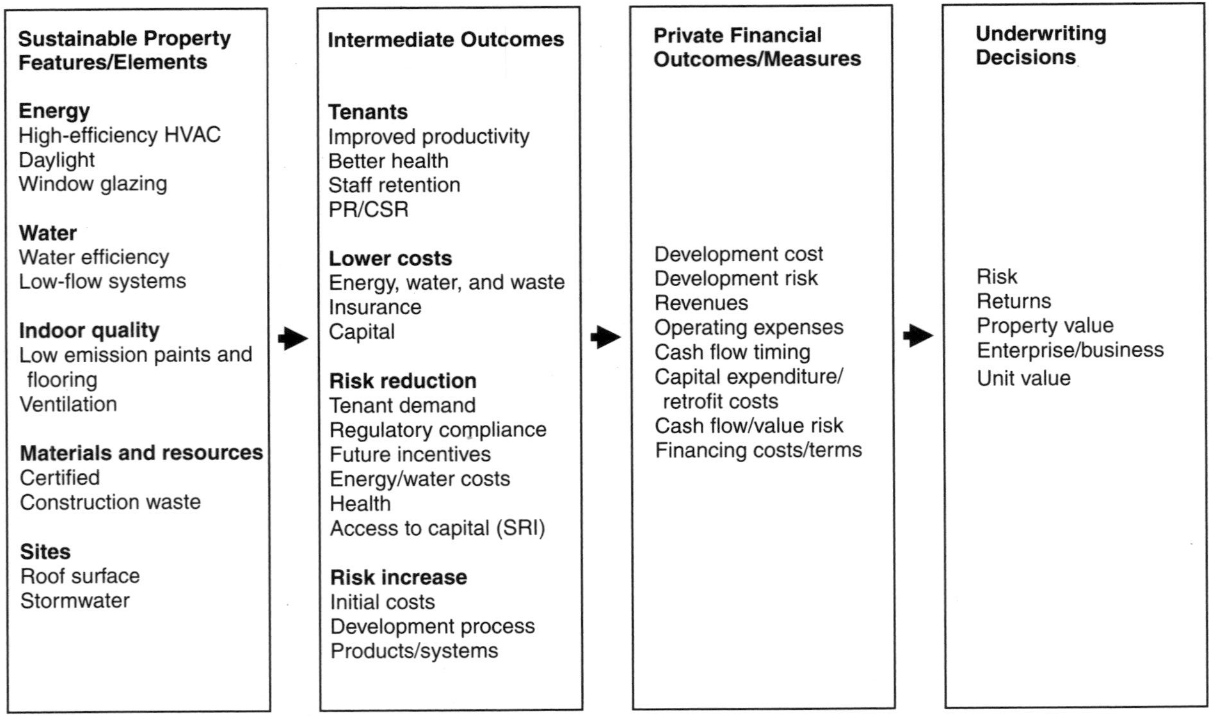

Sustainable Property Features/Elements

Energy
High-efficiency HVAC
Daylight
Window glazing

Water
Water efficiency
Low-flow systems

Indoor quality
Low emission paints and flooring
Ventilation

Materials and resources
Certified
Construction waste

Sites
Roof surface
Stormwater

Intermediate Outcomes

Tenants
Improved productivity
Better health
Staff retention
PR/CSR

Lower costs
Energy, water, and waste
Insurance
Capital

Risk reduction
Tenant demand
Regulatory compliance
Future incentives
Energy/water costs
Health
Access to capital (SRI)

Risk increase
Initial costs
Development process
Products/systems

Private Financial Outcomes/Measures

Development cost
Development risk
Revenues
Operating expenses
Cash flow timing
Capital expenditure/ retrofit costs
Cash flow/value risk
Financing costs/terms

Underwriting Decisions

Risk
Returns
Property value
Enterprise/business
Unit value

FIGURE 12.4 Linking sustainable features to financial assessment of sustainable property. Adapted from S. Muldavin, *Financial Assessment of Sustainable Properties' in Europe Real Estate Yearbook* (2008).

predominate in all markets, this is not surprising because of the relative scale issue: in the United Kingdom, for example, new build commercial property is equivalent to about 2 percent of total stock, with the balance comprising existing or legacy stock, which potentially may be hard to retrofit. So far, the vast majority of activity has been undertaken by private funds, often with the partnership of public pension funds. In contrast, investments from public equity markets have been very limited, held back by the paucity of certified sustainable properties for purchase, and the lack of common green product definitions throughout the industry (Nelson 2008).

It is also fair to say that the strategic and tactical decisions made by pension funds, corporation boards, and others (in alliance with their real estate or portfolio advisors) regarding whether to invest in sustainable buildings and in which properties to invest, are more highly evolved than the frameworks for detailed property-specific decisions (Miller et al. 2008). This is again because of the lack of rental and return data in what is essentially an emerging, innovative market, although if unified standards for green mortgages could be developed, this could also help drive the market.

Despite this, there is a growing set of investment opportunities in the commercial property market, both in terms of direct property investment by institutions and through specific property funds.

In the United Kingdom, for example, a number of commercial property investors have led by example through their commercial property activities, including Prudential Property Investment Managers (PRUPIM), Igloo, Hermes, Land Securities, British Land, and Hammerson; in Australia, they include Investa, Lend Lease Mirvac, and Martin Place Trust (MPT) (Dixon 2007; Newell 2008; UNEPFI 2008).

In the United States, favorable real estate yields and property borrowing conditions, declining green building costs, increasing technical sophistication, and rising confidence in green property performance have led to the "first generation" of sustainable real estate investment offerings. Initial institutional forays into green real estate have so far typically taken place in the context of additions to established portfolios (Tobias, 2007b), and examples include the Multi-Employer Property Trust and the Liberty Property Trust.

In the United Kingdom, PRUPIM's investment strategy for sustainable real estate has been driven by the need to minimize the business risks and maximize the market opportunities that the company faces, including (PRUPIM 2007):

- Legislative compliance and a tightening planning framework.
- Cost-cutting through eco-efficiency measures, including utilities efficiency, resource management, and waste management.

- Market advantages through increased competitiveness and increasing brand value.
- Future-proofing our investments to perform strongly under changing market conditions.

PRUPIM launched its Improver Portfolio of 25 properties in April 2007 to examine how best to reduce carbon emissions across its portfolio, but still maintain high returns on property, and is also working within the U.S. Climate Leaders Programme to help achieve consistency in standards across its international real estate portfolio. During 2007, PRUPIM improved its energy efficiency in shopping centers and managed offices by 17 percent and 16 percent, respectively, with consequent reductions in carbon emissions of 18 percent and 12 percent, respectively (Sketchley 2008).

In the United Kingdom, Hermes Real Estate's overall strategy is to implement sustainability improvements that will add value to their assets (Hermes 2008). In order to consider the financial impact of sustainability the company has developed what they claim is the property industry's first comprehensive sustainable investment evaluation tool, the Sustainability Rating System (SRS). The aim of the SRS is to rate the sustainability of each of the properties in each subsector of Hermes' directly managed portfolio, and the SRS is based on a suite of questions which address the issues Hermes believes are important with regard to sustainability and total returns. The list, which includes a range of factors measuring energy efficiency and other factors against benchmarks, establishes how each asset performs against estimated market conditions, and is designed to maximize financial performance by "identifying sustainability improvements to be achieved through management behaviour, service charge recoverable costs and capital expenditure" (Hermes 2008).

Leaders have also emerged elsewhere (Pivo and McNamara, 2008; UNEPFI 2008). In the United States, for example, California's two largest pension funds (California Public Employees' Retirement System [CalPERS] and California State Teachers' Retirement System [CalSTRS]) have set goals to reduce their property energy use by 20 percent in the next five years and have also increased their investment in urban inner city areas (UNEPFI 2008) in a move designed to underpin their RPI credentials.

These examples represent institutional responses in both the United Kingdom and United States to the broader sustainability and ESG agendas, and therefore to the holding and sale or leasing of sustainable property assets, but other investment opportunities in sustainable property are also expanding. A recent example of a specialist sustainable property fund is Climate Change Capital, based in the United Kingdom. This is, to date, the only sustainable property investment vehicle in the United Kingdom, which has succeeded in achieving its capital target (in this case, £50m). Stanhope, SNS Reaal, AIG, and Alliance Trust are among the investors. The aim is to secure further capital (55 percent debt and 45 percent equity) of £250m and to buy between 15 and 20 assets spread across the United Kingdom with a minimum environmental standard of BREEAM "very good," which is designed to offer the potential for higher returns to investors in a growing market.

Similar vehicles are developing elsewhere globally. In the United States, there have been several REITs, which have been developed to invest in sustainable property. These include the Liberty Property Trust, Rose Smart Growth Investment Fund, and the Revival Fund. More recently the first "green REIT" has been formed with the development of the Green Realty Trust (Dempsey 2008). According to its prospectus, targeted investments include green properties that (1) are certified under the LEED Green Building Rating System; (2) satisfy criteria for energy and environmental design under other established environmental rating systems; or (3) are properties that it intends to develop, redevelop, or renovate for subsequent certification as green properties. Similarly, Forward Progressive Real Estate Fund, a publicly traded fund holding the equities of publicly traded REITs, was repositioned in late 2006 to give greater emphasis to sustainability. The fund screens its investments on the basis of real estate market data, as well as energy conservation and efficiency, environmental impacts, the use of green development principles and effective management of natural resources and waste, and the fund's holdings are also screened on socially responsible investment criteria (Tobias 2008b).

BREWERY BLOCK II: Case Study in Successful Responsible Property Investing (RPI)

By Angelo A. Calvello, PhD

Brewery Block II, a LEED-Gold certified project that consists of a 10-story Class A office tower (completed in 2002) and two historic structures that were built in 1908 as a brewhouse for Blitz-Weinhard Brewery in Portland, Oregon. The project contains 219,695 square feet (sf), including 168,273 sf in the new office tower and 51,692 sf in the historic buildings.

Brewery Block II embodies many core RPI characteristics. The project is located in Portland's Pearl District and is transit oriented and pedestrian focused, offering bike storage, changing rooms, and a unique blend of historic preservation and modern amenities desirable to today's environmentally conscious tenants. Brewery Block II was designed with high levels of energy and water efficiency in mind, offering operating expenses well below national and local averages for comparable office buildings. The project promotes sound indoor environmental quality through use of low-emitting materials in tenant build-out, operable windows, and significant day-lighting of tenant spaces. During construction at Brewery Block II, more than 90 percent of construction waste was diverted from local landfills and the project's design sought to utilize locally sourced, sustainable materials. Estimated LEED-related costs associated with the project represented less than 1 percent of additional expense and were almost completely offset by the sale of state energy efficiency tax incentives. Additionally, the exclusive use of responsible contractors (i.e., those who pay living wages and provide appropriate worker benefits) for construction and tenant improvement work at Brewery Block II created over 2.2 million job hours and more than $52 million in personal income for local building trades members.

Brewery Block II was originally leased up during a challenging market cycle and has maintained a level of occupancy above the Portland CBD average. The project is currently 95 percent leased and has an attractive mix of national and local tenants. The project's green features and prime location enable it to remain highly competitive and a top tenant draw, resulting in strong tenant retention and interest—even compared to new "green" properties recently delivered to market. Brewery Block II remains a market leader with rental rates currently 8 to 10 percent above asking triple net rates of comparable Portland Class A office buildings.

Brewery Block II is owned by the Multi-Employer Property Trust (MEPT), a $6.0 billion open-end commingled real estate equity fund that invests in a diversified portfolio of institutional-quality real estate assets in major metropolitan markets around the United States. Kennedy Associates is MEPT's exclusive investment advisor and a national leader in responsible property investing (RPI). One of MEPT's top performers within the Kennedy Associates' portfolio, Brewery Block II has produced attractive historical returns that are above industry benchmarks. Moreover, in the current challenging market environment, the property continues to have strong cash flow despite the market's current downturn.

FUTURE TRENDS AND RISK: A PERFECT STORM?

As we enter a global economic downturn, however, what are the prospects for sustainable property? Sceptics might argue that just as in the 1990s recession when green politics became marginalized, sustainability could well drop down the priority list within corporate agendas. However, there are a number of reasons why this is unlikely. First, there is, as we saw earlier in this chapter, already a raft of legislation which is designed to underpin commitments to tackle climate change in the property sector. Second, it is likely that occupiers and institutions will want to maintain their competitive edge by continuing the push toward mainstreaming sustainable property as part of their corporate culture. Third, as energy costs continue to rise there will be increased urgency to invest in new renewables technology. For example, President Obama's $150 billion "Apollo project" to bring jobs and energy security to the United States through a new alternative energy economy, is likely to help underpin the United States' commitment to helping tackle climate change.

These views are supported by a recent U.K. Green Building Council poll, which found that 85 percent of respondents believed that sustainability would be a growth area in the economy despite the recession (UKGBC 2008b). Essentially, if the conditions for a "perfect storm" to help drive a step change in the property sector are not now with us, then it is unlikely they will occur again for some time, given that by some estimates we have only eight more years to deal with the problem of climate change. A key challenge will, however, be the fact that some 98 percent of buildings comprise existing stock (and about 75 percent of our buildings will still be with us in 2050), and therefore finding cost-effective ways of retrofitting commercial property will be of paramount importance, alongside the requirement for greater consistency in property certification globally so that investors and occupiers can subscribe to a unified system.

We can also expect to see further changes in fiscal regimes to persuade property developers and other stakeholders to behave more sustainably, and create a virtuous circle rather than a vicious circle, and we are also likely to see stronger mandatory standards applying to retrofitting as well as new build. As Nicholas Stern (2008) wrote recently:

> *The next few years present a great opportunity to lay the foundations of a new form of growth that can transform our economies and societies. Let us grow out of this recession in a way that both reduces risks for our planet and sparks off a wave of new investment which will create a more secure, cleaner and more attractive economy for all of us.*

The continued growth of the sustainable commercial property investment market can play a vitally important role in achieving this vision.

Useful Web Sites

Green Building Finance Consortium: www.greenbuildingfc.com/

Institutional Investors Group on Climate Change (IGCC): www.iigcc .org/

Oxford Institute for Sustainable Development (OISD): www.brookes .ac.uk/schools/be/oisd/

Responsible Property Investing Center: www.responsibleproperty.net/

United Nations Environment Programme Finance Initiative: www. unepfi.org/

Sustainable Buildings and Construction Initiative: www.unepsbci.org/

World Business Council for Sustainable Development: www.wbcsd.org/

NOTES

1. In Europe, about 5 percent of the value of investing institutions' portfolios are invested in property.
2. Criticisms of BREEAM, including the lack of a minimum threshold for each category of "score" have led to tighter standards in the relaunched BREEAM 2008 and a new "Outstanding" category.
3. LEED is due to be revamped in 2009.
4. The Energy End-Use Efficiency and Energy Services Directive are also expected to have important indirect impacts on the sustainable building market.
5. Business rates in England and Wales are the way in which businesses and other occupiers of nondomestic property contribute toward the cost of local authority services.

REFERENCES

ACE. 2003. *Energy efficiency in offices: Assessing the situation*. London: Research undertaken by J. Wade, J. Pett, and L. Ramsey for the Association for the Conservation of Energy funded by the Carbon Trust.

All Party Urban Development Group (APUDG). 2008. *Greening UK cities' buildings*. London: All Party Urban Development Group.

Austin, P., Y. Rydin, and M. Maslin. 2008. *Climate change: The risks for property in the UK*. London: University College London Environment Institute.

BRE. 2008. Accessed from web site www.bre.co.uk in August 2008.

CA Chevreux. 2008. *The lean way: Efficient buildings*. Courbevoie, France.

Cadman, D. 2000. The vicious circle of blame. Cited in M. Keeping. 2000. *What about demand? Do investors want "sustainable buildings"?* London: RICS Research Foundation.

Carbon Trust. 2008. *Low carbon refurbishment of buildings*. London: Carbon Trust.

Charles Schwab. 2007. *Global real estate securities: An attractive investment to diversify a portfolio*. New York: Author.

Chin, H., E. Topinzi, and P. Hobbs. 2007. *Global real estate investment and performance: 2006 and 2007*. London: RREEF Research, March.

Commission for Environmental Co-operation. 2008. *Green buildings in North America*. Quebec City: Author.

Committee on Climate Change. 2008. Building a low-carbon economy—the UK's contribution to tackling climate change. London: Climate Change Committee.

Compass. 2007. *Towards a green building and infrastructure investment fund*. Vancouver, BC: Author.

CoStar. 2008. *CoStar study finds Energy Star, LEED buildings outperform peers*. Press release and Powerpoint presentation at www.costar.com/News/Article.aspx?id=D968F1E0DCF73712B03A099E0E99C679 (accessed November 2008).

CORENET. 2004. *Corporate real estate 2010: Sustainability & corporate social responsibility*. Atlanta: Author.

Dempsey, B. 2008. Green Realty Trust: Nation's first green REIT is on the horizon. *Green Buildings NYC* (accessed November 2008 at www.greenbuildingsnyc. com/2008/02/06/green-realty-trust-nation%E2%80%99s-first-green-reit-is-on-the-horizon/.

Dixon, T. 2007. *Measuring social sustainability: Best practice from urban renewal in the EU 2007/02*. Oxford, UK: EIBURS Working Paper Series, OISD.

Dixon, T., A. Colantonio, D. Shiers, P. Gallimore, R. Reed, and S. Wilkinson. 2007. *A green profession? RICS members and the sustainability agenda*. London: Royal Institution of Chartered Surveyors (RICS).

Dixon, T., G. Ennis-Reynolds, C. Roberts, and S. Sims. 2009. *The demand for sustainable offices in the UK*. London: Investment Property Forum (IPF).

Drummer, R. 2007. REITs buying into 'green premium.' *CoStar Green Report*, October (accessed November 2008 at www.costar.com/News/Article. aspx?id=6E37F3E92F40A2AF38F72D9659791FCE,).

Edwards, M., and C. Marsh. 2001. Sustainable property development. In A. Layard, S. Davoudi, and S. Batty (eds.), *Planning for a sustainable future*. London: Spon Press.

Eichholtz, P., N. Kok, and J. M. Quigley. 2008. *Doing well by doing good? Green Office Buildings*. California: Institute of Business and Economic Research, University of California, Program on Housing and Urban Policy Working Paper Series.

Elkington, J. 1997. *Cannibals with forks: The triple bottom line of 21st century business*. Oxford, UK: Capstone.

Ellison, L., S. Sayce, and J. Smith. 2007. Socially responsible property investment: Quantifying the relationship between sustainability and investment property worth. *Journal of Property Research* 24(3) (September): 191–219.

Fridley, D., N. Zheng, and N. Zhou. 2008. *Estimating total energy consumption and emissions of China's commercial and office buildings*. Berkeley, CA: Ernest Orlando Lawrence Berkeley National Laboratory.

Fuerst, F., and P. McAllister. 2008. Does it pay to be green? Connecting economic and environmental performance in commercial real estate markets. Available from http://papers.ssrn.com/sol3/papers.cfm?abstract_id=1140409 (accessed November 2008).

Galbraith, K. 2008. Debating the green building premium. *New York Times (Green Inc.)* (accessed November 2008 at http://greeninc.blogs.nytimes. com/2008/11/20/debating-the-green-building-premium/.

Green Building Council of Australia. 2006. *The dollars and sense of green buildings GBCA*. Sydney: Green Building Council of Australia.

GVA Grimley. 2006. *Survey of property trends* (Summer). London: Author.

GVA Grimley. 2007. *Sustainability: Towards sustainable offices*. London: Author.

GVA Grimley. 2008. *From green to gold* (Autumn). London: Author.

Hermes. 2008. *Responsible property investment: Annual report*. London: Author.

Hinnells, M., Bright, S., Langley, A., Woodford, L., Schiellerup, P., Bosteels, T. 2008. The greening of commercial leases. *Journal of Property Finance and Investment*, 26(6):541–551.

HM Treasury. 2005. *Energy efficiency innovation review: Summary report*. London: Author.

Hobbs, P., and H. Chin. 2007. *The future size of the global real estate market* (July). London: RREEF Research.

Intergovernmental Panel on Climate Change (IPCC). 2007. *Fourth assessment report: Climate change 2007*. Geneva: Author.

Investment Property Databank (IPD). 2008. *Environment code*. London: Author.

Investment Property Forum (IPF). 2003. *Understanding commercial property investment: A guide for financial advisers*. London: Author.

Investment Property Forum (IPF). 2007. *The energy performance of buildings directive: A situation review*. London: Author.

Kats, G. H. 2003. *Green building costs and financial benefits*. Massport: Massachusetts Technology Collaborative.

Kremers J. 2006. Defining sustainable architecture. *Architronic*. Available at http://architronic.seed.kent.edu: cited in Williams and Lindsay (2007).

McGraw-Hill. 2008. *Commercial and institutional green building*. New York: Author.

McGraw-Hill Construction Analytics. 2008. *Global green building trends*. New York: McGraw-Hill.

McNamara, P. 2008. *The fundamentals of sustainable real estate: Towards responsible property investment*. MISTRA Workshop, June 12, Stockholm.

Miller, N., J. Spivey, and A. Florance. 2008. *Does green pay off?* San Diego: Working paper, University of San Diego.

Muldavin, S. 2008. Financial assessment of sustainable properties. In *Europe real estate yearbook, 2008*. The Hague: Europe Real Estate Publishers BV.

Nelson A. 2007. *The greening of U.S. investment real estate market fundamentals, prospects and opportunities*. San Francisco: RREEF Research, No. 57, November.

Nelson, A. 2008. *Globalization and global trends in green real estate investment*. San Francisco: RREEF Research, No. 64, September.

Newell, G. 2008. The strategic significance of environmental sustainability by Australian-listed property trusts. *Journal of Property Investment and Finance* 26(6):522–540.

Office of the Federal Environmental Executive (OFEE). 2003. *The Federal Commitment to Green Building: Experiences and Expectations*. Washington: Author.

Pivo, G. 2008. Responsible property investment criteria developed using the Delphi Method. *Building Research and Information* 36(1):20–36.

Pivo, G., and J. D. Fisher, 2008. *Investment returns from responsible property investments: Energy efficient, transit oriented and urban regeneration office properties in the U.S. from 1998–2007* (accessed November 2008 at www.responsibleproperty.net/assets/files/pivo_fisher_10_11_08.pdf).

Pivo, G., and P. McNamara. 2005. Responsible property investing. *International Real Estate Review* 8(1):128–143.

Pivo, G., and P. McNamara. 2008. Sustainable and responsible property investing. In C. Krosinsky and N. Robins (eds.), *Sustainable investing: The financial challenge of the 21st century*. London: Earthscan.

Prudential Property Investment Managers (PRUPIM). 2007. *Framework for sustainability management* (accessed November 2008 at www.PRUPIM.com/about/sustainability/ourapproach).

Rapson, D., D. Shiers, and Roberts, C. 2007. Socially responsible property investment (SRPI): An analysis of the relationship between equities SRI and UK property investment activities. *Journal of Property Investment and Finance* 25(4): 342–358.

Roberts, C., D. Rapson, and D. Shiers. 2007. Social responsibility: Key terms and their uses in property investment. *Journal of Property Investment and Finance* 25(4):388–400.

Royal Institution of Chartered Surveyors (RICS). 2005. *Green value: Green buildings, growing assets*. London: Author.

Sayce, S., L. Ellison, and P. Parnell. 2007. Understanding investment drivers for UK sustainable property. *Building Research and Information (BR & I)* 35(6): 629–643.

Sketchley, P. 2008. PRUPIM phases sustainability push beyond the UK. *IPE Magazine*, October (accessed at www.ipe.com/realestate).

Stern, N. 2008. Green routes to growth. *Guardian*, October 23.

Stern, N. 2007. *The economics of climate change: The Stern review*. Cambridge, UK: Cambridge University Press.

Tobias, L. 2007a. *Towards sustainable financing and strong markets for green building*. San Rafael, CA: Paper 2b, Green Building Finance Consortium.

Tobias, L. 2007b. Green builds a head of steam. *IPE Real Estate*, June, 44–46.

UKGBC. 2008a. *UK Green Building Council consultation: Code for Sustainable Buildings Task Group*. London: Author.

UKGBC. 2008b. UK Green Building Council Webinar poll (accessed November, 2008 at www.ukgbc.org).

United Nations Environmental Progamme (UNEP). 2006. *Sustainable buildings and construction initiative: Information note*. Paris: Author.

United Nations Environmental Programme Financial Initiative (UNEPFI). 2007a. *Responsible property investment: Property workstream* (accessed February 2008 at www.unepfi.org/work_streams/property/responsible_property_investment/index.html).

United Nations Environmental Programme Financial Initiative (UNEPFI). 2007b. *Building responsible property portfolios: A review of current practice by UNEP FI and PRI signatories*. Geneva: Author.

United Nations Environmental Programme Initiative (UNEPFI). 2008. *Responsible property investing: What the leaders are doing*. Geneva: Author.

U.S. Green Building Council. 2008. *Green building facts*. Washington, DC: Author.

Williams K., and M. Lindsay. 2007. The extent and nature of sustainable building in England: An analysis of progress. *Planning Theory and Practice* 8(1) (March):31–49.

World Business Council for Sustainable Development (WBCSD). 2007. *Energy efficiency in buildings: Business realities and opportunities.* Geneva: Author.

World Economic Forum. 2005. *Mainstreaming responsible investment.* Geneva: Author.

CHAPTER 13

Liquid Alpha

The Case for Investing in Water

Rod A. Parsley and Hua Liu

Water is the world's most important natural resource, a unique commodity for which there is no substitute. It is essential for human survival and an input into almost all facets of our economy, from growing crops to generating electricity to producing pharmaceuticals and semiconductors. However, while water is indeed pervasive on our planet, less than 1 percent is readily available for human use (Fry 2005). Moreover, water is inherently a local resource, given the costs associated with conveyance over long distances. This can lead to large discrepancies between where supplies of water are available and where they are needed most. Today, as Figure 13.1 illustrates, an estimated 1.1 billion people are without clean drinking water and 2.6 billion people are without water for sanitation (World Health Organization [WHO]/Unicef 2005).

This crisis is expected to significantly worsen in the decades ahead. Forty-seven percent of the world population will be living under severe water stress by 2030 (OECD 2008). The result could be even more challenging for countries like India, which the World Bank estimates could exhaust available water supplies by 2050 (Morrison, Morikawa, Murphy, and Schulte 2009). Rising standards of living and rapid urbanization will increase the demand for water, with some forecasting a 40 percent increase in consumption by 2025 (Siemens Water Technologies). Additionally, the supply of water, long plagued by incongruent local resources, pollution, and an aging

Rod A. Parsley is partner at Perella Weinberg Partners. Hua Liu is senior analyst at Perella Weinberg Partners. The views expressed by Mr. Parsley and Mr. Hua are personal and do not reflect the views of Perella Weinberg Partners.

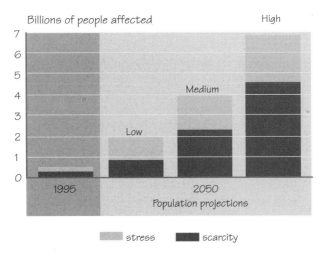

FIGURE 13.1   People suffering from water stress and
scarcity.
Source: UNEP/GRID-Arendal, Freshwater Stress 1995 and
2025, UNEP/GRID-Arendal Maps and Graphics Library,
http://maps.grida.no/go/graphic/freshwater_stress_1995_and_2025
(accessed March 11, 2009).

distribution infrastructure, now has to contend with perhaps its greatest
challenge—climate change.

The confluence of these issues is expected to have a dramatic impact on
the global economy. The World Bank estimates that water shortages already
negatively impact Chinese gross domestic product (GDP) by 2.3 percent
(Xie 2009), or in excess of $100 billion based on 2008 GDP figures. Climate
change is expected to be an additional burden, with studies indicating additional costs worldwide of close to $1 trillion per annum by 2100 (Ackerman
and Stanton 2008).

As discussed in Chapter 1, increased greenhouse gas (GHG) emissions
are resulting in warmer air temperatures. The ability of warmer air to
hold more water vapor than colder air serves to accelerate the hydrologic
cycle via increased evaporation. Scientists expect higher evaporative loss to
have an asymmetric impact geographically, with dry areas getting drier and
wet areas getting wetter. In dry areas, surface water will evaporate more
quickly, but will fail to saturate the warmer atmosphere enough to generate precipitation. Surface water and groundwater availability will decrease,
exacerbating shortages and the potential for severe droughts. Conversely,
in areas of abundant surface water, atmospheric saturation can take place
more quickly, generating enhanced levels of precipitation.

Ultimately, this could increase occurrences of extreme rainfall and flooding, causing a greater degree of nutrient runoff and therefore rising levels of water pollution. A rise in global temperatures will also accelerate the melting of snowpack and glaciers, a seasonal water storage medium and the source of drinking water for billions of people. Climate change experts fear that many Himalayan glaciers, the source of Asia's biggest rivers, may largely disappear by 2035 (Rai 2005). Juxtaposed with a greater demand for water given rising temperatures, the already fragile supply-and-demand balance will be severely challenged.

Unlike other categories of environmental investing, the investment thesis for water is less about mitigating GHG emissions than about *adapting* to the impact of climate change. Policies will need to account for the inevitability of greater demand for water in an increasingly constrained resource environment. Furthermore, solutions to our water problems will be significantly more localized than other challenges created by climate change. This is because of the geographic and physical restrictions of water, as positive externalities do not accrue from solutions implemented elsewhere. Boston, for example, does not benefit from a new treatment plant in Shanghai; the benefits stay local. Conversely, a cap-and-trade system like Kyoto can mitigate the risks of climate change precisely because the positive effects of reducing air pollutants locally will accrue *globally*. With water, the costs and benefits tend to occur proximate to the end user. As there is neither a spot nor futures market for water, investment opportunities are most often found in the companies that can provide solutions to address these local issues.

Key areas of focus will be alternative sources of water, improving an aging infrastructure, and the conservation of water through greater efficiency. These opportunities are reflected primarily in the public and private companies that source, treat, filter, distribute, and monitor water and wastewater, as well as in the companies that design and build the infrastructure and products that harness this essential resource. Changes in supply, demand, and regulation will give rise to an expected $22.6 trillion in capital investments in the water sector between 2005 and 2030 (Doshi, Schulman, and Gabaldon 2007), presenting investors with the potential to participate in alpha-generating opportunities for decades to come.

THE IMPACT OF ACCELERATING DEMAND ON WATER SUPPLIES

To date, the most glaring drain on water resources has been population growth and increasing usage per capita. Growth in worldwide water usage has markedly outpaced population growth over the last 100 years due to

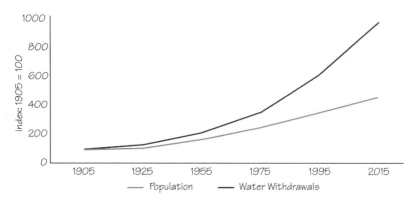

FIGURE 13.2 Water consumption outgrowing world population.
Original publication source: GWI and the International Desalination
Association (IDA).

increased access to fresh water, industrialization, and improving hygienic
conditions (Figure 13.2). Part of this disproportionate increase in water
usage is a result of rising standards of living and an increased demand for
food, which has a significant water footprint of its own. The next time
you go to your favorite steakhouse, consider that the ribeye on your plate
required the same amount of water as 66 showers.[1]

Migration patterns also have a significant effect on water supplies. The
percentage of people living in cities has grown from 29 percent in 1950 to
50 percent this decade, and poses a significant challenge to the water and
wastewater infrastructure in those locations (United Nations Department of
Economic and Social Affairs 2007). The burden of more densely populated
regions can drain local water supplies beyond their natural renewal rate, as
in the case of Lake Powell and Lake Mead, the primary reservoirs of the
Colorado River. These critical reservoirs serve already large and growing
southwestern cities such as Phoenix and Los Angeles, and may run dry by as
early as 2021 (Barnett and Pierce 2008). Situations like these are not likely
to improve anytime soon. The year 2008 was the first year that more than
50 percent of the world's population lived in cities, a figure that is expected
to rise to 70 percent by 2050 (United Nations Department of Economic and
Social Affairs 2007). The majority of this growth will occur in urban areas
in Asia and Africa, where more than 80 percent of all future population
growth is expected, but where the infrastructure is particularly ill equipped
to handle additional demand for water (United Nations 2009).

Compounding the stress on water resources is the nonlinear impact of
larger urban populations, as evidenced by the nexus between water and
power. By way of example, a typical family migrating from rural China
to Shanghai will purchase new appliances, many of which are unique to

an urban lifestyle. Some of these appliances (e.g., dishwashers, washing machines) require water, though all require power. This requisite power is, in turn, extremely water intensive as well. The operation of hydroelectric power plants or the cooling of nuclear and thermal power-producing equipment requires enormous amounts of water, as does extracting and processing the additional carbon-based energy sources such as oil, natural gas, and coal. All of that additional water requires power for extraction, treatment, and distribution, with water processing and distribution estimated to account for between 12 and 16 percent of total power use today (World Resources Institute [WRI] 2000). Given the close relationship between power and water demand, increasing urban populations will continue to exacerbate already acute water deficiencies in many major cities.

STRUCTURAL IMPACTS ON WATER SUPPLY

Unlike coal, oil, and other natural resources—which, once extracted and consumed, are essentially irreplaceable—water resources are continually replenished. The Earth serves to balance our demand via the hydrologic cycle, which naturally recharges the planet's freshwater supplies (Figure 13.3).

Ocean, lake, and river evaporation, combined with plant transpiration, is stored and transmitted by the atmosphere. Atmospheric water vapor condensation and precipitation returns water to the earth's surface, where it runs off or is stored in ice, snow, surface reservoirs, and underground aquifers. The runoff from the surface returns to existing bodies of water, later evaporating to begin the cycle again. As a result, there is roughly the same amount of freshwater on Earth now as there has been for tens of millions of years.

Despite the abundant supply of water on our planet, very little is readily available for human consumption, agriculture, or industrial use. Ninety-seven percent of the planet's water is located in the oceans, and of the 3 percent that is fresh, more than 80 percent is frozen in glaciers and polar icecaps or is located deep underground and difficult to access. Thus, while seemingly plentiful, less than 0.5 percent of the planet's water is fresh and readily useable (Fry 2005). Adding to this limited accessibility is the concentrated abundance of these freshwater resources, which are generally local in nature. This regional concentration can lead to incongruencies between the available supply and demand for water, especially as water is both difficult and expensive to transport over long distances. For example, China has 20 percent of the world's population and only 7 percent of the available water, and while Canada has approximately the same amount of water, it is home to less than 0.5 percent of the world's population (Food and Agriculture Organization of the United Nations [FAO] 2003).

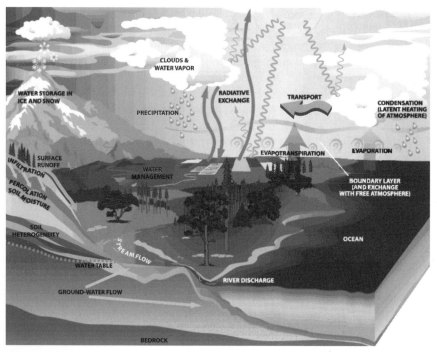

FIGURE 13.3 Hydro cycle.
Source: Paul Houser (George Mason University) and Adam Schlosser (MIT).

Due to these incongruencies, growing communities around the world are using freshwater resources at a faster rate than the natural hydrological cycle can replenish them, resulting in a net decrease of available water. In other situations, fossil water reserves, or groundwater that is not replenished through the hydrological cycle, are being drawn down. Unlike other fresh groundwater resources, the supply of fossil water is finite. The Ogallala aquifer, one of the largest sources of underground water in the world, has both rechargeable and fossil water resources that will potentially be depleted in the next few decades (BBC n.d.). This could present a dire situation for the American economy, as the Ogallala aquifer provides water for 82 percent of the people who live within its boundaries (spanning the Great Plains region from Texas to South Dakota) and for 30 percent of all agricultural production in the United States (U.S. Geological Survey [USGS n.d.]). Stressed water supplies can be further exacerbated by antiquated policy, as in the case of the Colorado River Compact of 1922. This agreement allows the seven basin states of the river to draw more than 15 million acre feet annually from the river and its reservoirs, allocations that do not jibe with recent drought flows of less than two-thirds this amount (McCabe, Betancourt, and Palecki 2004).

The dilapidated state of water distribution systems across the globe is a substantial source of water loss and further exacerbates the drain on water supplies. Much of the existing piping network in developed countries was installed more than five decades ago and is in need of substantial repair and upgrade. The American Society of Civil Engineers, in their 2009 Report Card for America's Infrastructure, gave both the water and wastewater infrastructure systems a D grade (American Society of Civil Engineers [ASCE] 2009). This compares to the much maligned bridge system in the United States, which received the second highest grade of C. The Environmental Protection Agency (EPA) estimates that over 240,000 water main breaks occur every year in the United States, a casualty of some pipes dating back 100 years or more (EPA 2007). An average of 15 percent of water can be lost between the treatment plant and household in developed markets, while physical water loss due to substandard infrastructure averages 35 percent in developing countries (Kingdom, Liemberger, and Martin 2006). The USGS puts the volume of water lost in the United States alone at 1.7 trillion gallons per annum (almost 5 billion gallons per day), costing $2.6 billion each year in lost sales (EPA 2007). In addition, Global Water Intelligence (2006b) estimates that the cost of physical and commercial losses worldwide has reached $18 billion. However, the capital cost for replacing the volume of water lost, either through long-distance transport, desalination, or water reuse, is exponentially greater.

The supply of readily available water is also heavily impacted by the vast amounts of pollution created by industry. The agriculture sector is the largest culprit, with fertilizers, pesticides, and processing chemicals causing 40 percent and 54 percent of total water pollution in high-income and low-income countries, respectively. The impact of industry is particularly startling in developing countries, where an estimated 70 percent of industrial wastes are dumped untreated into the water supply (UNESCO n.d.). Wastewater discharges in China, including heavy metals, solvents, and toxic sludge, rose to 53.7 billion tons in 2006. As a result, 60 percent of the rivers in China are unsuitable for human consumption (Xie 2009).

IMPACT OF CLIMATE CHANGE

These threats to our water supply are likely to be exacerbated by the consequences of climate change. As Dr. Betts explains in Chapter 1, global average temperature has risen by more than 0.7 degrees Celsius since the start of the twentieth century, and scientific models project warming of between 1.0 and 6.4 degrees Celsius by the end of the twenty-first century. This change in temperature intensifies the hydrological cycle, as the warmer air increases atmospheric water storage, accelerating evaporative loss and shifting precipitation patterns.

As a result, the geographic incongruity of water availability is expected to become more profound. Wet areas will get wetter due to higher levels of precipitation recycling, while water-scarce areas will get drier due to higher evaporative rates and a lower likelihood of saturating the atmosphere enough to cause precipitation. The result is likely to be an increase in extreme conditions such as floods and droughts, which in turn can increase the potential for polluted water in wet areas and severe water shortages in dry areas.

Additionally, increased evaporation, coupled with higher atmospheric water storage capacity, enhances the potential for water vapor to be transported *horizontally* across the atmosphere. This atmospheric impact can further exacerbate extreme conditions of wetness, causing additional surface runoff, pollution, and flooding. Shifts in temperatures can also create rising sea levels, leading to potential land loss and endangering coastal cities (Bates, Kundzewicz, Wu, and Palutikof 2008). But predicting the future impact of rising temperatures on climate change is a delicate matter, as the models are imperfect and there continues to be significant disagreements in the scientific community with regard to the magnitude of these events.

The *Stern Review* attempts to quantify the potential aggregate impacts of climate change on water. For example, it forecasts that with a 1 degree Celsius increase in global average temperature, small glaciers in the Andes disappear, jeopardizing the water supply for 50 million people. At a 3 degrees Celsius increase, an additional 1 to 4 billion people suffer water shortages, whereas 1 to 5 billion people experience greater water resources, putting them at risk for floods. At a 5 degrees Celsius increase, an additional 30 to 50 percent reduction of large glaciers from the Himalayas is possible, risking the permanent impairment of the water source for several hundred million people in China and India (*Stern Review* 2006).

There is evidence that some of this is occurring already. Studies by researchers have noted shifts toward less severe winters and earlier thaw periods in colder climates that have resulted in significant changes in water availability. Glaciers and snowpack—essentially above-ground freshwater storage facilities that provide a steady supply of water to populations through rivers—have been receding at a startling rate, causing water shortages from California to Pakistan during the crucial dry seasons. In addition, multidecade data on Africa shows decreases in water availability on the order of 20 percent between 1951 and 1990 for both humid and arid zone basins that discharge into the Atlantic (Millennium Ecosystem Assessment Board 2005).

The impact of climate change is also likely to have second derivative impacts beyond the water sector. Agriculture, as one of the largest single users of water, is in a particularly precarious situation. Higher evaporative rates will result in less water available for irrigation, while warmer temperatures will simultaneously increase the water stress levels for crops. Increased water stress levels would necessitate even larger volumes of water to sustain the same crop yields, raising the likelihood of desertification.

The *Stern Review* (2006) suggests that with the rise in temperatures, entire regions of the world will experience major declines in crop yields. This would create the potential for severe food shortages, particularly in emerging markets where agriculture is a disproportionately large user of water. Power-thirsty nations are also at risk, as evidenced by past shutdowns of baseload nuclear plants in Spain and France due to water shortages and drought. Rising temperatures can reduce the amount of water available for cooling purposes and drain nearby freshwater sources for effluent disposal. Additionally, for countries such as Brazil, which generates in excess of 80 percent of its power through hydroelectricity (EIA 2008), the loss of significant water resources would be devastating.

To address these escalating issues, world leaders are becoming more proactive in adapting to the current and future impacts of climate change on water. As with previously implemented policies such as the Clean Water Act of 1972 and Safe Drinking Water Act of 1974 in the United States, governments are attempting to change behavior toward crucial issues such as conservation and water pollution. Using mechanisms like the state revolving funds in the United States, the large-scale wastewater directives in places such as China, and government-supported desalination and water reuse projects in the Middle East, resources can be made available for the battle against water scarcity and an aging infrastructure. Through the movement toward the full-cost pricing of water, governments can provide the incentives that will encourage private enterprises to research, develop, and utilize cutting-edge technology to maximize the value of this precious resource. It will be through understanding the development, evolution, and consequences of these transformative events that investors can uncover the opportunities to generate environmental alpha.

INVESTING IN WATER

Investors looking to gain long or short exposure to natural resources and other commodities typically can do so in one of three ways:

1. Purchase (or sell) the commodity outright. For example, a gold investor might purchase gold bullion and hold the bars in a safe or with a local bank.

2. Access movements in the commodity's price via futures or other derivatives markets, thereby avoiding transportation and storage costs.
3. Purchase (or sell) an equity or fixed-income security whose price is closely correlated to the price movements of the underlying commodity. A gold investor, for example, might purchase Newmont Mining, while an oil investor might purchase Schlumberger or Exxon Mobil.

These options do not exist for water. Because it is expensive to store and transport, and because it has relatively little value on a volumetric basis, water does not trade on a global spot market or futures exchange. A barrel of oil, for example, is worth about $40 in early 2009, while a barrel of municipal water in the United States can cost as little as 6 *cents*. Compounding this problem of volumetric value is that water is heavy and cannot be compressed, making it extremely difficult and expensive to ship via rail, truck, or tanker. Even long-distance pipelines, with their power-intensive pumping requirements, are expensive to operate and maintain. Thus, moving water efficiently to a spot market hub would require transportation costs that far exceed the current value of the commodity itself. This contrasts markedly with the other primary utility commodity, electricity, which can be transmitted long distances simply by stringing transmission cables.

CASE STUDY: The True Price of Water

Given its seemingly abundant supply, there is a common misperception that water is like air—"It's free, and no one owns it." Unlike air, however, water's physical properties allow for it to be collected, stored, treated, and distributed. Thanks to the physics of dispersion, air has none of these characteristics; the availability of clean air can be addressed only through the ex ante mitigation of emissions. Water's quality and quantity issues, however, are typically dealt with through localized processing and distribution. Tremendous value can be added to the resource by removing dirt, pollutants, and harmful microbes and then delivering this treated water to an end user—for a price.

In the past, the true price of water has been masked by government subsidies, underinvestment in infrastructure, and a lack of awareness around water quality and pollution. Today, the proliferation of investor-owned water utilities and the build-out of water infrastructure have driven participants toward pricing based on economic fundamentals. The true price reflects the cost of treating and distributing water, combined with a premium that compensates utility owners for deploying large amounts of capital. However, this price can vary greatly depending on location. Desalination, recycling, and importation are relatively expensive ways of extracting or producing water, and water quality differences can meaningfully impact the cost of water treatment.

Nonetheless, substantial subsidies still exist, particularly in the agriculture sector. In California, where agriculture is a $36.6 billion industry (California Department of Food and Agriculture 2009), farmers can pay as little as $30/acre-foot (Burke 2008) for heavily subsidized water, while households typically pay in excess of 10 times this amount. Municipalities in California are forced to bear the cost of expensive new water sources such as the proposed desalination plant in Carlsbad, which is expected to produce water for upwards of $1,000 per acre-foot (Poseidon 2009). This has led to distortions in the economics of water usage. For example, water-intensive crops such as alfalfa, cotton, rice, and irrigated pasture now consume 54 percent of all agricultural water used, yet produce only 17 percent of the state's agricultural revenue (Gleick, Loh, Gomez, and Morrison 1996).

The transition from the subsidy of water as a public good to a full-cost pricing model will take time. The politics surrounding food and agriculture, which account for 70 percent of global water usage (World Water Assessment Programme [WWAP] 2003), are complicated. Maximizing the value of water resources by planting less water-intensive crops or potentially outsourcing production to areas of abundant water can be a challenging decision that goes beyond economics. Nonetheless, the increasing supply-and-demand imbalance (offset to some degree by advances in technology) increases the financial burden on governments that currently elect to subsidize water rates for users. The cost of water is thus expected to evolve over time to become a more meaningful input cost for most households and industries. The availability of water resources will increasingly become a competitive advantage for some and a disadvantage for others, much like oil and gas are today. Ultimately, this could cause geographic shifts in manufacturing capacities and migration patterns, as well as be a source of potential conflict.

Given the lack of these traditional investment options, those looking for water exposure must focus on direct or indirect investment opportunities across the water supply chain. Typically, this involves investing in the companies that source, treat, distribute, monitor, and test water for municipal, residential, industrial, and agricultural end users. This investment universe of water companies is large, diverse, and global (Figure 13.4), with over $1 trillion in market capitalization of companies with varying degrees of water exposure.

The water industry includes a variety of subsectors, ranging from utilities to cutting-edge desalination technologies. Given the sector's burgeoning popularity with investors in recent years, a plethora of options exist beyond direct investments in equities and fixed-income securities. These include investments in exchange-traded funds (ETFs), mutual funds, hedge funds, and venture and private-equity funds.

ETFs like the PowerShares Water Resources Portfolio (PHO) provide exposure to the water sector and are typically linked to an underlying basket of publicly traded companies. All of the currently available water ETFs are long-only and offer sector exposure with considerably lower fees than actively managed investment vehicles. However, many of these ETFs have held equities like General Electric that derive only a small percentage of their overall revenues from water. As a result, investors can end up with unintended economic exposures—in the case of General Electric, exposure to financial services, entertainment, and aviation—and diluted exposure to

Water/Wastewater Utilities

Regulated Utilities
- American Water Works
- American States Water
- Aqua America
- California Water Service
- Pennon Group
- Severn Trent
- South West Water
- United Utilities Group

O&M Operators
- American Water Works
- Suez Environnement
- Veolia Environnement

Concessionaires
- Acciona
- Befesa
- COPASA
- China Everbright International
- Manila Water
- Puncak Niaga
- SABESP
- Suez Environnement
- Veolia Environnement

Water Source

Desalination
- Consolidated Water
- HyFlux
- IDE Technologies*
- Qatar Electricity & Water
- Suez Environnement
- Veolia Environnement

Raw Water Supply
- Eastern Water Resources
- Guangdong Investment

Water Rights
- PICO Holdings
- Pure Cycle

Engineering & Construction
- Aecom Technology
- Black & Veatch
- C2HM Hill*
- Impregilo
- Mitsubishi Heavy Industries
- Orascom Construction industries
- Tetra Tech

Treatment Equipment

Filtration & Desal
- Asahi Kasei
- Doosan Heavy
- Dow
- General Electric
- Layne Christenson
- Pall
- Kurita Water Industries
- Nitto Denko
- Siemens
- Suez Environnement
- Toray Industries
- Veolia Environnement

Disinfection
- Calgon Carbon
- Danaher
- ITT

Chemicals
- Ashland
- Dow Chemical
- General Electric
- Kemira Chemicals
- Nalco

Infrastructure Equipment

Pumps, Valves & Other
- Ebara
- Energy Recovery
- Flowserve
- ITT
- Mueller Water Products
- Pentair
- Watts Water Technologies

Pipes
- Ameron International
- Insituform Technologies
- Mueller Water Products
- Northwest Pipe
- Underground Solutions
- Westlake Chemical

Meters
- Badger Meter
- Itron
- Roper Industries

Irrigation Equipment
- Lindsay
- Valmont Industries

*Denotes private company.

FIGURE 13.4 Sample industry landscape.

water. This can result in the performance of the ETF being driven by various exogenous variables in addition to the macroeconomic tailwinds in the water sector. Moreover, the mechanical approach of ETFs does not allow them to capture the fullness of the opportunity. A quarterly "rebalance," for example, does not necessarily coincide with the time at which a stock trades above its intrinsic value.

Actively managed mutual funds and hedge funds are less constrained in evaluating and accessing opportunities in the water sector. By performing fundamental analysis on the various macroeconomic factors impacting the sector, in addition to analyzing company-specific attributes for companies with water exposures, active managers are better able to uncover valuation disparities and time-sensitive opportunities. Additionally, hedge fund managers can short excessively overpriced stocks and bonds and have the ability to hedge out undesired exposures, thereby establishing a mechanism to approximate pure water exposure. By combining this fundamental analysis with the opportunity to capitalize on short-term catalysts, including regulatory changes, earnings announcements, and large project awards, actively managed portfolios offer an alternative to ETFs for investors seeking to capitalize on alpha-generating opportunities in water.

Earlier-stage and nonpublic water companies can be accessed via venture capital and private-equity funds. The risk-reward profile of these investments offers the potential for high returns offset by illiquidity, technology risk, and adoption risk. Adoption risk is typically associated with venture-stage companies that are subject to the risk-averse nature of their end customers. More often than not, utilities are hesitant to test new and unproven technologies as the downside risk of failure can far outweigh efficiency benefits. The manager of a municipal water utility, for example, rarely earns a bonus for successfully introducing a new filtration technology. However, he may face considerable downside risks, including job loss, if the new technology fails. Private investment opportunities range from interests in small equipment and technology companies to owning water and wastewater utility assets, pipelines, and water rights.

CASE STUDY: Water Rights (Not as Liquid as They Seem)

In many countries, including the United States and Australia, individuals and corporations can purchase the right to an annual volumetric draw of water from an aquifer, stream, river, or lake. These water rights have historically been associated with farmland real estate and are often purchased and sold with the land. As legal frameworks have slowly developed around these rights, some municipalities, utilities, and industries have turned to purchasing water rights to supplement their water needs.

These water rights are the closest thing to a spot market for water and can be accessed through publicly traded entities like Pure Cycle and PICO Holdings that specialize in such transactions. However, the market is highly illiquid and region specific, and monetizing these rights can be onerous and problematic. Water Strategist (2009) shows that 194 transactions were completed in the United States in 2008, the majority in the West and Southwest. Transactions can vary greatly in size, ranging from a few thousand dollars to tens of millions of dollars.

The endgame for investors in water rights is to find another party with an insufficient supply, most commonly a real estate developer or a municipality. Developers in Arizona, for example, must prove sufficient water resources to meet the needs of their development for a minimum of 100 years (Arizona Department of Water Resources 2006). Often, water rights owners will look to build conveyance systems that allow their water to be distributed tens if not hundreds of miles away from the source. This piping infrastructure can run into the hundreds of millions of dollars, making financing a critical variable in the execution of a transaction. Other difficulties include the cost and potential use of the property that is often inextricable from the right to the water. These factors, combined with an often arduous and opaque governmental approval process and legal difficulties in moving water across state lines, introduce significant layers of complexity when analyzing investments in water rights.

CAPTURING LIQUID ALPHA

The water sector, in its most basic form, is fairly straightforward (Figure 13.5). Water utilities source untreated "raw" water from wells, reservoirs, lakes, and other sources, and then filter and purify it to meet regulated quality standards. Treated, potable water is then delivered to commercial, residential, and industrial end users through a complex distribution system consisting of a variety of pumps, valves, pipes, meters and other equipment. End users are typically connected to a separate sewage network that transports wastewater to a wastewater utility, where it undergoes several

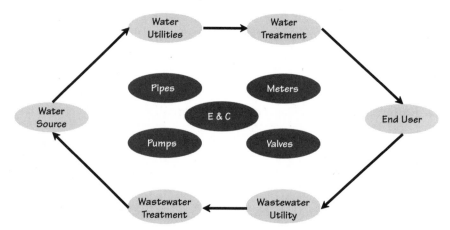

FIGURE 13.5 Simplified water industry map.

rounds of further treatment before being returned to nature or recycled for reuse.

The water industry is uniquely capital intensive, with water and wastewater utilities requiring two to three times more capital than electric and gas utilities to generate $1 of revenue (National Association of Water Companies [NAWC] n.d.). A 2002 report by the U.S. EPA estimates that between *$485 and $896 billion* needs to be spent on water and wastewater in the United States alone by 2020 (EPA 2002). The opportunity for investors to capitalize on these spending trends starts with water and wastewater utilities.

Water and Wastewater Utilities: The Engines Driving Infrastructure Spending

Much of the responsibility for addressing this required spending lies on the shoulders of water and wastewater utilities, and the organizations that establish the relevant policy and regulatory frameworks. The majority of water assets continue to be owned by governmental entities, with 84 percent of U.S. water utilities and 98 percent of wastewater utilities controlled by municipalities (American Water Works 2009). However, more onerous capital and regulatory requirements have resulted in governments looking increasingly toward the privatization of water assets. Facing a choice between raising taxes or debt to fund water infrastructure investments or selling/partnering their system with a reputable private operator, resource-constrained municipalities are increasingly opting for the latter. Doing so allows them to take advantage of the operational expertise, economies of scale, and financing capabilities of private operators. Privatized assets generally operate under two types of regulatory frameworks:

1. *"Rate base" or "regulated asset base" regimes.* Developed countries tend to have regulated monopoly frameworks where a single regulator approves capital expenditures and sets the allowed return on that capital base, also known as rate base or regulated asset base. In effect, these utilities grow by making capital expenditures and then earning a return on this capital via fees to customers. In the United States and the United Kingdom, companies such as American Water Works, Aqua America, California Water, and United Utilities are regulated by utility commissions and are given the exclusive right to treat and distribute water in specific territories. They must, however, apply for fair rates of return and get approval for capital investments in order to charge consumers additional fees or "tariffs." This limits the potential for abuse by the regulated entities but ensures a clear framework for capital investment decisions.

2. *Public-private partnerships (PPPs)*. Operators in both developed and developing countries can engage in PPPs that rely on concession-type contracts. Under structures such as Build-Own-Transfer (BOT) and Transfer-Own-Transfer (TOT), operators are given the right to collect future cash flows for a finite period of time, based on an assumed capital investment profile. The counterparties on these contracts are typically highly rated governments and municipalities, which can mitigate the service provider's risk of investing large amounts of capital. The winners of these concessions are usually influential local companies, such as Puncak Niaga in Malaysia and Manila Water in the Philippines, although large international water companies with successful track records (e.g., Veolia Environnement and Suez Environnement) have also had substantial success in developing countries. PPPs can also be structured as asset-light operation and management contracts, where the government is responsible for asset ownership and capital investment but outsources the day-to-day operations of the assets to an independent third party.

The key difference between the rate base regime and the PPP models is in asset ownership structure. Rate base regimes are structured as perpetual monopolies that own their assets outright, whereas PPPs are generally priced with a hypothetical termination date and can be viewed as temporary stewards of the assets with responsibility for operations. The risk profile can vary as well, as rate base regimes tend to benefit from more stable, developed-market regulatory frameworks, whereas PPPs can be subject to substantial developing market risk and regulatory ambiguity. These factors aside, both models seek to augment growth through territorial expansion via asset acquisitions or additional PPP contracts. Both models are also structured to generate significant operating cash flows and, combined with secure and generally high dividend payouts, are considered to be a defensive way to gain exposure to water.

Engineering and construction firms like Aecom, Mitsubishi Heavy Industries, and Tetra Tech are involved in many of the capital expenditure decisions made by public and private utilities. They provide governments and utilities with the critical front-end consulting, engineering and design, procurement, project management, and construction expertise necessary to build and upgrade their systems. Correspondingly, the market for these services is substantial, with *Engineering News Record* (Rubin 2008) sizing the market at $18 billion per year. This market should continue to grow in line with the increased capital spending requirements in the water industry.

Adapting to a Water-Constrained Future

The significant amount of capital expenditure necessary to adapt to an increasingly water-constrained future must address three major categories: (1) developing alternative sources of water; (2) upgrading an aging infrastructure to cut water losses; and (3) achieving conservation through efficiency-enhancing technologies.

Addressing the Need for Alternative Sources of Water   One of the most critical issues facing the water industry is access to safe and reliable water supplies. To address the shortage of clean water in certain areas, governments have been forced to develop new ways to sustain water supplies, including desalination, water reuse, and water importation.

Desalination   Desalination is the process of removing the salt and other minerals from water. Desalination technologies potentially make 97 percent of the world's water that lies in oceans available for agriculture and human consumption. Globally, there are approximately 12,500 desalination plants in operation, with 60 percent located in the Middle East. These technologies can also be used to treat salinated or "brackish" groundwater in aquifers deemed unfit for potable use. Of the 250 desalination plants in the United States, all 38 located in Texas were built inland to treat brackish groundwater (Texas Water Development Board 2009).

There are two general methods for removing salt from water. Thermal distillation applies heat to saline water in order to generate water vapor, which is then condensed into fresh water. Reverse osmosis (RO) is the process of using applied pressure to force saline water through a semipermeable membrane, separating salt from water. Seawater RO has gained significant traction in recent years, growing at 55 percent per annum since 2002 (Global Water Intelligence [GWI] 2006a), as rising energy costs and technology developments have improved its efficiency relative to thermal distillation techniques. Desalination plants are often co-located with power plants to maximize the energy efficiency of the combined system.

While desalination can be considerably more expensive than traditional potable water treatment processes, it is often a viable solution in areas of water stress. Energy can constitute up to 50 percent of total operating costs for a desalination plant (Energy Recovery 2008), by far the largest contributor to operating costs. Therefore, desalination tends to make the most sense in areas of extreme water scarcity and cheap energy resources, such as in the Middle East and North Africa. Companies with desalination expertise include Qatar Electricity & Water, which operates concessions under long-term, take-or-pay contracts in the state of Qatar, and Hyflux, which won the tender to build, own, and operate the world's largest desalination plant in Algeria in 2008. On the backs of these trends, Global Water Intelligence (2006a) expects desalination capacity to double by 2015.

Beneficiaries will also include manufacturers of equipment that improve the cost-competitiveness of desalination. Advances in membrane technology by Dow Chemical, General Electric, and Nitto Denko, among others, along with energy efficient pumps and pressure exchangers made by companies such as Flowserve and Energy Recovery, have contributed greatly to increasing the efficiency and increasing adoption of desalination.

Water Reuse   Water reuse, or recycling, is another option for enhancing water supplies. Reuse is the process of applying advanced technologies to treat residential and industrial wastewater for the purpose of using it again, typically for irrigation and other nonpotable water needs. Recycled or "reclaimed" water is usually not used for potable water purposes, given the political sensitivities associated with the "toilet-to-tap" concept. The one notable exception is the NEWater program in Singapore that utilizes a combination of filtration and disinfection technologies to turn sewage water into clean and safe drinking water. This treated water is piped to a variety of industries for manufacturing purposes as well as blended with raw water for potable use.

Water reuse greatly reduces the amount of wastewater discharged back to the environment and decreases strains on traditional water supplies. It does, however, require a separate distribution network to operate, typically represented by distinctive purple pipe. Many countries that had previously focused on desalination are now considering water reuse as an alternative, principally because of the reduced energy requirements and lower costs associated with water reuse. For example, in California, where 19 percent of total electricity demand is for the movement and treatment of water, water reuse can reduce energy requirements by up to 85 percent (GWI 2008). An interesting case study is Orange County's Groundwater Replenishment System, which treats wastewater for the purposes of recharging vast local aquifers. This system combines water reuse with storage capabilities, simultaneously increasing water availability and replenishing dwindling water reserves. With these compelling advantages, GWI (2007) expects a fourfold increase in water reuse capacity over the next decade.

CASE STUDY: Orange County, California, Groundwater Replenishment System

In 2008, Orange County, California, opened the world's largest and most advanced wastewater reclamation plant, costing $500 million with daily capacity of 70 million gallons. The system was designed to protect against dwindling water resources available to Californians by treating sewage with a droughtproof process including microfiltration equipment from Siemens, a reverse osmosis system from Nitto Denko/Hydranautics, and an ultraviolet and hydrogen peroxide disinfection system from Danaher/Trojan. The ultrapure water is then injected into California's vast underground aquifer system. The state of the art system covers 20 acres of land, with production costs estimated to be in the $550/acre-foot range (Royte 2008).

Water reuse utilizes proven water and wastewater treatment technologies ranging from nanofiltration membranes to ultraviolet disinfection and chemical treatment to achieve its purposes. While the design of an individual plant can vary, a typical plant will utilize a combination of these technologies and processes to treat wastewater, as shown in Figure 13.6.

Companies such as Calgon Carbon, Danaher, General Electric, ITT, Nalco, and Siemens provide much of the equipment and technology that helps utilities and industries produce, treat, and recycle water to meet increasingly stringent rules and regulations.

Water Importation   Importing potable water over long distances to areas of scarcity is an expensive water replacement option because of the required infrastructure and associated costs. The water-constrained country of Cyprus has responded to its water supply crisis by importing water from Greece in tankers. China's $59 billion South-to-North Water Diversion Project directs water from southern China to water-scarce regions in the north. To provide water to coastal cities such as Tripoli, Libya has spent a quarter century constructing a $25 billion system of reservoirs and pipeline to pump water from an underground aquifer beneath the Sahara desert. In North America, more of these types of projects are being considered as a method to deliver water to arid regions. Several projects that are being discussed are hundreds of miles long and can cost in excess of a billion dollars, a trend that will benefit large-diameter pipe manufacturers such as Northwest Pipe and Ameron International.

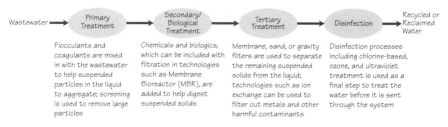

FIGURE 13.6   Typical water reuse treatment process.

CASE STUDY: South-to-North Water Diversion Project in China

China faces a unique challenge with its water resources: Northern China, which accounts for 65 percent of agricultural production, has access to only 35 percent of the country's water resources. In 1952, Chairman Mao Zedong proposed a water diversion project that today, over 50 years later, is known as the South-to-North Water Diversion Project. The largest project of its kind ever undertaken, the project calls for three routes (western, central, and eastern) of canals and raised aqueducts that will supply 400 million people with 44.8 m^3 of water annually from the water-rich south of the country to the parched northern parts. The construction of the eastern and middle routes is already under way. The western route, which will replenish the Yellow River with the water from the top of the Yangtze River through tunnels in the mountains of western China, is still at the planning stage. The project has an estimated total price tag of $59 billion ($3 billion investment in 2009) and has raised many environmental concerns, mainly the loss of antiquities, the displacement of people, and the devastation of pastureland (Jakes 2006; Ministry of Water Resources P.R. China [MWR] 2009).

Upgrading an Aging Infrastructure   Upgrading old and dilapidated water and wastewater infrastructure is a priority for many systems, considering that every ounce of water saved from leakage represents direct treatment cost savings. A significant number of the pipe, pumps, and valves that comprise an area's water distribution network can suffer from corrosion and leaks. The cost of repairing and replacing this network is substantial; the American Society of Civil Engineers estimates at least an $11 billion annual shortfall in spending in the United States alone for the next 20 years (ASCE 2009). A variety of companies supply this equipment and provide services aimed at improving water infrastructure. For example, Mueller Water Products and Westlake Chemical manufacture small- and medium-diameter ductile iron and polyvinyl chloride (PVC) pipes used for local distribution infrastructure. Companies such as Insituform and Underground Solutions have developed trenchless pipe rehabilitation technologies that help utilities to replace dilapidated pipes without digging up roads and stopping traffic. Products such as pumps, backflow preventers, water softeners, and valves made by Flowserve, General Electric, ITT, Pentair, and Watts Water help manage and treat the flow of water all the way from the source to within the home itself.

Achieving Conservation through Increased Efficiency   The third way to alleviate stress on the water supply is to improve the way water is utilized. Substantial efficiencies in water conservation can be achieved by improving and updating technologies embedded in water infrastructure. The adoption of irrigation, water meter, and sensing technologies increases the efficiency of water usage and provides simple return-on-investment calculations for customers. Center-pivot irrigation equipment manufacturers such as Valmont Industries and Lindsay Manufacturing provide products that can irrigate farmland more precisely and generate labor, energy, and water savings. Likewise, installing new meters in previously unmetered locations and replacing outdated meters can increase both the breadth and accuracy of metering. This can impact consumer behavior as households are charged directly for the water they use as opposed to a flat fee. New advances in water meter technologies can also provide a method of increasing the efficiency of water systems by allowing municipalities to eliminate on-the-ground meter readers. Traditionally, water meters have been read manually, a time-consuming, labor-intensive way of recording water usage. Developments in radio technologies by such companies as Badger Meter, Itron, and Roper Industries have increased the viability of automated meter reading (AMR) and fixed network systems. Combined with sensing equipment made by companies such as Danaher, utilities now have the tools to not only accurately track usage real time but to easily detect leaks and quality issues.

RISKS TO INVESTING IN WATER

Investing in water is not without its risks. For starters, the industry is extremely capital intensive, largely dependent on the capital markets for funding. Municipalities as well as private utilities and concessionaires depend heavily on the credit markets, as infrastructure projects are typically heavily leveraged. As a result, weakness in the credit markets can have a dramatic impact on capital expenditures, limiting demand for the products and technologies in the water supply chain. A source of credit that has stood out in the current poor market conditions are government export credit agencies, which have provided financing globally for many water projects that involve their country's companies.

Water's crucial role in human life makes it a highly controversial subject. Governments and municipalities are constantly trying to strike a balance between fair economic returns and the need to provide a life-sustaining product to its people at a reasonable price. Regulated utilities, concessionaires, and operations and maintenance companies typically operate under legally binding contractual frameworks but they can frequently be subject to political interference. Water is still viewed as an essential public good (i.e., some still think it should be free), and politicians occasionally use it as a tool to gain favor with their constituencies. In many countries, water has been subsidized by the government for so long that it can be difficult to implement tariff increases, even if they are contractually based. This potential for regulatory risk can put the return parameters of private investors in jeopardy and demonstrates the need for a keen sense of the relevant political environment.

This political sensitivity also leads to risk aversion in the business, causing the slow adoption of new technologies that address water supply, quality, and efficiency. Utilities are reluctant to put their water quality at risk by using new technologies, unless compelled by law. The adoption of new technologies is uncertain and can make it difficult for small companies with unique products to gain traction.

Finally, many water equipment companies have cyclical components that have exposure to the general economy, particularly residential and commercial construction. While the repair and replacement cycle provides a solid foundation for many of these businesses, much of the equipment necessary to move, control, and measure water is subject to incremental capital investment that can vary with the business cycle. Municipal exposure can also be impacted, as a slowing economic environment and expectations of lower tax receipts can temporarily halt capital outlays. Additionally, the often significant nonwater businesses of many multinational conglomerates can introduce exposure to the broader economy that can impact earnings power.

SUMMARY

The case for investing in water is simple. Water is an essential resource across multiple dimensions, for which no substitute exists. Demand is increasing, and supply is facing pressures on multiple fronts. The demand curve for water is highly inelastic and with the rapid population growth and the industrialization of emerging economies, it is shifting farther and farther to the right. However, these growing populations are facing an increasing shortage of local freshwater resources. Consumption continues to grow faster than the renewal rates of fresh water, and a crumbling infrastructure and rampant water pollution are limiting the supply to end users. Finally, the impact of climate change could potentially be devastating, with billions of people facing severe water shortages by the middle part of this century.

To adapt to these challenges, public and private sectors alike will be required to invest substantial amounts of capital, driving growth and profits for the companies that serve this sector. The requisite capital requirements and the importance of technology development make the water sector one of the key areas of focus for environmental investing for the foreseeable future.

Despite strong macroeconomic drivers, the water sector is not well covered by sell-side research analysts. Given the highly diverse nature of the companies in the water universe, most firms allocate these companies across a traditional coverage spectrum as opposed to ring-fencing the water names as a unique category. Also, few analysts are knowledgeable of the policy and regulatory issues that drive many investment opportunities. The disparate nature of the universe of water companies, complexity of the regulatory environment, lack of focused research coverage, and uncertain impact of climate change creates an environment ripe with asymmetric information that knowledgeable water investors can exploit to their advantage.

ACKNOWLEDGMENTS

The authors gratefully acknowledge Dr. Paul R. Houser, the Director of Atmospheric and Hydrological Sciences at EarthWater Global, LLC and former Director of Hydrology at NASA for his valuable insight and consultation into the challenges our planet faces with water. The authors would also like to extend special thanks for the hard work and dedication of their colleagues Kristen Kelly, Marty Cheatham, Marci Guillerme, and Amy Treat, integral members of our team at Perella Weinberg Partners, all of who were instrumental in getting this chapter across the finish line.

NOTES

1. One kilogram of beef requires 15,000 to 70,000 liters of water over time. At roughly 15,000 liters per kilogram, or 5,000 liters per 12 ounces, this approximates 1,320 gallons of water, or 66 showers at 20 gallons a shower.

REFERENCES

Ackerman, F., and E. A. Stanton. 2008. *The cost of climate change.* Natural Resources Defense Council.

American Water Works. 2009. *Water industry dynamics.* http://ir.amwater.com/about.cfm.

American Society of Civil Engineers (ASCE). 2009 report card for America's infrastructure. www.asce.org/reportcard/2009/grades.cfm.

Arizona Department of Water Resources. 2006. *Assured/adequate water.* www.adwr.state.az.us/dwr/WaterManagement/Content/OAAWS/default.asp.

Barnett, T. P., D. W. Pierce. 2008. When will Lake Mead go dry? *Water Resource Research,* March 29. www.agu.org/pubs/crossref/2008/2007WR006704.shtml.

Bates, B. C., Z. W. Kundzewicz, S. Wu, and J. P. Palutikof (eds.). 2008. *Climate change and water.* Technical Paper of the Intergovernmental Panel on Climate Change, IPCC Secretariat.

BBC. n.d. *Ogallala aquifer.* http://news.bbc.co.uk/1/shared/spl/hi/world/03/world_forum/water/html/ogallala_aquifer.stm.

Burke, G. 2008. *Water shortage: Some farmers in California would rather sell their water than grow something.* Associated Press, January 25.

California Department of Food and Agriculture. 2009. www.cdfa.ca.gov/.

Doshi, V., G. Schulman, and D. Gabaldon. 2007. Lights! Water! Motion. *Booz Allen Hamilton: Strategy + Business.*

Energy Information Administration (EIA). *Country Analysis Brief: Brazil.* 2008. www.eia.doe.gov/emeu/cabs/Brazil/Electricity.html.

Energy Recovery, Inc. 2008. IPO prospectus.

Environmental Protection Agency (EPA). 2002. *The clean water and drinking water infrastructure gap analysis.* Office of Water.

Environmental Protection Agency (EPA). 2007. *Addressing the challenge through innovation.* EPA Aging Water Infrastructure Research Program.

Food and Agriculture Organization of the United Nations (FAO). 2003. *Review of world water resources by country.*

Fry, A. 2005. *Water: Facts and trends.* World Council for Sustainable Development.

Gleick, P., P. Loh, S. V. Gomez, and J. Morrison. 1996. *California water 2020: A sustainable vision.* www.sarep.ucdavis.edu/NEWSLTR/v8n1/sa-13.htm.

Global Water Intelligence (GWI). 2006a. Desal's double digit future. *GWI* 7(10), October.

Global Water Intelligence (GWI). 2006b. Turning losses into gains. *GWI* 7(12), December.

Global Water Intelligence (GWI). 2007. Reuse revolution. *GWI* 8(11), November.

Global Water Intelligence (GWI). 2008. End of the road for big pipe? *GWI* 9(10), October.

Jakes, S. 2006. China water woes. *Time Asia*, October 2.

Kingdom, B.L, R. Liemberger, and P. Martin. 2006. The challenge of reducing non-revenue water (NRW) in developing countries. *Water Supply and Sanitation Board Discussion Paper Series, Paper No. 8.* December.

McCabe, G., J. Betancourt, and M. A. Palecki. 2005. *Drought: Historical Context.* http://co.water.usgs.gov/drought/workshop200501/pdf/mccabe.pdf.

Millennium Ecosystem Assessment Board. 2005. *Ecosystems and human well-being: Current state and trends,* Volume 1. Washington, DC: Island Press.

Ministry of Water Resources P.R. China (MWR). 2009. *China to pump 21.3b yuan into water diversion program.* www.mwr.gov.cn/english/20090202/94596.asp.

Morrison, J., M. Morikawa, M. Murphy, and P. Schulte. 2009. *Water scarcity & climate change: Growing risks for businesses & investors.* Thousand Oaks, CA: Ceres.

National Association of Water Companies (NAWC). n.d. *Price, cost, value.* www.nawc.org/pdf/nawc_brochure_web.pdf.

Organization for Economic Co-operation and Development (OECD). 2008. *OECD Environmental Outlook to 2030.* http://www.oecd.org/document/20/0,3343,en_2649_34305_39676628_1_1_1_1,00.html

Olmstead, S. M., and R. N. Stavins. 2007. *Managing water demand, price vs. non-price conservation programs.* Boston: Pioneer Institute.

Poseidon. 2009. *The Carlsbad desalination project.* www.carlsbad-desal.com/faq.aspx?id=2.

Rai, S. C. 2005. An overview of glaciers, glacier retreat, and its subsequent impacts in Nepal, India and China. *WWF Nepal Program.*

Royte, E.. 2008. A tall, cool drink of . . . sewage? *Times Magazine*, August 8.

Rubin, D. K., P. Hunter, B. Buckley, and T. Illia. 2008. Top 200 environmental firms. *Engineering News Record (ENR)*, June 30.

Siemens Water Technologies. 2007. *Fast facts fiscal year 2007.* Munich: Siemens. www.siemens.com/water.

Stern, N. 2007. *The economics of climate change: The Stern review.* Cambridge, U.K.: Cambridge University Press.

Texas Water Development Board. 2009. *Desalination.* www.twdb.state.tx.us/iwt/desal/faqgeneral.html#14.

United Nations Department of Economic and Social Affairs. 2007. *World urbanization prospects: The 2007 revision population database.* United Nations, http://esa.un.org/unup.

United Nations Educational Scientific and Cultural Organization (UNESCO). n.d. www.unesco.org/water/wwap/facts_figures/water_industry.shtml.

United Nations. 2009. *The United Nations World Water Development Report 3: Water in a Changing World.* http://www.unesco.org/water/wwap/wwdr/wwdr3/pdf/WWDR3_Facts_and_Figures.pdf

U.S. Geological Survey. n.d. *High plains regional ground-water study.* http://co.water.usgs.gov/nawqa/hpgw/factsheets/DENNEHYFS1.html.

World Health Organization (WHO) and UNICEF. 2005. *Water for life: Making it happen.* World Health Organization and UNICEF.

World Water Assessment Programme (WWAP). 2003. *Water for people, water for life.* The United Nations World Water Development Report.

World Resources Institute (WRI). 2000. *People and ecosystems: The fraying web of life.* http://pdf.wri.org/world_resources_2000-2001_people_and_ecosystems.pdf

Water Strategist. 2009. Analysis of water marketing, finance, legislation and litigation. Stratecon, Inc.

Xie, J. 2009. *Addressing China's water scarcity.* The World Bank.

Practical Considerations

A Collaborative Response to Climate Change

Danyelle Guyatt, PhD

The importance and threat of climate change to our way of life, our long-term savings, and our investments—as well as the opportunities that it produces—has been examined in earlier chapters of this book. This chapter sets out the rationale for a collaborative response to climate change and a framework with possible actions for pension funds and other institutional investors to take. One of the difficulties for investors in considering the implications of climate change for their investments is that modern portfolio theory—namely, efficient frontiers, diversification, and mean-variance analysis models—fail to capture the complexity of most real-world investment issues, especially the impact of systemic risks such as climate change. Indeed, the current credit crisis and its impact on the banking system, the real economy, and pension schemes' funding positions highlight our vulnerability to systemic risk. Without meaning to sound alarmist, climate change could very well be the next elephant in the room that the financial sector is not taking seriously and, if left unchecked, the consequences could potentially be far greater than the credit crisis. The challenge is therefore to widen the umbrella of investment theory beyond traditional portfolio management to bring the theory closer to the real-world realities of how decisions are made and how economies actually function.

One school of thought that offers insights and practical tools for action are the various theories related to collaboration, such as the theory of cooperation, game theory, and the formation and evolution of conventions

Danyelle Guyatt is a principal in Mercer's Responsible Investment and a visiting fellow at the University of Bath.

(Axelrod 1984; Boyer and Orlean 1992; and Samuelson 2002). These theories are particularly useful for considering appropriate action to redress systemic issues (such as climate change) in those situations where individuals might not feel powerful or, indeed, personally responsible; but when left unchecked, everyone suffers the consequences of inaction. Economists call these issues "externalities" and the textbook response is to handball such matters to governments. But, as the saying goes, "many hands make light work," and the degree of urgency needed to respond to climate change means that everyone will benefit from playing an active role in moving to a lower-carbon economy.

This chapter begins with a few home truths about the lack of action among institutional investors in responding to climate change and sets out some possible reasons for that. The conditions for a collaborative framework for pension schemes and other institutional investors will then be presented, where opportunities exist both at the industry level and in implementing specific investment strategies. Some of the collaborative initiatives that have emerged over recent years in relation to climate change at the industry level will be discussed, before moving on to examine some specific actions that like-minded institutional investors could take to benefit from a shift to a lower-carbon economy.

DIARIES OF AN INVESTMENT CONSULTANT

We have come a long way in terms of raising awareness about climate change and its potential consequences over the past decade, owing to a raft of compelling scientific evidence and, of course, the comprehensive coverage of the issue by Al Gore in his documentary and book *An Inconvenient Truth* (2006). In the investment world, we have witnessed the birth of a new range of alpha strategies designed to capitalize on the opportunities produced by climate change (see Chapters 7 through 13 for a comprehensive discussion of the different investment themes). Indeed, there are many options available for institutional investors in responding to climate change, including:

- Integration of climate change policies into voting and engagement activities of listed equity investments.
- Review of portfolio holdings to include carbon footprint analysis and evaluation of investee companies' response to climate change in adapting their corporate strategy and operations.
- Review of investment manager agreements, where trustee expectations regarding climate change issues are clearly specified and written into agreements with fund managers across all asset classes.

- Modification of selection criteria for new investment mandates across all asset classes, such that climate change considerations are given a higher priority.
- Expansion to the way portfolio performance is reported and measured on an ongoing basis to include carbon footprint analysis, corporate strategy regarding climate change, engagement outcomes, and specific actions taken by investment managers.
- Investing in climate change–themed products, which can include passive or active strategies with opportunities across all regions and asset classes.
- Engaging with public policy makers on issues surrounding climate change that may impact on the investment backdrop and long-term risk/returns.
- Raising the priority of climate change across all mainstream investment mandates (and asset classes) such that climate change considerations are embedded in the way all assets are managed.

These options have been set out in detail in a trustee guide to climate change (Mercer 2005). So the possible actions for fiduciaries have been defined, strategies are being developed, and general advice from investment consultants and other specialist advisors on how to respond to climate change is becoming more readily available. Progress has undoubtedly been made.

But let's be honest—apathy still reigns supreme toward climate change in many aspects of our lives as consumers, as policy makers, as corporate executives, and as investors. Speaking from the perspective of an investment consultant specializing in providing advice on environmental, social, and governance (ESG) issues to pension schemes, I can still count on two hands the number of pension schemes that have tackled environmental issues in a very serious way—that is, with subsequent and meaningful changes to their investment strategy. There is increased awareness and increased willpower among some, but there is still little real action among mainstream institutional investors. The following two stories might help to illuminate this observation and shed some light on why this is still the current state of play.

Story 1: Many Still Believe that Climate Change Is a "Soft" Issue

I recently attended an industry event with a small gathering of international pension fund trustees and executives, the purpose of which was to discuss the "big trends and issues" facing pension funds and how the executives might best respond. Many of the participants in the room represented organizations that had signed the UN Principles for Responsible Investment (PRI), which advocates the integration of ESG issues into active ownership activities and investment decisions. Naturally, one would expect the audience to be relatively sympathetic and open to considering how issues such as the environment might be integrated into the way they manage pension scheme assets.

Various investment topics were discussed over the course of a few days, including a panel discussion on active ownership that included a summary of the academic evidence supporting the materiality of ESG issues, how they might impact on equity and fixed-income investments, and how collaboration might be used as a tool to improve corporate practices. During the question-and-answer session, one of the workshop participants made the following comment in response to the presentations on how ESG factors can be integrated into investment processes:

> *All of these factors are soft issues. There are other more pressing concerns for pension schemes to think about.*

This viewpoint quickly became the consensus opinion around the room, with a lot of heads nodding in approval. In terms of environmental considerations, this seemed counterintuitive, given the broadly accepted reality that climate change is one of the greatest challenges facing modern civilization, and one that is already directly impacting the standard of living of many people, the bottom line of many corporations, and therefore pension scheme returns.

The quandary is that while most senior thought leaders in the investment world acknowledge that climate change is real, they fail to see it as a priority investment issue or how it is of concern for them *right now*. It is a question of timing and competing priorities. Pension schemes, endowments, foundations, and other institutional investors have a lot on their plate with the credit crisis damaging funding positions, so it's not surprising that there are "more pressing concerns" for schemes to think about. There always have been, and the events in 2008 have made this situation even worse. The problem, of course, is that if we wait too long before responding in any meaningful way, then the consequences will be far greater for our standard of living and for pension schemes' financial position. Unlike the credit crisis, government bailouts and interest rate cuts will be rendered redundant tools for responding to climate change events. Furthermore, there are potentially attractive returns to be generated from proactively seeking climate change–themed investments and raising corporate standards in adapting to a changing business environment that asset owners would directly benefit from *right now*. Why, then, the still low priority assigned to climate change as an investment issue by many institutional investors?

Here is where the behavioral economists can shed some light. First of all, this story demonstrates a classic case of myopia; that is, institutional investors' tendency to place more weight and importance on the here and now rather than proactively contemplating and responding to future uncertainties and opportunities (Black and Fraser 2000). The second observation is that individuals have a tendency to stick to what they know and what they have done in the past, otherwise known as the status quo bias (Samuelson and Zeckhauser 1988). For most pension schemes, the mean-variance analysis model and quest for diversification are still the default frameworks underpinning investment decisions to meet their long-term funding obligations. In times of extremities, such as the credit crisis, these default mechanisms become even more powerful drivers of investment decisions, rendering climate change and other systemic issues to the ancillary bucket. Finally, I observed a classic case of herding at this workshop, where the consensus opinion around the room enveloped everyone; after the "soft issue" comment was made, it quickly became more comfortable and acceptable to think of climate change as an issue that is a low priority when developing a core investment strategy (Grinblatt, Titman, and Wermers 1995).

Story 2: Intergenerational Shift in Attitude Toward the Environment

The second story draws from an informal conversation I had with a trustee who sits on the board of a number of pension schemes. This trustee struck me as being a very experienced and knowledgeable individual who was actively engaged in the debate around current investment issues and challenges. After speaking with him for some time, I casually asked for his view on climate change and the pension scheme's response to the changes it brings, to which he replied:

> *Climate change? No we haven't done anything about that yet. Too many other things to worry about and we don't have the money to spend on those sorts of things. . . .*

After a pause, the trustee went on to say:

> *But my daughter follows me around the house turning all the lights out and is constantly reminding me about how I can reduce my carbon footprint.*

The first comment struck me as a very honest depiction of the perceived reality of being a trustee of a pension scheme, where resources are constrained ("we don't have the money") and other priorities take precedence over systemic issues such as climate change ("other things to worry about"). The impression I was left with after our conversation is that he believes that climate change is a "nice" thing to consider and he will probably get around to thinking about "those sorts of things" one day, but that it is not seen as a core function of his role as a fiduciary of a pension scheme (or else presumably he would have responded already).

The second comment is particularly insightful as it demonstrates the complexity of individuals in carrying out their multiple functions in society (Elster 1986). The clear acknowledgment that climate change is an issue has been compartmentalized by the trustee into his personal life rather than his professional capacity as a fiduciary. By making reference to his daughter and his carbon footprint at home, he demonstrates an awareness of the issue and what it means for him in his personal life (such as turning the lights out!). But in his professional capacity, any connection between his investment decisions as a fiduciary and the reality of climate change has been lost. The challenge is therefore to develop a framework for responding to climate change that resonates with an individual's self-interest in both their professional and personal capacities; could collaboration be one of the tools to help bridge that gap?

The happy ending to this story is the intergenerational shift in attitude toward the environment, where small steps and actions are more commonly accepted and natural, as indicated by his daughter's awareness of his "carbon footprint" and her eagerness to take action in day-to-day life. Economists call this "adaptive efficiency," and this has been applied to study how conventional modes of behavior change over time by the likes of Sugden (1989), Boyer and Orlean (1992), and Young (1996). The nugget to take from these theories is that people behave according to conventions, modes of behavior, or "rules of the game" that are well entrenched, often unquestioned and reaffirmed by the fact that everyone else follows them (Guyatt 2006). Then, after a period of time and often out of necessity, these conventions change and we develop new norms and behavioral patterns. This can happen in a sudden way in response to a major catastrophic event, or it can gradually evolve over time in response to changing conditions. The intergenerational shift in attitude toward the environment is a classic demonstration of this adaptive process in practice.

So what have these stories taught us about human behavior, and how can these observations be built into investment theory in a practical way for institutional investors, in direct response to climate change? We know that people focus more on the here and now than the future (myopia), that they prefer the safety of others (herding), that there are competing priorities and limited resources, and that there is a reluctance to change course (status quo bias) but that these patterns of behavior can and do change over time (evolution of conventions).

WHY MIGHT A COLLABORATIVE APPROACH TO CLIMATE CHANGE BE APPEALING?

Collaboration could help to redress some of the aforementioned financial and behavioral obstacles in responding to climate change; as the process of coordinating actions with others taps into our herding instincts, it helps to speed up the process of change as new norms develop, and it could also be cheaper and more effective for asset owners than acting individually (Guyatt 2008).

It's Cheaper

Collaboration to promote system-wide change can be more cost-effective than acting individually owing to the economies of scale benefits of pooling resources, sharing costs, sharing knowledge, and power/influence. For institutional investors, formulating a coordinated response to climate change might therefore appear to be less onerous, in terms of both money and time spent.

Share the Risks

There are benefits in terms of sharing the risks associated with introducing change and innovation, or what some refer to as the "first-mover risks" (Hoffman et al. 1999). This is particularly true when the innovation is aimed at promoting industry-level change, where there are perceived risks of being different from the crowd that may encourage some coordinated level of effort (Keynes 1936). Sharing the risk is an important factor for institutional investors seeking to alter their investments to position for a lower-carbon economy, in terms of both investing in clean technologies and promoting adaptation to the way corporations conduct their business.

It Signifies "Buy-In" and Commitment

By engaging institutional investors in the change process, collaboration can mobilize intrinsic motivation in a way that externally imposed rules/laws cannot. This is not to suggest that collaboration is a replacement for changes to government regulation or guidelines in responding to climate change, but rather that investors have a role to play in being proactive. Voluntary collaboration among interested parties is not for everyone, but those that do commit themselves to climate change considerations will have more willpower to creatively pursue risks and opportunities as they arise than those who are only meeting minimum standards as required by regulations.

It Bolsters Power and Influence

In many situations, a coordinated effort could be a more powerful and effective means of driving change and improvement than going alone. Not only would a group of institutional investors have more power in terms of dollar value and representation at the industry level, but they would also be perceived as a legitimate voice that carries more weight in influencing the behavior of others than could be achieved individually. This is especially true for redressing systemic issues such as climate change, where the potential externalities (positive and negative) are significant.

It Taps into the Herding Instinct of Investors

Models of herding show that investors are driven by interpretation of a similar set of information used in their analysis (investigative herding; Froot et al. 1992) and also to protect their reputations in the event that their analysis proves to be incorrect (reputational herding; Scharfstein and Stein 1990). Climate change is still not regularly included in mainstream investment analysis, hence collaborating to improve the quality and use of the metrics will encourage more investors to follow suit without fearing the consequences for their reputation.

It Speeds Up the Evolution of Conventions

Building on the notion that there is a deep set of norms or conventions that guide behavior and investment decision making, a collaborative approach to change can help to speed up the pace of adaptation and responsiveness of agents to the changing environment. By proactively managing this process of evolution in the norms that underpin our behavior, we reduce the risk of responding at the 11th hour to major catastrophic (climate change–related) events. Restating this in investment terms, for pension funds and other long-term investors, this means being better prepared for unforeseen events and risks that emanate from climate change.

It Can Better Meet Stakeholders' Needs and Expectations

Many collaborative initiatives raise public awareness and generate positive publicity from which participating institutions (and their stakeholders) can benefit without incurring all the expenditure from such activities. For many pension funds the reputational benefits of such an association can be important, especially in a defined contribution world where individuals have more choice as to where they place their retirement savings pool. For defined benefit schemes, the membership base/beneficiaries may also view such activities favorably (e.g., charities, endowments, some industry funds, and corporate schemes, particularly where the corporation prides itself on its sustainability credentials). Public pension schemes will also have responsibilities under their mandate from the government agencies on whose behalf they manage funds, which often place a high value on sustainability initiatives. Indeed, based on our knowledge of the actions that pension funds around the world have taken, the public schemes have undoubtedly been the first movers in responding to climate change.

WHAT ARE THE PITFALLS OF COLLABORATION?

As with all good ideas, there are some pitfalls associated with collaboration to be aware of. Collaboration for collaboration's sake could be a costly and wasteful exercise. A number of factors need to be taken into account in formulating and sustaining a successful collaborative initiative. First, an honest and careful appraisal of the self-interested motivation of all agents must be factored into the design of any collaborative initiative to maximize its chance of success. Each participant must genuinely believe that collaboration will be beneficial for them and/or for the institution that they represent (Figure 14.1), where self-interest is a function of the value placed on

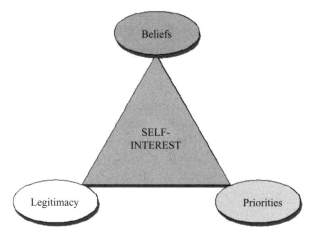

FIGURE 14.1 Motives for collaboration.
Source: Guyatt, D. 2008. Pension collaboration:
Strength in numbers. *Rotman International Journal of Pension Management*, www.rotman.utoronto.ca/icpm
doi: 10.3138/rijpm.1.1.46.

protecting the environment (beliefs), expectations of the financial benefits of action (priorities), and/or the perceived reputational gains from taking proactive steps in responding via collaborative action (legitimacy). If at least one of these basic motivations does not exist, then collaboration is unlikely to succeed in responding to climate change. [Details of the theoretical framework underpinning the collaborative model are provided in Guyatt (2007).]

Second, it is necessary for any collaborative initiative to have a measurable objective with clearly defined responsibilities, monitoring of performance, and reporting of progress. Some of the potential downsides of collaboration also need to be taken into consideration, such as the risk of increased bureaucracy of decision making, the time/effort required, the long lead times to implementing change and free riding of noncollaborative agents (Sullivan and Mackenzie 2006, p. 336).

Finally, collaboration should not be overly demanding in terms of prescribing specific actions or reduce participants' autonomy, nor should it be considered something that is permanent; as and when behavior and industry-wide change emerges, the need for collaboration will subside. These factors should all be considered and built into the design of any collaborative endeavor.

A COLLABORATIVE FRAMEWORK APPLIED TO CLIMATE CHANGE

So the first condition is that an institutional investor will need to believe that responding to climate change is in their best interest, be it to enhance returns, better manage risks, improve their reputation publicly, because of strongly held convictions among board members, and/or to meet beneficiary expectations on environmental standards (or a combination of all of these factors). This sets the foundation for finding suitable partners and considering opportunities for collaboration with other investors that share similar long-term goals, priorities, and a belief that climate change is a legitimate issue that they want to respond to proactively.

The next step is to consider what specific actions institutional investors might wish to collaborate on with respect to climate change. I have divided these opportunities into the macro (industry-wide) and the micro (investment specific).

Macro Collaborative Opportunities

At the macro or industry level, there are a number of tools available for institutional investors to collaborate with others in responding to climate change. For example, they can engage with corporations in which they are invested to improve disclosure and develop best practice standards (either directly or through their investment managers); they can engage with governments and policy makers who are responsible for designing new mechanisms for responding to climate change; and/or they can engage with representative industry bodies on developing suitable standards and codes of conduct.

There are some excellent examples of collaborative initiatives that focus on promoting industry-wide change specifically in response to climate change, such as the Carbon Disclosure Project, Ceres, Institutional Investor Group on Climate Change, Investors Group on Climate Change, and the Investor Network on Climate Risk, to name but a few. To varying degrees, these initiatives focus on raising industry standards on the environment such as advocating improved corporate reporting and transparency, engagement on the formulation of corporate strategy and government policy, and development of codes of conduct and best practice standards together with industry bodies, as set out above. These initiatives have some level of representation from the investment community, either at the board level through relevant committees or as fee-paying members, although typically the day-to-day management is carried out by an independent secretariat.

A MODEL OF COLLABORATION: Investor Network on Climate Risk (INCR)

By Christopher Fox and Erica Scharn

In November 2003, major pension funds and other leading institutional investors joined forces at the first Investor Summit on Climate Risk at the United Nations Headquarters. The Summit enabled investors to discuss the risks posed by climate change, as well as the investment opportunities that lie ahead as the world transitions to a low-carbon economy. The meeting was cohosted by Ceres President Mindy Lubber, Connecticut Treasurer Denise Nappier, and United Nations Foundation President Timothy Wirth.

At the Summit's conclusion, 10 U.S. institutional investors issued an "Investor Call for Action on Climate Risk," urging corporations, Wall Street firms, and the U.S. Securities and Exchange Commission (SEC) to take specific actions to address the risks posed by climate change. Investors also announced that in order "... to further promote investor and corporate engagement and understanding of the range of risks posed by climate change, we will support the creation of an Investor Network on Climate Risk (INCR). We have asked Ceres, a U.S.-based coalition of investment funds and public interest groups, to serve as secretariat to INCR." (Investor Call for Action on Climate Risk, released at United Nations press conference on November 21, 2003.)

Christopher Fox is director of Investor Programs at Ceres. Erica Scharn is program associate of Investor Programs at Ceres.

Rather than having each investor conduct its own assessment of the risks and opportunities posed by climate change, the founding members of INCR recognized that a collaborative knowledge management system would be more efficient, cost-effective, and would result in higher-quality information for all investors. In addition to gaining access to expertise and training, another major motivation for joining INCR for the original members—and for the other fiduciaries who have since joined—is that INCR provided them with staff support on coordinating engagement with fund managers, companies, and government agencies, including staff support on shareholder letters, proposals, and petitions to the SEC.

Following a second UN Summit in 2005, and a third in 2008, INCR's membership has grown from 10 institutional investors with $600 billion in assets under management to 81 investors with over $7 trillion in assets in early 2009. Members include public and labor pension funds; state and city treasurers, comptrollers, and controllers; asset managers; foundations; and other institutional investors. A complete list of INCR members is available at the INCR web site (www.incr.com).

INCR has catalyzed four significant changes in the investment community since its founding. First, institutional investors have dramatically sharpened their focus on the climate risks and opportunities in their portfolios by making new investments, engaging with companies in their role as shareholders, adopting disclosure standards, and sharing information. INCR members have deployed more than $4 billion in clean technologies, filed dozens of climate-related shareholder resolutions with U.S. companies, and developed the Global Framework for Climate Risk Disclosure to make it easier for companies to learn how best to disclose climate risks.

Second, fund managers and financial advisors have improved the quantitative analysis of climate risks. Investor action has encouraged many financial firms to release dozens of detailed reports about the bottom-line implications of climate change. In addition, some INCR members are beginning to ask their fund managers and consultants how they are factoring climate risks and opportunities into their investment processes.

Third, companies in industries with the highest greenhouse gas (GHG) emissions have increased their focus on climate risk and opportunities in product plans, capital investment decisions, and public policy stances. Pressure from investors, regulators, and consumers has driven major corporations to develop plans to deal with climate change throughout their businesses. Investors have also increased their corporate engagement efforts, urging companies to set GHG emissions reduction targets, increase spending on renewable energy, and improve corporate disclosure and governance on climate risk, among other actions.

Finally, a growing number of investors have called on Congress, the SEC, and other policy makers to address climate change. In September 2007, a subset of INCR members managing more than $1.5 trillion in assets, along with state officials, petitioned the SEC to require all publicly traded companies to assess and disclose their financial risks from climate change, leading to INCR's testifying before the U.S. Senate Banking Committee on October 31, 2007. And on November 11, 2008, more than 130 leading investors with $6.4 trillion in assets released a joint investor statement calling for a strong global agreement on climate change, warning that clear and long-term policy signals are essential if investors are to allocate the huge amounts of private capital required to fund the transition to a low-carbon economy.

While INCR has achieved significant results over the past five years, there is still much to be done. In 2009, INCR members are focused on the following goals:

- Investing in clean technology, with a goal of deploying $10 billion collectively over the next two years.
- Improving energy efficiency of their real estate portfolios, including aiming for a 20 percent reduction in energy used in core real estate investment holdings over a three-year period.
- Pressing companies to disclose climate risk, such as the business impacts of foreseeable carbon costs—particularly on carbon-intensive investments such as new coal-fired power plants, oil shale, tar sands, and coal-to-liquid projects.
- Persuading the U.S. government to swiftly enact a mandatory national, market-based, economy-wide policy to reduce national GHG emissions to levels called for by scientists to prevent the most dangerous effects of climate change. INCR is also calling on the U.S. to realign national energy and transportation policies to achieve climate objectives, support and enact policies to maximize energy efficiency throughout our economy, and support a strong global agreement on climate change.

These results achieved by INCR are possible because of INCR's collaborative structure. From its founding in 2003, INCR has been interweaving investor relationships, understanding, and action so that each contributes to the other. Through information sharing and coordinated engagement with companies and policy makers, INCR has become a powerful new community. The INCR community is deploying its collective resources on behalf of its common interests: limiting the downside risks of climate change and capturing investment opportunities posed by the transition to a low-carbon global economy.

This case study is based on a report by Ceres titled *Investor Progress on Climate Risks & Opportunities: Results Achieved Since the 2005 Investor Summit on Climate Risk at the United Nations, February 2008.*

While this is not the time or place for analyzing the pros and cons of each of these initiatives [see Guyatt (2007) for an overview], it is fair to say that the focus of their collaborative efforts is on changing the external environment that has an impact on the investment institutions, rather than on changing the direct behavior of investment decision makers. These macro/industry changes are crucial for creating the building blocks for shifting to a lower-carbon economy, but these can be complemented by activities that focus specifically on investment decisions. By way of illustration, in my conversations with investment managers and pension schemes who signed the CDP letter—requesting that companies improve their corporate reporting standards on the environment—it seems that very few of the signatories have used this information when carrying out their investment analysis or making actual investment decisions. The CDP is currently embarking on a series of initiatives to tackle the apparent gap between collaborative action to promote systemic change and the lack of follow-through in terms of investment decisions (and therefore shifts to capital allocation). One possibility for institutional investors and their agents who share a desire to respond to the climate change challenge is to participate in specific initiatives focused on investment process changes (at the individual firm level, rather than industry-level change), as the following section will set out.

Micro Collaborative Opportunities

Micro collaborative opportunities refer to those that go to the heart of the investment strategy formulation and implementation, such that climate change gets embedded into investment actions and outcomes. One notable example of collaboration at the investment level is the arrangement between Dutch funds ABP and PFZW, who together mandated Alpinvest to manage a portion of their private-equity investment with a special focus on clean technology. Generally speaking, there are far fewer examples of such initiatives, I suspect because the industry has been focused on getting the building blocks in place first (through the macro initiatives) before being ready to move on to consider specific actions. Also, it takes some time and effort to find the suitable partners for specific collaborative outcomes and to design how these might work in practice. Many schemes are also more accustomed to considering investment strategy and implementation as something that you do alone, rather than through a collaborative model. In many instances this is indeed the case, but a few examples of areas where a collaborative response to climate change might be beneficial at the investment strategy level are set out next (this list is by no means meant to be exhaustive).

Environmental Performance Audit via Industry Body The idea here is that institutional investors could pool their resources to negotiate a reduced fee for portfolio auditing services that measure and evaluate the environmental performance of their underlying investments. This might include measuring the carbon footprint of their investments, the performance of investee companies in terms of strategy and adaptation to climate change, and the environmental footprint associated with alternative assets such as infrastructure projects, private-equity investments, and real estate. An example of one such initiative was the venture between the Australian Institute of Superannuation Trustees and Trucost to analyze the carbon exposure of 14 superannuation funds (producing a report entitled *Carbon Counts 2008: The Carbon Footprints of Australian Superannuation Investment Managers*).

The benefit of pooling resources in this way for fiduciaries is that they obtain the information about the performance of their investments against environmental criteria at a lower cost than would otherwise be the case because the service provider(s) will likely be more flexible with fees as an introduction to the tools available for a larger pool of clients. Moreover, the recipients of this information will benefit from sharing and comparing their environmental portfolio performance with others in a way that would not be possible if they undertook the analysis individually. Follow-up strategies could be developed to collaboratively engage with laggard companies to improve their environmental performance. On the downside, of course, there is less chance to tailor the analysis to meet the specific needs of a pension scheme which would be advantageous to develop follow-up actions where investment holdings differ. Nevertheless, this collaborative approach could be a useful and cost-effective starting point for assessing the environmental performance of a scheme's investments.

Modify Investment Manager Selection Criteria   Another area where asset owners could potentially benefit from collaboration is to jointly develop a new industry standard that integrates climate change considerations into the selection of new investment products and fund managers. While it may appear on the surface that this is one aspect of the investment process that institutional investors would carry out independently, in reality there are a narrow set of norms that tend to be applied in selecting new investment strategies in the traditional asset class. These tend to cluster around views around the expected risk/return (and tracking error from the benchmark for active strategies) which in turn, is a function of the robustness of the investment process (particularly its performance track record and expectations), the quality of the people, the reputation and strength of the organization, and the ability for the manager to meet the administrative requirements of the institution. Many of these norms have evolved from the industry's belief and attachment to portfolio theory and the drivers of risk/return, with some practical additions added to tailor the criteria for each scheme's particular needs (especially on the administration side).

At present, there are only a handful of asset owners that have attempted to widen the selection criteria to actively consider and evaluate systemic issues, such as climate change. Why not collaborate, perhaps via a representative industry body or advisor, to consider how this fairly standard set of selection criteria would need to be expanded to include systemic risks (like climate change) as a priority across different asset classes? The signal sent to asset managers would certainly be stronger if carried out collaboratively, the industry would move to a new set of norms more quickly, and, of course, it would be more cost-effective than for each individual asset owner trying to tweak its criteria independently.

Modify Investment Manager Agreements (IMAs) Similarly, the standard wording for agreements between asset owners and asset managers could be modified in a way that gives higher credence to climate change and other systemic risks. One thing I have observed as a consultant is that if an asset owner contemplates changes to portions of what can be fairly standard IMAs in this direction, it pretty quickly becomes apparent that the cost of changing documentation and the likely resistance from asset managers makes this a fairly unappealing option. For small and medium-sized schemes with less negotiating power, there is a lot of merit in considering joining forces to formulate new standard wording on climate change to build into IMAs with asset managers, perhaps orchestrated via a representative industry body, legal group, or set of advisors. Of course, this is not to suggest that IMAs become standardized across all investments as that would be completely impractical, but rather the idea is to agree on new standards for including climate change as a factor worthy of consideration.

This may sound like a fairly basic idea, but to design the appropriate wording that is well accepted, legally robust, and encourages the desired change in behavior is no small feat—and, certainly, my experience suggests that this has tended to stop many schemes from taking that important step. A collaborative approach means that asset managers will be more inclined to have the financial incentive to respond, and again, the signal would be clearer on the expectations of asset owners with respect to climate change.

Strategic Research and Scenario Analysis One of the challenges when considering systemic issues is that the cost of conducting sophisticated analysis and modeling of scenarios at the individual scheme level might outweigh the benefits in terms of practical (short-term) implications. This is one of the reasons for underinvestment in longer-term research of this kind, although the credit crisis reminded us of the consequences of not digging a little deeper in considering the risks to our investments. Again, the limitations of traditional finance theory as it stands today doesn't provide a lot of room for thinking the unthinkable in a way that investors could justify financially. It is for this reason that collaboration on strategic research, such as scenario analysis, might be an option worthy of further consideration. A few examples of groups interested in long-term strategic research are the Network for Sustainable Financial Markets (NSFM) and Rotman International Centre for Pension Management (ICPM), both of which bring together industry experts as well as academics with the aim of improving institutional investment decision-making processes. The following section discusses a possible application of collaboration on strategic research and how this might work in practice when applied to climate change.

COLLABORATION IN ACTION: ASSET/LIABILITY MODELING AND SCENARIO ANALYSIS

The Intergovernmental Panel on Climate Change (IPCC) was set up in 1998 as a scientific body set up by the World Meteorological Organization and the United Nations Environment Programme to provide decision makers and others interested in climate change with an objective source of information about climate change. In addition to this, the U.K. government commissioned a report on the economic implications of climate change scenarios, the so-called *Stern Review* (2006). These projects are large scale (and ongoing) endeavors, involving hundreds of people and taking many years to produce and disseminate. The translation of the implications of climate change for the real economy was considered in the *Stern Review,* with some industry reports also taking up the challenge of measuring the impact of climate change at the company level. In addition to this, the insurance industry has evolved its risk models and moved some way toward integrating climate change events into risk assessments and underwriting policies.

One of the puzzles that remains unsolved is what climate change means for institutional investors at the strategic level—in other words, what impact could it have on the risks surrounding asset allocation? How might different climate change events impact on property investments in different parts of the world? How will different emerging market regions fare, and what countries are better prepared and advancing their policy response more quickly than others, and, therefore, how will that feed back through to their long-term growth prospects? All of these issues are central for considering the assumptions underpinning long-term risks and return opportunities of investment portfolios. Indeed, the U.K. Institute of Actuaries members agree that climate change is becoming of increasing importance. A survey conducted in 2007 yielded a response rate of over 1,000, with the results showing that more than 50 percent of respondents believe that model assumptions regarding pricing and reserving of nonlife products may need to be modified and that there are scenarios that could lead to large losses under some investment products. Yet no one has started the work in earnest. Why not?

Well, there are a few reasons, and they are behavioral and financial. First, it's expensive to do the analysis, especially if it is going to be done properly. Second, it's not a current priority for the investment world as the credit crisis and immediate funding shortfalls are of higher importance. Third, the wider perception across the finance industry is that climate change effects are not imminent and are therefore a problem to consider in the future, not right now. Finally, there is a misplaced sense of security that it will all work out okay in the end and that the governments around the world will fix the mess before it gets out of hand. So the response of the financial world to date has largely focused on the investment opportunities that climate change produces, such as clean tech and low-carbon investments across asset classes (as set out in Chapters 7 through 13). These are valid responses and crucial for fostering the research and commercialization of new technologies in our shift to a lower-carbon economy globally. However, for long-term investors to start and finish here would be a missed opportunity, and certainly one that underestimated the risk side of the climate change equation as laid out in the *Stern Review* and IPCC reports.

Let's bring some theory into the picture at this point, to see how a collaborative response might work to fill this research gap. Figure 14.2 sets out the collaborative framework that was developed by Guyatt (2007) as

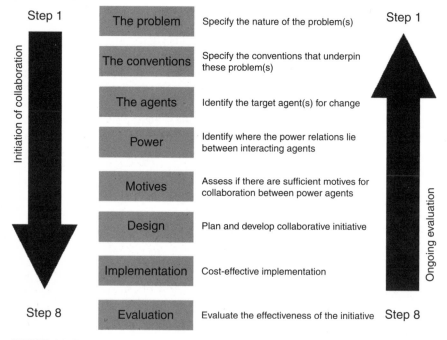

FIGURE 14.2   The collaborative framework.

Source: Guyatt, D. 2008. Pension collaboration: strength in numbers. *Rotman International Journal of Pension Management*, www.rotman.utoronto.ca/icpm doi: 10.3138/rijpm.1.1.46.

part of research funded by Rotman ICPM. In brief, some key observations regarding climate change and strategic research:

- *The problem.* Starting at the beginning, we have identified a need for strategic research on the implications of climate change for pension schemes' long-term assets and liabilities. The problem is that this research is not being conducted owing to the aforementioned financial and behavioral obstacles.
- *The conventions.* We are dealing with a short-term mind-set across the finance industry that does not lend itself to thinking the unthinkable on systemic issues, as illustrated by the credit crisis and climate change. Climate change is not part of the institutional investors' armory in building and implementing investment strategy. New research needs to be carried out to fill this gap.
- *The agents.* The people who need to change are primarily the actuaries and consultants, although they won't change unless they are provided clear signals from their clients (or perhaps industry bodies/regulators) that climate change is of sufficiently high priority for their clients to build into scenario analysis.
- *Power.* The key agents with the power are the asset owners who pay the actuaries and consultants to provide advice on strategic asset allocation, including asset liability analysis. Relevant government agents and industry bodies also have some sway over industry standards that could be utilized, particularly given the widespread potential impact of climate change scenarios on future retiree savings.

This leads on to the next stage of building a collaborative framework, namely the motivation for collaboration. There is no doubt that some of the leading asset owners might benefit in pooling their resources to commission research that considers the strategic implications of climate change. As a basic starting point, they would need to hold the shared belief that climate change is of sufficiently high importance to take a closer look at the strategic implications. A collaborative approach would not only be cheaper, but there is the added benefit of legitimacy when accompanied by other investors in such an endeavor. Adding to that is the benefit of staying ahead of government regulatory changes and meeting the wider goals and expectations of stakeholders in considering not only the opportunities but the risks that climate change produces in safeguarding the savings pool of future retirees.

This final example is a "live" project on which we at Mercer are currently embarking. I hope this discussion has shown how a collaborative research project on climate change could be one example of a collaborative initiative that has direct implications for pension schemes in considering their asset/liability risks, but one where there are natural synergies across asset owners to collaborate in making it happen.

CONCLUSION

While there is no doubt that awareness surrounding climate change among institutional investors has increased over recent years, the actions taken are still patchy and indicate a sense of complacency. Indeed, this chapter has argued that there are still some barriers hindering climate change from being truly integrated into core investment processes, including behavioral and financial considerations. It also has been argued that collaboration could be an effective tool for fostering change and consideration of climate change, provided a few basic conditions are met. First, an institutional investor will need to believe that responding to climate change is in their beneficiaries'/stakeholders' best interest, be it to enhance returns, better manage risks, improve their reputation publicly because of strongly held convictions among board members, and/or to meet beneficiary/stakeholder expectations on environmental standards (or a combination of all of these factors). Second, there is the question of finding suitable partners with whom to coordinate activities, who share similar priorities and long-term goals with respect to climate change, be it to change the mechanisms at the industry level (macro collaborative initiatives) or specific enhancements to the core investment decision-making process at the firm level (micro collaborative initiatives).

In sum, the notions of collaboration and behavioral biases in decision making are not new, but their relevance and significance for extending the toolkit used by long-term investors in their role as fiduciaries still has a way to go to complement and fill the gaps of traditional finance theory that the industry has come to rely on. In considering systemic risks such as climate change, extending these tools is a crucial part of creating a more sustainable capital market system.

REFERENCES

Axelrod, R. 1984. *The evolution of cooperation.* New York: Basic Books.

Black, A., and P. Fraser. 2000. Stock market short-termism—an international perspective. *Journal of Multinational Financial Management* 12(2):135–158.

Boyer, R., and A. Orlean. 1992. How do conventions evolve? *Journal of Evolutionary Economics* 2(3):165–177.

Elster, J. 1986. *The multiple self: Studies in rationality and social change.* Cambridge, UK: Cambridge University Press.

Froot, K.A., Scharfstein, D.S., and Stein, J.C. 1992. Herd on the street: Informational inefficiencies in a market with short-term speculation. *The Journal of Finance* 47(4):1461–84.

Gore, A. 2006. *An inconvenient truth.* New York: Rodale Press.

Grinblatt, M., S. Titman, and R. Wermers. 1995. Momentum investment strategies, portfolio performance, and herding: A study of mutual fund behaviour. *American Economic Review* 85(5):1088–1105.

Guyatt, D. 2006. Identifying and overcoming behavioural impediments to long-term responsible investing. Unpublished PhD thesis, University of Bath.

Guyatt, D. 2007. *Identifying and mobilising win-win opportunities for collaboration between pension fund institutions and their agents.* Unpublished research paper funded by the Rotman International Centre for Pensions Management. www. rotman.utoronto.ca/userfiles/departments/icpm/File/2_Danyelle%20Guyatt_ Identifying%20and%20Mobilising%20Collaborative%20Opportunities_ Overview_Key%20Recommendations_10%20pages.pdf.

Guyatt, D. 2008. Pension collaboration: Strength in numbers. *Rotman International Journal of Pension Management* 1(1):46–53.

Hoffman, A. J., J. J. Gillespie, D. A. Moore, K. A. Wade-Benzoni, L. L. Thompson, and M. H. Bazerman. 1999. A mixed-motive perspective on the economics versus environment debate. *American Behavioral Scientist* 42(8):1254–1276.

Keynes, J. M. 1936. *The general theory of employment, interest and money.* New York: Cambridge University Press.

Mercer. 2005. A climate for change: A trustee's guide to understanding and addressing climate risk. A report commissioned by IIGCC and Carbon Trust.

Samuelson, L. 2002. Evolution and game theory. *Journal of Economic Perspectives* 16(2):47–66.

Samuelson, W., and R. J. Zeckhauser. 1988. Status quo bias in decision making. *Journal of Risk and Uncertainty* 1:7–59.

Stern, N. 2007. *The economics of climate change: The Stern Review.* Cambridge, U.K.: Cambridge University Press.

Scharfstein, D.S., and Stein, J.C. 1990. Herd behaviour and investment. *American Economic Review* 80(3):465–479.

Sugden, R. 1989. Spontaneous order. *Journal of Economic Perspectives* 3(4):85–97.

Sullivan, R., and C. Mackenzie. 2006. *Responsible investment.* Sheffield, UK: Greenleaf Publishing.

Young, H. P. 1996. The economics of convention. *Journal of Economic Perspectives* 10(2):105–122.

BIBLIOGRAPHY

Guyatt, D. 2005. Meeting objectives and resisting conventions: A focus on institutional investors and long-term responsible investing. *Journal of Corporate Governance* 5(3):139–150.

Guyatt, D. 2009. Beyond the credit crisis: The role of pension funds in moving to a more sustainable capital market. *Mercer's Responsible Investment* newsletter Q1 2009, www.mercer.com/referencecontent.htm?idContent=1332305.

Kahneman, D., J. L. Knetsch, and R. H. Thaler. 1991. Anomalies: The endowment effect, loss aversion, and status quo bias. *Journal of Economic Perspectives* 5(1):193–206.

CHAPTER 15

Corporate Responsibility and Environmental Investing

Tony Hoskins and Martin Batt

The earlier chapters of this book examine the issues and consequences of climate change and how it affects investment decisions. In this chapter, we look at the way in which climate change is handled within organizations.

At the outset, it is important to define the different roles that organizations (or individuals) may have in regard to the environment. In this chapter, we have used the terms *asset owner, asset manager,* and *asset user:*

- **Asset owner:** The person or institution (e.g., a pension fund) with the original rights to the assets, and holding the ultimate authority to switch the ways in which these assets are invested.
- **Asset manager:** The institution managing investment of the assets under the direction of the asset owner.
- **Asset user:** The corporate entity or enterprise using the assets (in the form of cash invested as equity or loaned, such as in the form of corporate bonds) and has authority to use them as defined by the corporate memoranda of association.

We start by looking at how asset users address environmental responsibility (ER), often as part of their corporate responsibility (CR) approach. In this respect, asset users may be both corporations (including privately held companies) and organizations within the public sector (where the assets would be government bonds) and voluntary sector, such as

---

Tony Hoskins is chief executive of the Virtuous Circle. Martin Batt is managing consultant of the Virtuous Circle.

nongovernmental organizations (NGOs). For the sake of simplicity in this chapter, the term *corporations* represents all asset users. We then move on to how asset managers can evaluate the nature of a corporation's ER. It should be borne in mind that a corporation may be both asset user and asset manager—this would be the case for banks and investment houses managing funds for clients. We will look at some examples of such conjoined entities and inquire about the extent to which their asset manager ER matches their role as an asset user.

WHAT IS CR AND ER?

For most large corporations, environmental impacts such as climate change are addressed within the corporate responsibility function and strategy. There is no universal definition for corporate responsibility. Generally, it is defined as how corporations consider the interests of society by taking responsibility for the impact of their activities on their internal and external stakeholders, as well as their impact on the environment. This obligation can be addressed on a basis of statutory requirements (compliance approach) or by voluntarily building on the compliance approach and taking further steps to reduce these impacts (proactive approach). The stakeholders considered include the corporation's community, workplace (covering employees and contractors), and marketplace (suppliers and customers). A corporation's impact on the environment can include aspects such as biodiversity loss and greenhouse gas (GHG) emissions. The latter arise from the "environmental aspects" of its activities, including use of fossil fuels in buildings and transportation, use of nonrenewable and hazardous materials, and water consumption.

The relative corporate priority of each of these impact areas depends on the nature of the corporation's business, geographic location, and the related culture of the society in which it operates.

For support service businesses, where typically there is a low asset base and the key business driver is employee performance, the CR priority will tend to be the relationship between the corporation and its employees. In contrast, corporations such as those operating in heavy manufacturing or power generation will be more focused on environmental impacts.

Historically, office-based business sectors have considered their environmental impacts to be relatively small. However, there is a growing realization that data centers and call centers (usually linked to office-based businesses) are large users of energy, with a consequent significant impact on GHG emissions. A recent McKinsey article (Forrest, Kaplan, and Kindler 2008) suggested that between 2000 and 2006, the amount of energy used to store and handle data had doubled. It is estimated that the average data facility uses energy equivalent to that used by 25,000 households. It went on to state that the world's 44 million servers consume half a percent of the world's electricity. It estimated that emissions from the world's data centers now approach the total emissions of countries such as Argentina or the Netherlands. Projections show that without efforts to curb demand, carbon emissions from worldwide data centers will quadruple by 2020.

However, regardless of the nature of business sectors, different cultural attitudes result in differing priorities. This can be seen from Figure 15.1, which shows the attitudes of different geographic populations toward corporate activities.

Clearly, in 2004, environmental issues were one of the highest priorities for European respondents but of relatively lower priority for respondents from developing nations, who considered the protection of human rights to be of greater importance.

Nowadays, among corporations progressively developing their CR strategy, these differing priorities tend to reflect their perceptions of the risks relating to CR impacts. These corporations will prioritize CR activities based on the nature of the risks related to their impacts.

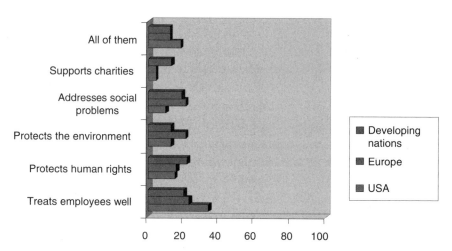

FIGURE 15.1 Corporate activities most are interested in learning about—percent strongly agreeing.
Source: Kate Fish, Managing Director, Europe, Business for Social Responsibility. 2004. *Are Americans from Mars? CSR in North America.* presentation given at the European Conference on CSR in Maastricht on November 8.

In the late 1990s, CR was considered to be driven by the NGOs' agenda—largely because corporations considered CR to be "nice to have," and NGOs were able to exert pressure through lobbying in what was essentially somewhat of a corporate vacuum. It was not necessarily part of their business strategy or business processes. Toward the end of the first decade of the twentieth century, this attitude has changed. More often, corporations now consider CR to be part of their reputation and risk management.

Corporations also have recognized that they can choose where they wish to be positioned on CR, on a spectrum covering leading edge, middle of the road, or laggard, against their sector peers. It is not only the overall position that needs to be considered but also that of the relevant CR elements. It is acceptable for a corporation to be at the leading edge for environmental initiatives (perhaps because it sees its reputation as being correlated to its perceived environmental management). At the same time, it could decide to be laggard or compliant for community initiatives (perhaps because it sees its impact on the community as very limited).

This risk-related approach is relevant also to asset owners and asset managers that invest in a corporation. We recently undertook investor and customer stakeholder surveys for two clients—one in the manufacturing sector and the other in the services sector. In the former, investors regarded the environment as a key priority for the corporation to address because these investors saw environmental impacts as having a significant potential influence on both future profit line and longer-term reputation. They considered that failure to manage environmental impacts effectively could have a detrimental effect on shareholder value. In contrast, investors considered the environment to be low risk for corporations operating in the service sector. Instead, because of the nature of this company's sector, investors saw the presence and effectiveness of a strong code of ethics and accompanying compliance system as a far higher priority.

Within this context of a corporation's risk management agenda, what is the role of the environment as part of a corporate entity's CR agenda? What is its ER? Inevitably, it has to relate back to the potential risks a corporation may face. These risks may include potentially some or all of the following:

- Environmental mishaps, leading to potential fines or litigation
- Environmental issues in the community, leading to possible closure of facilities
- Forthcoming environmental legislation requiring changes in business or product processes
- Publicized environmental mishaps or processes causing unacceptable reputational comment
- Government legislation addressing environmental dilemmas, causing the potential loss of the "license to operate"

In this respect, a corporation's ER can be described as:

At a minimum, to manage its environmental approach, business processes, products and services, to ensure the risks of environmental impacts have minimal injurious effect on stakeholders and are to the long-term benefit of society as a whole.

With this definition of ER in mind, we move on to the environmental footprint for a corporation.

WHAT COMPRISES AN ASSET MANAGER'S ENVIRONMENTAL FOOTPRINT?

As we have already established, there are likely to be differing corporate attitudes, and hence priorities, towards environmental impacts. The impact of influential advocates, such as former vice president Al Gore (with his documentary *An Inconvenient Truth*) and Sir Nicholas Stern (with his report to HM Treasury, *The Stern Review*), has been to raise both public awareness of the impact of climate change and the economic impacts of environmental issues. As a result, there are probably few major corporations not aware of the need to monitor their environmental impacts.

However, different corporations have different environmental priorities. The typical choice of environmental impacts would cover elements:

- *GHG emissions*—emanating from the consumption of energy, as well as the consumption of other GHGs, such as those used in refrigeration or air conditioning systems. The Greenhouse Gas Protocol (www.ghgprotocol.org) has helped guide corporations to account for these emissions and are referred to as being in one of three scopes:
  - *Scope 1 emissions* occur from resources owned or controlled by a company, such as combustion facilities (e.g., boilers), combustion of fuels in transportation (e.g., car fleet), and physical or chemical processes (e.g., refrigeration).
  - *Scope 2 emissions* are from the generation by another party of electricity that is purchased and consumed by the company.
  - *Scope 3 emissions* cover all indirect emissions (other than from purchased electricity) that occur from sources not owned or controlled by the corporation (e.g., purchased materials, business travel and supply chain transport).
- *Energy consumption*—clearly closely linked to GHG emissions across all three scopes, but in this respect, the question asked is the extent to which the corporation is managing its energy consumption (often described in a normalised manner, such as consumption per employee

or per production unit). In addition, corporations are being asked to describe their ability to generate some of the energy they consume through renewable energy sources they own.

- *Pollution*—which could be in the form of gases (such as nitrous oxide or sulphur dioxide), chemical spillages (including those released into water courses), as well as noise or light pollution that impact on the well-being of neighboring communities.
- *Water consumption*—reducing the amount of unnecessary water consumption, for example, by recycling "gray water" in activities not involving ingestion by human beings.
- *Resource management*—reducing materials created as waste unnecessarily in business processes, as well as ensuring that any waste generated is reused or recycled. In particular, corporations are being asked to ensure that their waste sent to landfill is minimized. This is particularly relevant since the methane that is emitted by landfill is a significant GHG.
- *Biodiversity*—the way in which a corporation maintains or enhances the biodiversity of habitats that surround locations where owned facilities are situated.
- *Environmentally friendly products*—the manner in which corporations modify products and services to ensure that they minimize environmental impacts in production and the ultimate use and disposal of their products by customers—either by modifying materials used, as is required by the EU Biocidal Products directive, or by ensuring that product consumption occurs in an environmentally friendly manner, as is the case for end-of-use disposal of electrical equipment under the EU's waste electrical and electronic equipment (WEEE) directive.

An emerging area is that of the environmental impacts in a corporation's supply chain. Such impacts may include the carbon emissions involved in long-distant freight transport, as well as the biodiversity issues, especially where raw materials are procured from the developing world. In the past this was a low priority for corporations, but now, particularly for carbon emissions, the subject of supply chain impacts is becoming more significant. The Carbon Disclosure Project initiated its CDP Corporate Supply Chain Programme in 2007. The initial participants included Cadbury, Dell, Hewlett-Packard, Imperial Tobacco, L'Oréal, Nestle, PepsiCo, Procter & Gamble, Prudential, Reckitt Benckiser, Tesco, and Unilever. These participants have now been expanded considerably. One of their actions is to put pressure on their own supply chain to commence reporting on carbon emissions, as a precursor to building these considerations into their own procurement selection procedures.

In the main, corporations mitigate these impacts according to their own standards and needs. While the trend toward independent verification of data by external bodies has helped, this is still in the minority. Hence, an investor often has to take a corporation's word regarding the extent of most of the preceding impacts, as well as the extent to which the corporation has attempted to mitigate them.

However, this is different for GHG emissions, where corporations are increasingly being asked to declare the level of emissions—often as a part of a program of trading carbon in an open market. Energy-intensive industries, such as power generators and steel and glass manufacturers, are already facing financial pressures to improve carbon emissions.

The EU's Emissions Trading Scheme (EU ETS) caps emissions in specific industries, by requiring corporations to trade their carbon emissions (below or above this cap) in an open market as an incentive to reduce emissions. The EU ETS was launched in 2005 and is the largest multicountry, multisector GHG emission-trading scheme worldwide. However, it is not the only scheme. Since 2006 the state of California has set caps on the emissions of power generators within the state. Similarly, New South Wales established its Abatement Scheme in 2003, which builds on an existing emissions benchmarking program in connection with electricity retailer licensing conditions.

But while the percentage of emissions covered by of these schemes was large, the range of business sectors and size of business establishments was relatively limited. This changed with the United Kingdom's Carbon Reduction Commitment (CRC), introduced in 2008 and effective from 2010. It is the first cap-and-trade scheme to impact across all sectors. Targeted at all organizations (both private and public) with total U.K. electricity consumption of more than 6,000 MWh (measured by half-hourly meters), it is expected to encompass more than 5,000 organizations—and many more individual establishments in these organizations.

Schemes such as the CRC will ensure for the first time that a wide range of organizations will internalize the external costs of carbon emissions. In the past, managers in organizations recognized the cost of energy consumption, but did not take the impact of GHG emissions into account in business decisions. Now, with the CRC, the cost of carbon will appear for the first time on the profit-and-loss account.

In addition, there are means other than trading carbon permits by which corporations declare their GHG emissions. The best example is the Carbon Disclosure Project (www.cdproject.net), which produced its sixth report in 2008 (CDP6), and is backed by 385 investment management signatories, representing over $57 trillion of assets at the launch of CDP6 in autumn 2008. The project is designed to provide a data source of carbon emissions for investors in major quoted corporations. CDP differs from the carbon trading schemes, since the information requested from corporations includes:

- *Strategic considerations*—the extent to which corporations are identifying risks and opportunities related to climate change and how they are managed.
- *Data reporting*—requiring emissions reporting across the three scopes (scope 1, representing bought-in electricity; scope 2, representing energy consumed by owned assets; and scope 3, representing emissions arising through the procurement of other corporations' services or products, including business travel, supply chain, and carbon embedded in purchased products), as well as energy usage and emissions intensity (the normalized value of emissions relating to the corporation's revenue).
- *Management procedures*—setting targets and action plans, emissions forecasting, and governance aspects.

The Carbon Disclosure Project has grown in coverage, both of data requested and corporations involved. For CDP6, more than 3,000 corporations were sent questionnaires requesting information, and 77 percent of the Global 500 responded. In addition, CDP6 included a Carbon Disclosure Leadership Index for the first time, measuring how much of the questionnaire is covered in a corporation's response.

However, as can be seen from the CDP, simply measuring a corporation's emissions is just a first step—there must be management strategies in place to develop improvement initiatives. Ways in which corporations attempt to mitigate environmental impacts vary from the passive to the proactive (and often creative). This differentiation is also visible in attempts to avoid responsibility for emissions, as can be seen from the following examples:

- *Passive*. Carbon offsetting is a means of mitigating a corporation's GHG emissions, by offsetting them against schemes that are seen to be GHG friendly. These can include programs such as planting trees. Typically, such schemes are not seen as an effective form of mitigation, with many stakeholders questioning whether they truly offer a net reduction in emissions.

  Similar views are held about corporations that attempt to mitigate their emissions by purchasing "green electricity" from power generators, often where a high percentage of electricity generation is based on fossil fuels. Indeed, purchase of "green electricity" is not permitted in the calculation of net GHG emissions for the United Kingdom's CRC scheme because the U.K. government states that existing evidence suggests that green tariffs deliver insignificant additional carbon savings from renewable energy. However, there are more acceptable offsetting schemes such as investing in wind power projects supplying renewable energy to the China national grid, but these tend to be few in number.

- *Proactive.* Such schemes demonstrate how mitigation strategies can produce both environmental improvements for a corporation and, very often, internal cost reductions. A good example is investment in energy efficiency schemes, which may range from better insulation to more capital-intensive combined heat and power plants. Another would relate to improvements in distribution planning, which could be achieved by a range of actions:
  - Equipping vehicles with GPS systems to reduce route miles.
  - Fitting vehicles with speed inhibitors to reduce maximum speeds.
  - "Backhauling," by which businesses such as supermarkets manage supply chain transportation to ensure that delivery vehicles collect waste packaging for return journeys via recycling depots.

A further example is property owners designing new buildings to take account of the local biodiversity, while at the same time offering a highly attractive facility for those working at the new premises. The result will ensure that rental values reflect the value invested in the biodiversity scheme.

However, the most creative schemes tend to be those that involve employees of corporations directly assisting in environmental impact improvements. Man Group, a leading financial services company, undertook a series of employee awareness seminars, designed to give their employees greater awareness of their environmental impacts, both at work and in their own personal lives. The benefit of such broad-ranging communications is that it sets a framework for changes within a working environment that may otherwise be challenging. In the case of Man, one output was a change from provision of bottled water for consumption by employees to a filtered drinking water system. In this case, the scene for such a change had already been set through the employee awareness seminars.

Another example is from BAA (which runs Heathrow airport, along with others around the world). At Heathrow, road traffic congestion is a significant issue, and for its new offices BAA liaised with employees to introduce a car-sharing scheme. After eight years, this is now the largest of its kind in the United Kingdom and has been effective with the cooperation of the employees, for whom parking permits are prioritized for those participating. It has also been extended to employees of other companies at Heathrow.

For employees to participate actively in mitigating the environmental impact of a corporation, they need to see the benefits of their participation.

A good example of how this may be undertaken is safer driving course for employees (either with delivery vehicles or company cars). The benefit to employees of attending such courses is improved personal safety, while for the corporation providing the course, the benefit comes in several ways—through reduced insurance claims as well as from reduced fuel consumption, because these courses usually include "eco-driving" as a way to improve driving standards. In addition, corporations increasingly focus product developments to reduce environmental impacts of products for users.

These developments occur in both business-to-consumer markets (an example includes the (Red) product range, devised by U2's Bono to provide funds for HIV/AIDS prevention, with the participation of private-sector corporations such as Dell, Gap, and American Express in the United Kingdom) and in business-to-business markets.

The latter is possibly where some of the greatest changes are occurring. Examples include Rentokil Initial's EnviroFresh products, which not only attack odors in men's washrooms, but also save up to 90 percent of water used (equivalent to 280 liters each year). Similarly, IMI's solution for a biofuel-powered turbine involved a system of 30 desuperheating and bypass valves to control the flow of steam, and resulted in twice the electricity output, halving emissions out of the turbine. In a more commercial manner, BT has been active in promotion of its video and teleconferencing services, which help customers to reduce business travel while operating efficiently and cost-effectively at the same time.

To achieve high levels of ER, corporations have to be proactive in both the inputs and outputs of their business processes. They have to consider both internal (direct) and external (including direct and indirect) impacts.

- *Avoiding responsibility.* Avoidance is not a mitigation strategy but is used by corporations where measurement of environmental impacts does not cover the full scope of operations. For those companies that do not attempt to declare the environmental impacts of their supply chain, transferring activities to outsourced supply chains or call centers is a form of migrating the company's environmental footprint to a third party, with no other physical process changes occurring. This is a form of moving GHG reporting from scope 1 or 2 to scope 3, where the levels of reporting are currently low. A more extreme example of this migration was suggested by U.K. chemical industries' representatives, who suggested that extension of the EU ETS to sectors such as agricultural fertilizer manufacture could lead to the move of production facilities from the United Kingdom to countries that do not fall under the scope of the ETS.

WHAT ROLE DOES ER (AND CR) PLAY IN A CORPORATE ENTITY'S BUSINESS MODEL?

The previous section focused on ER, which usually lies within a corporation's overall CR approach. The way the CR approach is handled within a corporation will affect the nature of that corporation's ER.

CR has a strong correlation to the reputation of, and the trust in, a corporation. CR at its most basic is about the way a corporation (and its employees) behaves toward stakeholders. Implementing CR effectively within a corporation requires the organization culture to match the CR approach being introduced. If there is a mismatch initially, then the implementation process takes on some of the problems of organizational change, which can often be both challenging and long term.

Depending on the corporation's culture, the role of corporate responsibility can vary from compliance to having CR integrated and embedded within business processes and corporate culture. Correlatively, where CR is positioned in a corporation typically determines the role of ER.

The management of an entity's environmental impacts will be a function of the quality of commitment of the corporation's employees to manage impacts in a positive and transparent manner. If employees see management adopting a cavalier attitude toward customers, suppliers, or themselves, then they will take these actions as an indicator of how they themselves should behave. No matter what a CR policy says about managing environmental impacts, if employees do not see it as part of their day-to-day activity, this policy will not be effectively implemented.

The issues to be considered in determining the effectiveness of a corporation's CR approach will cover several areas, such as:

- *Business integrity.* A key question is "Does the corporation's code of business conduct include how employees should behave toward stakeholders?" If the answer is yes, then the follow-up question is "How effective is this code and is there an effective management system in place to ensure that breaches are notified and employees trained in how to implement the code?" If there is no evidence of either a code or a business integrity management system, then it is highly likely that the CR approach is based on compliance, and therefore the role of ER would be largely passive within the organization.
- *Employee communications.* Here the question to be asked is whether there are effective forms of employee communication (over and above a "births and marriages" newsletter). This would ensure that employees are aware of CR initiatives and understand their roles in these initiatives. In addition, there is a need to review how employees' views are heard.

This may take the form of an employee survey or works councils or more informal processes such as regular management cascade briefings. There is a risk that if employees are not communicated with effectively, they will not understand their role in CR initiatives and will not behave in the manner expected of them.

■ *Customer and supplier initiatives:* In this area, the questions focus on how proactive the corporation is in its relationships with customers and suppliers. Examples of such initiatives could include improvements in customer service, actions taken to improve product labeling or reduce misselling or mismarketing to customers, as well as actions to assist suppliers to improve their own CR standards, or joint initiatives with suppliers to develop more environmentally friendly products. Each example demonstrates enhanced levels of relationships with customers and suppliers, over and above a price-based relationship. If relationships are based solely on limited commercial grounds, then it is unlikely that the corporation takes the needs and attitudes of stakeholders fully into account. As a result, the CR approach toward customers and suppliers is based more on a compliance approach—"Just do it"—than on a partnership or proactive basis.

The important message is that a corporation's attitudes and actions toward CR will affect how employees view both its CR messages and, as a consequence, the degree of commitment toward ER.

Employee attitudes are a critical issue for ER because developing individual employee responsibility is an important activity in establishing effective ER for a corporation. If we take climate change as an example, the attitude of employees toward basic actions such as switching off computer monitors when not in use may be significant in reducing GHG emissions, particularly for heavily office-based corporations. As an example, we recently visited a client with distribution warehouses and were shown around one at lunchtime. There were no employees present in the warehouse—and no productive activity—but conveyor belts were still running at maximum capacity and all the lights were blazing.

Corporations may tend to forget that their environmental responsibility is delivered by employees. Such employees are often at lower levels of the corporation, and their managers may have a primary focus based on delivering specific production targets, rather than having accountability for other aspects, including environmental management.

Given the increasing degree of risk attached to environmental issues, developing ER within the workforce is an important area of focus for corporations. It ensures that all members of the workforce become responsible for their own personal environmental impacts. In addition, it ensures that those making management decisions are fully aware of the environmental effect of their actions.

Managers may make key strategic choices that, depending on the criteria considered, may have significant environmental impacts. Examples could include decisions such as the development of environmentally friendly products, the choice of alternative modes of supply, selection of investments, mergers, or acquisitions.

The development of ER within the workforce involves some other elements of the CR agenda, such as the quality of employee training and development, employee communications, and ethical conduct. If employees consider that the corporation's management believes in its own CR agenda and if they are well informed and trained, they will understand the relevance of ER and required behavior for their own job.

WHAT ER STATEMENT SHOULD AN INVESTOR EXPECT FROM A CORPORATION?

Corporate ER statements need to be set in a relevant sector context. While reporting schemes such as the Carbon Disclosure Project have been effective in raising the issue of GHG emissions within both asset managers and asset users (corporations), they have tended to dictate reporting coverage, rather than allowing corporations to report on what they consider to be relevant to their business.

That said, environmental measurements included by corporate entities within annual or CR reports are notoriously difficult to benchmark. There is a need to develop a template for relevant ER reporting. We have developed the following model based on our experience with clients in projects involving corporate reporting and environmental impacts.

Our basis for the template is the corporation's business strategy and, flowing from this, its key business drivers. Relating ER to the business drivers ensures that the corporation's own management is able to understand their meaning and make this meaning relevant to shareholders and potential investors. The latter need to understand both current business performance and potential risks that may affect its continuing growth trend. The ER statement includes both business strategy and key business drivers. It relates to each business driver the key environmental aspects that will support its delivery and the key performance indicators (KPIs) that demonstrate the quality of performance management for each aspect.

Table 15.1 shows an example of the template in use with a client (based in the manufacturing sector).

TABLE 15.1 Environmental Responsibility Template Used in Manufacturing Sector

Business Strategy Objectives	Key Business Drivers	Information to Be Provided (description of initiatives, how they relate to the business drivers, and how they risks)	Relevant KPIs
Operational excellence and cost efficiency	Improving cost per unit of production	Energy efficiency initiatives	Energy consumption GHG emissions
		Waste management programs	Waste recycling and reuse percentages
	Operational excellence	Environmental management systems (EMS)	EMS coverage EMS actions—progress against plan
Strengthening customer relationships	Innovative product developments	Development of more environmentally friendly products	Pipeline of product development and the anticipated percentage of overall revenue
	Developing customer partnerships	Business process improvement discussions with customers	Improvements in logistics data
Recruitment and retention—having the right people in place	Employees participation in environmental decision-making process	Employees forum on environmental matters Employees training on environmental impacts	Number of meetings and numbers of key actions decided upon Numbers of training days per 100 employees
	Ethical conduct	Whistle-blowing encouraged on environmental issues	Numbers of whistle-blowing calls related to environmental matters recorded and resolved satisfactorily

Source: The Virtuous Circle.

This approach enables asset managers to judge a corporation's ER in the context of its own business, without having to benchmark across the sector. Sectoral benchmarks may be as desirable, but often the difficulty of cross-sector comparison makes their value less than may be perceived.

Our experience shows that there will be a great dispersion of performance reporting, even by companies within the same sector, and that comparing performance between companies is often like comparing apples and pears. While producing this ER statement is of value in its own right, the critical question is how (and where) it should be communicated. The most relevant method is to provide this statement, with regular updates, within the investor relations section of the corporation's web site or, if in the public sector, in an appropriate area of its web site.

This enables an asset manager to review the ER statement and reference it to relevant financial information from within the same source. Of course, if XBRL (which enables tagging of various components in an annual report for subsequent downloading into investors' spreadsheets) becomes an investor industry standard, then this would achieve the same objective, but until this is the case, there is a need to take a lower-technology approach. Other options include featuring ER as part of the annual report and accounts, but variations in reporting standards around the world limit the value of such approaches.

Indeed, the benefit for organizations of including the ER statement within the publicly available sections of their web sites as a voluntary disclosure means that they are able to construct it in a manner appropriate to their business context and strategy. One of the issues of having regulatory reporting is that it can be prescriptive, which often results in the publication of data that is not relevant to the organization's strategy or business performance.

Similarly, while the ER statement could be included in the corporation's CR report, this document tends to be produced for all stakeholders and includes a wider set of information than may be required by investors. As a consequence, including the ER statement within the CR report tends to diminish its value to an investor. However, some asset managers will need to compare environmental activity on a cross-sector basis. For this, there are several public sources of information, although most are limited in the coverage.

One source is the Dow Jones Sustainability Index (DJSI). This has been established on a sector basis and includes questions about environmental management, performance and reporting as well as sector-specific areas such as product impact and product stewardship. However, the DJSI includes only the top-performing 10 percent (320 by number) of quoted corporations in a sector, which limits it utility.

Another source is the Carbon Disclosure Project. While it covers more corporations than does the DJSI, CDP6 still represents only 1,550 responses worldwide. Its value is further limited because it captures only GHG emissions and excludes matters of some potential impact such as waste management, water consumption, and product stewardship. But these benchmarking sources may be of less relevance to an investor if they are able to gain a better assessment of risk and opportunity by having the corporation's ER statement, which relates environmental activities and performance to the delivery of its business strategy.

AN INVESTOR'S APPROACH TO ER

In this section, we discuss how investors (the original asset owner) can assess the quality of ER within the operations of an asset manager in which they are considering making an investment. For this purpose, it is necessary to distinguish the asset manager's role as a fund manager from its role as an asset user. The question to be asked is how often asset managers have both environmentally focused funds and an environmentally focused internal management approach.

To make this distinction, it is necessary to consider what should be the key signposts used as a means to assess the extent to which an entity's ER is managed successfully. For this, we separate out the signposts between those as an asset manager and those as an asset user. Examples of signposts that could be used by the asset manager and the asset user include:

- Asset manager:
  - *Signatory to international standards* would demonstrate that the asset manager takes ER and CR issues into account as part of its investment approach. Examples of standards would include the Equator Principles (www.equator-principles.com) or the UN Principles for Responsible Investment (UNPRI, www.unpri.org), which has just incorporated the Enhanced Analytics Initiative (www.enhancedanalytics.com).
  - *Utilization of external ER benchmarks* would demonstrate that the asset manager goes beyond the signatory approach or compliance level and is absorbing ER-related information into investment decision making. Examples of these benchmarks would include the use of CDP data in its investment process as well as offering investment funds based on the DJSI.
  - *Active socially responsible investment analyst teams* would demonstrate that asset managers adopt a proactive approach to reviewing the extent to which potential investment choices are assessed for their overall corporate responsibility approach and ER procedures.

- *An expanding range (or value) of ER-related funds* is a necessary requirement for an investor as a means of establishing a suitable investment entity. The task is to observe if the asset manager is able to demonstrate that there is commitment behind a statement of belief in ER.
- Asset user:
  - *An ER statement,* such as the template suggested previously, demonstrates that the asset user "walks the talk" when it demonstrates its commitment to ER.
  - *Examples of ER initiatives* demonstrate that the asset user takes a proactive approach to acting responsibly within this arena.
  - *ER and CR reporting* enable an investor to observe the extent the asset user performs in these areas, and whether its initiatives have an impact in improving its environmental footprint.

With these signposts in mind, we looked at eight asset managers from around the world—HSBC, Insight Investment (part of HBOS, subsequently the Lloyds Bank Group), Mercer, Merrill Lynch, Nomura, Shinkin Asset Management, Société Générale, and State Street Global Advisors—and reviewed their ER positions as asset managers and asset users. With the exception of Mercer and Nomura, all are CDP6 2008 signatories. Our review focused on the information publicly available via the companies' web sites.

Perhaps the first conclusion is that the majority of these companies make it hard for an investor interested in their ER to find the relevant information. Very few (as an asset user) had relevant information about their environmental stance visible on the first page of their web site—and it was even harder when it came to the asset manager pages relevant to the potential investors as asset owners. Merrill Lynch came out strongly in both respects, but perhaps this is not surprising given its high profile as sponsor of the CDP.

The second conclusion is that often a difference exists between asset managers that claim they offer socially responsible investment (SRI) vehicles and those that have a more proactive focus on development of environmentally responsible investments. Merrill Lynch (which can offer integrated environmental investing into client portfolios and has developed a series of six environmental investment indices) would meet the latter, more proactive criterion. For the others, we were surprised at how difficult it was to find the SRI section of their web site—even if, when found, it demonstrated how advanced they were. This may have been because some of the information was held within password protected sections, but if available, then surely it would make sense to include some form of communication about the asset manager's intent in this area.

The final conclusion is that if an asset user offers some form of ER, then it is more likely that the company, as an asset manager, would position itself as an environmentally responsible asset manager. For the asset owner, the message from this relatively low-scale research is that, when considering a range of potential asset managers, a request for information about their position on ER is essential. Some corporations (as an asset user) may not be as effective as they should be in communicating these positions via their web sites. For the asset manager, the research suggests that there is probably competitive advantage available if an ER statement is provided—and easily accessible—via its web site. But the ER statement will be seen to be stronger if the asset manager can demonstrate a proactive approach through significant environmental initiatives that not only mitigate its own impacts but also contribute to the benefit of wider society.

CONCLUSION

Environmental impacts resulting from climate change (including water usage and exposure to flooding) are likely to have significant effects on the valuation of many corporations over the next few decades. These impacts will arise from negatives such as new regulations (such as the expansion of emissions trading schemes to new sectors and new countries) to positives such as new technologies and customer behaviors.

If corporations do not actively manage their environmental impacts over the next decade, then their capital valuation will be affected. Since most asset user activity is hidden from the view of the asset manager or the original asset owner, there is a need for corporations to become better at both developing environmental initiatives and communicating them. The latter should not be, however, a case of "greenwash," as demonstrated by some corporations' CR reporting around the turn of this century. Instead, it should be targeted at the asset manager (as the intermediary for the asset owner) and demonstrate how the corporation's ER is bound together with the corporation's governance procedures and support the development of the corporation's business strategy.

In addition, there is a need to build trust in a corporation's ER position, especially employee trust. Employee behavior has an important role in determining whether these impacts are actively managed or passively observed. If a corporation does not "walk the talk," then employees will not be trusting, and passive observation will be the regrettable outcome.

A sound approach to ER will ensure that the corporation's performance—in terms of cost efficiency, market development, and risk management—will improve over time. Effectively communicating this performance, including the development of an open and transparent ER statement, will ensure that asset managers will be able to build this performance in their valuation models.

Now is the time for corporations to start explaining to investors how they perceive environmental impacts affect profitability and cash flows and how they will implement necessary actions to maintain a sustainable and competitive position. The pressure is likely to come from asset owners as they become better informed about the potential risks that could affect corporations in which they invest. This pressure will become stronger, with new regulatory regimes being introduced over the next five years, as the Obama administration begins to address the issues of climate change at a federal level, the EU moves into its 7th Environmental Action Plan in 2012, and the Kyoto Protocol is replaced by its Copenhagen successor in 2010.

APPENDIX: AN INTERVIEW WITH MICHELLE CLAYMAN OF NEW AMSTERDAM PARTNERS

Michelle Clayman is the managing partner and chief investment officer of New Amsterdam Partners (NAP), an asset management firm with about 30 employees, specializing in offering a blend of quantitative and fundamental investment research. Clayman founded the firm in 1986, and it remains 100 percent employee owned. NAP had $2.7 billion of assets under management at the end of 2008, with funds focusing on environmental, social, and governance (ESG) amounting to $679 million (about 25 percent of the total).

ESG Funds and Client Motivation

NAP entered the ESG fund market in 1997, when these funds represented less than 1 percent of the firm's assets. Since then, and particularly in the past five years, Clayman reports that more investors are considering ESG funds. NAP acts in subadvisory programs, including, in the ESG area, for Calvert investments. Their public fund clients include California Public Employees' Retirement System (CalPERS) and California State Teachers' Retirement System (CalSTRS).

In the past, investors tended to question whether ESG-based funds represented "real investing." They wondered, "Is it more about being a good guy, living to a moral code, and giving up investment returns as a result?" However, in Clayman's experience, NAP's Large Cap Socially Responsible Active Equity Fund has performed in line with equivalent large-cap active equity funds. But this performance measure is very time sensitive.

The motivation of ESG fund clients is not dissimilar to other categories of investors, Clayman believes, in that they are looking for "great market-beating returns from investments that meet their guidelines."

In this respect, "Clients love it when guidelines work in their favor, but may squeal when they go against it." In such circumstances, clients are reminded that investment guidelines are agreed at the outset, and Clayman's experience is that they will stick with these. Indeed, her opinion is that when investors return to the equity markets, there will be a focus on ESG funds, which will gain in market share.

Investment Research

NAP's investment teams are each charged with integrating ESG issues in the investment decision-making process. A Fundamental Checklist is completed for every stock purchased, regardless of whether it will be in the socially responsible portfolios. The checklist includes the target company's ESG track record (both positive and negative) and how that might affect the company going forward. ESG performance is integrated in the company assessment and is not viewed as "separate" from traditional financial criteria. However, rarely, if ever, are companies seen as being all "bad "or all "good."

In this context, environmental impact is one of the elements considered. Clayman's view is that investors have become more interested in environmental and climate change risks because they are very much in the public domain, and investors like to hear about companies' responses. However, companies like to "trumpet their good environmental activities all over their web sites," but are not as good at communicating some of the "less than good activities."

Nevertheless, Clayman believes investors' interest in environmental risk is unlikely to diminish. Aspects such as "peak oil" will result in investment attention on other energy sources, over the longer term (5 to 10 years). She foresees increased environmental regulation by the Obama administration. If these measures are stringent, companies will have to respond—but investors' decisions will depend on a wider range of factors.

In support of investment analysis, NAP uses public sources such as the CDP and the Global Reporting Initiative Register, as well as the commercial KLD Database and RiskMetrics. NAP is a signatory to the Principles for Responsible Investment.

While much environmental data is published by companies, the extent of this information—and most importantly, its quality—will vary. Corporate sustainability reports are one tool that NAP uses. However, it does not require that such reports be structured in any specific form. Clayman believes that companies will readily communicate "nice, warm, and cuddly" activity, but they are more reluctant to report on negative environmental factors.

Here, NAP's analysts will examine in detail the regulatory 10K and 10Q reports, where companies are obliged to comment on any matters that are material, such as to report on the need to clean up operational sites or pollution affecting public health.

Clayman gave one example where NAP found it necessary to dig deeper than the company report. In researching a manufacturing company, NAP noticed that it used a number of environmental monitoring companies to comment on the impact of some of its activities. In one case, NAP discovered that the company's claim that its use of boreal forests was acceptable was based on a rating from an entity that NAP deemed very lax, and viewed its performance with a more critical eye.

On the need for a more transparent ER statement, Clayman described this as being an ideal in the perfect world, but companies "promote the good stuff and downplay the bad stuff." "Greenwashing" is a constant threat, and requires the investment manager to analyze more deeply than corporate sustainability reports. There is a wealth of information in a company's regulatory filings, but these are often not examined in great detail.

The need for a detailed approach was reinforced by Clayman's comment that NAP tracks the differences between the statements made in investors' quarterly conferences and what they include in their regulatory reports. NAP's analysis probes differences between statements in the annual 10k filings, where material environmental matters should be disclosed. These issues should also be covered in the 10Q filings.

Clayman rates environmental management as being low on a company's board agenda, unless it is revenue or cost improving, and companies see a trade-off between acting progressively and addressing profit-and-loss (P&L) issues. Leadership from the CEO is key. Clayman quotes Duke Energy's ex-CEO, who led the company into a series of green initiatives, sometimes at a potential cost to the P&L.

From an investment perspective, ESG matters are significant. NAP's director of quantitative research has been exploring the time horizons affecting environmental and governance practices of companies. The findings indicate that better governance practices are reflected in stock outperformance in the nearer term, whereas better environmental practices result in stock outperformance in the long term. Other research by NAP's quantitative team completed a study in mid-2008 showing that companies with neutral or positive ESG scores outperform companies with negative scores.

Walking the Talk

Clayman was asked whether NAP practices what it preaches. As a relatively small firm, it does not have rigorous ESG procedures laid down. However, influenced by Clayman's own beliefs and values in this area (she is an active volunteer in her community), NAP has adopted practices on reducing and recycling paper and actively working in the community. In particular, the firm operates internship programs and supports after-school programs and mentoring. Clayman is particularly proud of the fact that 50 percent of the firm's employees volunteered as mentors in the East Harlem Tutorial Program, an education and enrichment center that works with children from early childhood through adulthood.

Explaining why NAP employees participate, Clayman said, "Being more complete human beings is one reason." But she went on to emphasize sound business benefits for the firm—40 percent of NAP's business is with state and local government bodies who value volunteerism. In her mind, walking the talk pays, but it starts with living your values first.

Our Conclusions

Clayman represents an investment firm that is active in ESG-related investment. However, the investment world encompasses a wide range of investment (and investor) attitudes.

Nevertheless, the way in which NAP integrates ESG considerations into their overall investment analysis is an indicator of a possible future trend. Clayman is sanguine about the extent to which investors consider environmental matters: "It will depend...."

On transparency in ER, while Clayman accepts the need "in a perfect world," she recognizes that the information is already out there, provided that analysts are willing to delve. A more transparent environmental report may be more important for analysts less interested in examining the detail.

Overall, Clayman's convictions came through strongly in the interview—and confirm her belief that a firm's values stem from its leadership.

REFERENCES

Fish, K. "Are Americans from Mars? CSR in North America." 2004. Presentation given at the European Conference on CSR in Maastricht on November 8, 2004.
Forrest, W., J. Kaplan, and N. Kindler. 2008. "Data centers: How to cut carbon emissions *and* costs." *McKinsey Quarterly*, http://www.mckinseyquarterly.com/Data_centers_How_to_cut_carbon_emissions_and_costs_2255.

CHAPTER 16

Beyond Best Practices

Angelo A. Calvello, PhD

Although it is neither practical nor necessary to think that you should become experts on climate risk, climate risk can have a real impact on portfolio holdings. There is a growing case for trustees to attain some level of knowledge around these issues, and to take steps to mitigate any negative consequences of not taking action.
—Mercer Investment Consulting, *A climate for change: A trustee's guide to understanding and addressing climate risk*

The original plan was to end the book with a chapter on the best practices institutional investors use in approaching climate change, but then a friend correctly pointed out the problem with this approach: it is backward looking and simply distills what actions have worked well. Because climate change is a new consideration for institutional investors, looking backward holds little value. Instead, I would suggest that general sets of best practices exist for institutional investors. These practices are well documented, and dozens of sources exist to help institutional investors properly discharge their fiduciary duty over the assets entrusted to them. As Paul Watchman makes clear in Chapter 5, fiduciaries should incorporate the consideration of climate into their best practices. I would like to build on Paul's work and take a forward-looking approach to suggest how institutional investors could best approach climate change within the existing framework of best practices.

These suggestions are tied more to common sense than to statutes. Some are reflections on a theme. In general, they represent thoughtful approaches to climate change and environmental investing.

STEP 1: UNDERSTANDING CLIMATE CHANGE AND THE VIRTUAL TEAM

The myriad effects of climate change bring unparalleled uncertainties to the investment landscape and forces investors to look for ways to gather and interpret disparate pieces of data.
                                        Carbon Disclosure Project (CDP), 2009

1. The critical element is understanding climate change. The goal is to understand "if there is the potential that climate change could have a material impact on the assets in our care" (Mercer Investment Consulting 2005).
2. Be as comprehensive as possible in building your understanding of climate change. For example, focus on the drivers of returns, fiduciary issues, risks, investment options, and operational issues.
3. Recognize that climate change is a cluster concept, but continually pull the research and discussion toward a center—understanding climate change and how it impacts our duties and actions as fiduciaries.
4. Build a virtual climate change team to achieve this understanding of climate change. As Danyelle Guyatt persuasively argues in Chapter 14, collaborative initiative by parties with different interests and skills is key to understanding and responding to climate change.
5. Assess the current competencies of your immediate resources (e.g., staff, board members, current service providers, and advisors). Identify other resources to supplement this knowledge base. Be creative in this search. Of course, explore relationships with other service providers and advisors but also include some nonconventional sources:
   a. Your host organization—the endowment, foundation, corporate or public sponsor—and see if it has a sustainability program or environmental responsibility program that you could leverage. Leaders of these programs tend to be quite well connected in the environmental community and could provide access to key resources.
   b. Managers and advisors specializing in environmental investing.
   c. Peers who have or are exploring climate change and investing.
   d. Industry organizations with expertise in climate change like Ceres, Carbon Disclosure Project, Investors Group on Climate Change Australia/New Zealand, Responsible Property Investment Center, Chicago Climate Exchange.
   e. Academics (climate scientists of all sorts, economists, public policy experts, etc.).
   f. Policy makers.

g. Governmental and quasi-governmental organizations (e.g., the Environmental Protection Agency and the United Nations)

h. Nongovernmental organizations such as Global Canopy Programme, the Nature Conservancy, the World Wildlife Federation, and The Pew Center on Climate Change.

i. Think tanks.

j. The media.

6. Create a specific climate change learning agenda and designate explicit goals as part of your overall agenda.

7. Be global, not provincial, in your search for knowledge and virtual team members. Great ideas can be found near and far.

8. Use technologies like videoconferencing and networking sites to connect and interact. (You don't have to fly people in for meetings to build relationships and acquire information. Skype and videoconferencing are effective and environmentally friendly ways to hold meetings.)

9. Be patient. It will take time to create the virtual team and build the knowledge base. But be diligent.

10. Expect to add and remove members from this team as your understanding will mature over time and your needs and interests change.

11. Understanding comes with the right mind-set. Skepticism is healthy; dogmatism is stultifying.

12. Articulate your current view of climate change and share it with stakeholders. For example: "We believe that climate change poses a real and material risk to the financial performance of our investments (particularly over the long term), and therefore the returns that the fund will make" (Mercer Investment Consulting 2005).

13. Integrate climate change considerations into your processes.

STEP 2: ASSESSMENT (ENVIRONMENTAL STRESS TEST)

> *Whilst it is one thing to state "I can see that climate change has the potential to affect materially the assets held in our fund's investment portfolios," it is quite another to be able to add "and I know what to do about it"*
>
> Mercer Investment Consulting, 2005

14. Recognize climate change as a bona fide and mainstream risk.

15. Perform an environmental stress test on your portfolio to determine climate change's possible impacts. Model possible changes in science, economics, policy, and technologies to see how individual strategies

and the aggregate portfolio might be impacted. Consider different time horizons. Keep in mind your position as a universal owner.

16. This stress test could be developed in concert with your peers.
17. Evaluate the available tools for measuring climate risk. Recognize the limitations of the current risk models and quantitative tools but make an earnest attempt to quantify the types and kinds of risks. (See the following case study for an example of one set of tools.)
18. Use your virtual team to build a process to identify, monitor, and measure climate risks in the portfolios.

Carbon Profile Case Study: Tools for Measuring Climate Risks

By Lisa Hayles

Investors rank climate change as one of the most financially significant risks facing companies today. Many investors are requesting specific climate-related information from companies in order to understand the carbon exposures of their portfolios. However, a lack of a robust methodology for capturing consistent, comparable data has made it difficult for investors to determine which companies are responding effectively to climate change or what to do about those companies that are not.

EIRIS, a leading global provider of independent research into the social, environmental, governance, and ethical performance of companies, has developed one approach to assist investors to accurately understand and measure the climate risk in their portfolios. The EIRIS Carbon Profile allows investors to quantitatively assess the climate change impact and environmental policy of public equity and bond holdings by comparing the specific carbon performance of an investment strategy or portfolio to a relevant index or benchmark. This assessment of performance combines a measure of carbon impact with an analysis of each holding's or company's response to managing the challenges and opportunities of climate change.

In order to measure the climate change performance of a portfolio, EIRIS constructs an assessment framework based on two broad questions:

1. What is the "carbon impact" of each constituent?
2. How is each firm addressing the issue of climate change?

What Is "Carbon Impact"?

EIRIS calculates the carbon impact of a company according to the direct GHG emissions associated with its industrial sector. Each sector is assigned to one of four categories: very high, high, medium, or low climate change impact based on their direct, indirect, and product emissions. The impact classifications also factor in projected growth of emissions in the sector, net impact of the sector (e.g., benefits of public transport), the allocated

Lisa Hayles is a senior client relationship manager at EIRIS based in London.

share of upstream and downstream emissions, and the strategic importance of the sector in contributing to solutions to climate change.

How Is Each Company Responding to the Issue of Climate Change?

A company's specific strategy for dealing with climate change is assessed via qualitative indicators such as how it manages its operational emissions, involvement in the development of new low-carbon products or services, and disclosure on performance. Based on internal expertise and comprehensive engagement with subject matter experts from nongovernmental organizations and companies (including the World Wildlife Fund, Climate Group, Carbon Trust, and the Institutional Investor Group on Climate Change), EIRIS develops a series of indicators to assess how well companies are addressing their climate change impacts and risks. These indicators cover:

- Governance
- Strategy
- Disclosure
- Performance

The assessment of impact and environmental strategy is combined to produce a company-specific "Carbon Risk Factor." The Carbon Risk Factor ranks each company's climate change performance on a 100-point scale. (See Figure 16.1.)

The higher the Carbon Risk Factor score, the better positioned (and less exposed to climate risk) a company is. Companies providing carbon "solutions" such as alternative energy firms will obtain the highest possible scores—from 75 to 100. These firms pose the lowest risks in a portfolio as their products are designed to reduce or replace typical

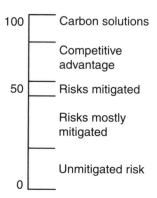

FIGURE 16.1 Carbon Risk Factor scale.
Source: EIRIS.

carbon-intensive products. Energy-intensive companies that have failed to translate concern for climate change into action will score between 0 and 30 points. EIRIS has set 50, the midpoint, as the score that identifies companies that have begun to effectively address the issue of climate change.

How an Asset Manager Uses Climate Change Data

MACIF Gestion, a leading provider of sustainable and responsible investment strategies, commissioned EIRIS to create a comprehensive Carbon Profile of one of its socially responsible investment (SRI) funds. MACIF believes that the more a firm integrates sustainable development into its core business strategy, the better are its long-term prospects. In order to test its philosophy, MACIF asked EIRIS to analyze the impact of climate change on its Croissance Durable Europe Fund. EIRIS designed an objective assessment of the carbon exposure and environmental policy of the constituent companies in this Eurozone equity fund and produced a Carbon Profile report that concluded:

- The average market capitalization-weighted outperformance of the Fund's holdings was 2.34 points (on the EIRIS proprietary Carbon Risk Factor scale) better than the benchmark.
- The Fund had a lower carbon impact than the benchmark (26 percent of its holdings are rated "high" or "very high," compared with 41 percent for the benchmark).
- Sector allocation more so than security selection was responsible for most of the carbon outperformance.

MACIF used the Carbon Profile results to:

- Identify which energy-intensive constituents of their reference index were better at managing their responses to climate change then their sector peers.
- Determine if and how its investment process adds value through the consistent selection of both less energy-intensive companies and companies with strong corporate environmental policies.
- Provide unit holders with an objective assessment of the environmental impact of the MACIF sustainable investment approach.

Buoyed by this success, MACIF Gestion is exploring how it might further use climate change metrics to enhance performance returns by possibly tilting sectors' weights toward lower-impact sectors and identifying companies that are well positioned to benefit from the transition to a low-carbon economy.

STEP 3: FROM THEORY TO PRAXIS

Factoring climate change data into investment decisions requires both science and art—this is the struggle the industry faces today.

CDP, 2009

19. With understanding comes the possibility of prudent action. (Or, as a trustee once told me, "I recycle at home—isn't that enough?")
20. Determine what steps—over specific time periods—should be taken to mitigate the risks and adapt to possible investment opportunities. Codify these decisions.
21. Realize that even if you decide not to pursue environmental investments, you still must confront climate change and its risks to your portfolio. Moreover, from this point on, you should acknowledge that every new investment you make has exposure to climate change.
22. If you do decide to pursue environmental investments then:
    a. Evaluate your manager selection process to ensure it accommodates environmental investments. Supplement as needed.
    b. Recognize that there are both beta-centric and alpha-centric environmental investment opportunities.
    c. Paraphrasing Russell Read and John Preston, rather than being a fringe area of investment, environmental investing is taking a central role *within* virtually every asset class.
    d. Just as you think broadly about the risks, think broadly about the opportunities. Do not succumb to the tyranny of the policy allocation.
    e. If you decide to pursue environmental alpha, acknowledge that environmental investing is not necessarily a subset of socially responsible investing. Do not be distracted by socially responsible investment (SRI) issues; it's about alpha, not absolution.
    f. Mark out the practical investment challenges associated with environmental alpha. For example:
        i. *Manager due diligence.* Does your current manager due diligence process support environmental investing?
        ii. *Access.* The size and scale of some environmental investments are so large that they might not be available to smaller institutional investors.
        iii. *Direct or indirect?* As with other alpha-centric opportunities, you need to determine when to go direct and when to go indirect.
        iv. *Time horizons.* While climate change is a long-term theme, individual environmental investment opportunities will come with their own time frames. Some opportunities will be enduring and match up well with your liability structure; others will prove to be only transitional.
        v. *Fees.* Environmental alpha, if it is truly the result of manager skill, will be expensive. Be clear on the fees because net alpha is all that matters.
        vi. *Operations.* Be sure your custodian can accommodate new strategies and instruments.

    k. Consider creating a consortium of peers who could pool assets to improve the access and reduce fees, sort of a Commonfund model for environmental investing.

23. Enact the appropriate changes but remain mindful of associated transaction costs.

STEP 4: GENERAL OBSERVATIONS

The trouble with our times is that the future is not what it used to be.
Paul Valery

24. Climate change is a challenge that requires short-term and long-term responses.

25. Invest with a clear mind and a strong stomach (Calvello 2009). Sure, you might want to save the world, but keep your emotions in check.

26. The uncertainty surrounding just how we'll move to a low carbon economy will cause high volatility. There will be blood.

27. Be mindful that exogenous events can occur almost out of nowhere and materially impact your investments.

28. A corollary: avoid falling in love with an environmental investment. Remember ethanol.

29. Climate change makes engagement a part of your fiduciary responsibility. "In order for investors to exercise appropriate judgment and for fiduciaries to act responsibly, disclosure of the potential economic risk posed by climate change is essential" (Social Investment Organization 2004). There will be times when you will need to use your role as a shareholder or investor to get companies to provide you (and your managers) with relevant information so you could properly understand material risks associated with your ownership. This engagement is not about social causes; it's about having the necessary information to properly do your job.

30. Think globally. Climate change risks might manifest themselves locally (i.e., company- and deal-specific risks) but the investment opportunities are global.

31. Don't be overwhelmed by the enormity of the problem and the scale of the response. Design and implement a process that can be habituated into your investment and governance structures.

32. Doing nothing is not an option—and not doing enough is not an option either.

Case Study: Two Meanings of Stewardship

The University of Minnesota endowment office, the Office of Investment and Banking, with over $1.5 billion in assets under management, might not seem like the organization that would be setting the standard for an institutional investor's response to climate change. But, through its partnership with the university's Facilities Management Group's Energy Management Group (EMG), it is part of a game-changing energy conservation program that could serve as the prototype for fiduciary behavior.

The conservation program has three parts:

1. The university efficiently and resourcefully manages its energy needs in part through the use of renewable energy sources and a campus-wide conservation program.
2. As part of its commitment to be a leader in sustainability, the University joined the Chicago Climate Exchange (CCX) and accepted the Exchange's binding but voluntary emission reduction standards. By complying with or exceeding the CCX's standards, the university is able to monetize its energy conservation programs in the form of carbon credits.
3. The Office of Investment and Banking (OIB) uses investment knowledge and CCX membership to strategically trade these credits.

This innovative collaboration between the EMG and the OIB is a striking example of a how a fiduciary—in this case, an endowment—can use its core competencies to extend its stewardship beyond the oversight and management of its financial assets to include the enterprise's use and management of natural assets. The OIB's engagement could serve as a prototype for how fiduciaries could use their investment expertise to directly and effectively help their sponsoring organization respond to the challenges of climate change.

The University's Twin City Energy Conservation Program

The university's energy conservation activities are formally managed by its Facility Management Group's Energy Management Group and covers two major areas:

1. **Use of renewable sources of energy in the heating plant**
   The university needs to generate steam for heating, humidification, sterilization, and air conditioning. The primary steam plant in Minneapolis is a "circulating fluidized boiler" that is capable of burning just about any type of fuel: coal, gas oil, wood, oat hulls, and other biomass. The EMG chooses to use fossil-fuel and non-fossil-fuel sources of energy. For example, they offset 5 percent of natural gas use with oat hulls or wood chips from pallets.
2. **Campus-wide conservation program**
   The university has had a robust campus-wide conservation program since 1994.

It was started by a group of academics and eventually brought into the Facilities Management Group. The program has been a success on a number of fronts.

Over the past few years, the university has built a number of energy intensive or "heavy research" buildings, growing the Twin Cities campus from 17 million to 23 million square feet, but because of its conservation program, the university's carbon footprint has actually decreased by 1 percent. Significant energy savings opportunities have come from aggressive adoption of energy efficiency techniques, such as the transitioning away from incandescent lights to higher-efficiency lighting options and aggressive optimization of building control schedules. Such optimization of the Art Building on the campus's West Bank has resulted in an annual electricity savings of over 1 million kWh and a corresponding CO_2 emission reduction of over 830 tons.[1]

The conservation program also includes an innovative $6 million revolving loan–type arrangement used specifically to fund energy-efficient projects. The EMG typically spends about $1 million of this loan every year on energy efficiency projects that have an expected payback of about five years or less. The principal and interest is paid out of utility rates the university charges its clients.

It is also important to understand that this conservation program is about more than just adopting energy-saving technologies. Since 2006, university instructors have integrated this program into their curricula. For example, students in a senior engineering design class identify university buildings in need of energy efficiency upgrades and share their ideas with the EMG. The EMG vets the feasibility of the ideas and attempts to implement the most promising before the end of the course (or at least include the ideas into the capital budget). The EMG also works with the University's Initiative on Renewable Energy and the Environment (IREE) and Environmental Institute to build visibility for the conservation program. In 2008, their efforts resulted in an orientation session on energy reduction to 5,500 incoming freshman.

Monetizing the Efficiencies

Other universities certainly employ similar energy saving techniques, practices, and programs. What differentiates the University of Minnesota is its explicit and public commitment to reducing its GHG emissions.

First, it joined the Chicago Climate Exchange (CCX) in 2004. As an exchange member, the university makes a voluntary but legally binding commitment to meet annual greenhouse gas emission reduction targets. (As of this writing, eight universities were members of the CCX.) Each CCX member is allocated an annual emission allowance with its emissions baseline and reduction schedule. If it emits less than the target, then it earns surplus allowances—tradable carbon financial instruments (CFI) issued by the CCX—that it can sell or bank. Of course, if it emits more than the target, it must buy credits or other instruments.

Second, as a CCX member, the university registers its energy conservation projects as offset projects—schemes that sequester, destroy, or reduce a member's GHG emissions. These projects are verified by approved third parties, and all verification reports are inspected for completeness by the Financial Industry Regulatory Authority (FINRA, formerly the National Association of Securities Dealers [NASD]).

To date, the university's use of renewable energy sources and energy-efficient technologies have met the CCX standards for accredited offset projects and earned the university about 1,150 CFIs, or carbon credits, representing over one million metric tons of carbon emissions reductions.

The Role of the Endowment Office

Another differentiating feature of the university's energy conservation program is the role played by the OIB. The EMG and OIB have had a close working relationship since 2005 when the OIB began hedging the University's heating oil and natural gas exposure in the futures markets. This relationship has been extended to include the OIB's active management of the university's CCX carbon credits. Since 2005, the OIB has used its investment insights and acumen to strategically trade these credits on the CCX. This activity has resulted in a profit of about $500,000—all of which is intended to fund additional energy reduction efforts.

These trading profits are a financial windfall, as the university originally foresaw its CCX membership as an extension of its overall commitment to sustainability and its initial goal was to use any carbon credits to cover the cost of its membership (around $5,000). But the OIB's ability to consistently add value to the energy efficiency and conservation program through its trading has generated significant benefits to all of the university's stakeholders.

This collaboration is the first step in an evolving sustainability partnership between the EMG and OIB. For example, the university is extending its energy conservation program by making its Morris campus, located 160 miles from the Twin Cities, the first carbon neutral university campus in the country. The campus already uses a wind turbine to produce about 30 to 60 percent of its electrical load. This capacity will increase with the addition of a planned second turbine. The university also plans to use a biomass gasification system that uses corn stover as its primary fuel source. Once completed, up to 75 percent of the heating and cooling loads for the campus will come from alternative energy sources. The university already has made significant lighting modifications and incorporated energy control technologies to further reduce their energy consumption and plans to use a steam production boiler to efficiently meet its (considerable) heating needs. The university fully expects these efforts to comply with the CCX standards and earn additional carbon credits that the OIB will trade.

CONCLUSION

Perhaps this ethic of conservation is in the university's DNA. The Cedar Creek Ecosystems Science Reserve, considered by many to be the birthplace of modern ecology, was started at the University of Minnesota in the 1940s.

In any case, it certainly is manifest in the collaboration between the EMG and OIB. What is truly groundbreaking is how the endowment staff exercises its stewardship not only over its investment assets (which include a variety of alpha-centric environmental investments) but also over the university's use of natural assets. This is an exceptional model of stewardship that other institutional investors should evaluate as they develop their response to the challenges and opportunities associated with climate change.

NOTES

1. The university also manifests its energy efficiency in its management of transportation. The university is one of the largest users of E85 in the state and is now using B20 in its diesel vehicles. As a result, in both 2005 and 2006, University Fleet Services was named among the 100 Best Fleets in North America.

REFERENCES

Calvello, A. 2009. *Green trends and greenbacks*. Portfolio.com, January 7. www. portfolio.com/news-markets/national-news/portfolio/2009/01/07/Environmental-Investing.

Carbon Disclosure Project (CDP). 2009. *Investor Research Project—investor use of CDP data*. Carbon Disclosure Project, March.

Mercer Investment Consulting. 2005. *A climate for change: A trustee's guide to understanding and addressing climate risk*. The Carbon Trust, www. thecarbontrust.co.uk/trustees.

Social Investment Organization. 2004. *Climate change and investment risk: Best practices for Canadian pension funds and institutional investors*. A report on the Climate Change and Investment Risk Workshop. Toronto, Ontario, March 11. Original: Investor Network on Climate Risk. 2003. *Investor call for action on climate risk*. New York, November 21.

About the Author

Angelo A. Calvello, PhD, is the founder of Environmental Alpha, a consulting firm that helps institutional investors develop climate change–based risk management and investment strategies. He has worked in the investment business for over 25 years.

Dr. Calvello earned a PhD in Contemporary European Philosophy from DePaul University and a postdoctoral certificate in business from New York University Stern School of Business. He writes and speaks extensively about environmental investing, alpha-centric investing, and alpha/beta separation. Recent publications include *Green Trends and Greenbacks*d (for portfolio.com) and *Build, Buy, Lease: Three Approaches to Alpha Generation* (AllAboutAlpha.com).

Dr. Calvello has been the keynote speaker at numerous environmental, portable alpha, and asset allocation conferences and regularly provides board education to pension funds and nonprofits on environmental investing. He is on the Charter Alternative Investment Analyst (CAIA) Association Advisory Board, a member of the Chicago Quantitative Alliance, a member of the Chicago Climate Futures Exchange, and president of the Illinois Chapter of U.S. Lacrosse.

Together with his wife, Lisa, and three children, Giana, Joe, and Michael, he resides in suburban Chicago.

About the Contributors

Martin Batt has extensive experience in the financial services sector, having been responsible for financial marketing, internal communications, and environmental management number of U.K. building societies, ultimately at the Halifax. As corporate responsibility manager for HBOS from 2001, his key achievements included the development and launch of a new ethics statement for the company, and gaining a top 10 placing in the Business in the Community Corporate Responsibility Index. In 2008, Martin worked as group environment manager for Prudential Group, before returning to the Virtuous Circle. His particular interests lie in the areas of sustainable development, climate change, energy efficiency, and renewable energy. He has an MSc in environmental decision making.

C. Shawn Bengtson, PhD, CFA, is a senior portfolio manager for Woodmen of the World Life Insurance Society's Investment Division. Shawn is committed to bridging actuarial, accounting, and finance theory and practice, and publishes applications in these areas. Her focus is on integrating enterprise risk management into investment decision making.

Shawn's work on the boards of nonprofits has provided the opportunity to participate in assessing the potential impact of carbon credit trading on Nebraska-domiciled businesses. Her objectives include responsible stewardship of our natural resources.

Martin Berg is a vice president and a carbon emissions originator at Merrill Lynch, a wholly owned subsidiary of Bank of America. Based in London, Martin is responsible for the bank's global transaction in the flexible mechanisms of the Kyoto Protocol (CDM and JI) and in the voluntary carbon markets. Martin has worked on climate change and the carbon markets for over 10 years. Prior to joining Merrill Lynch, he was a carbon finance specialist at RNK Capital LLC, a New York–based hedge fund. Before his time at RNK Capital LLC, Martin worked at the climate change unit of the Organisation for Economic Co-operation and Development (OECD) in Paris and at conferences of the United Nations Framework Convention on Climate Change (UNFCCC) in Bonn, Lyon, and the Hague.

Martin holds an MPA in international energy management and policy from Columbia University and an MA in political sciences from the University of Bonn.

Richard A. Betts, PhD, is head of climate impacts at the Met Office Hadley Centre. He leads a team of research scientists, operational forecasters, and consultants in understanding and predicting the impacts of climate change and variability for a wide range of customers both in government and the private sector.

Richard has a bachelor's degree in physics and a master's and PhD in meteorology. He has worked as a climate modeler for 17 years, and has pioneered a number of key developments in the extension of climate models to include biological processes. He has published over 50 peer-reviewed scientific papers and other articles. Richard was a lead author on the *Fourth Assessment Report of the Intergovernmental Panel on Climate Change* (IPCC), which shared the 2007 Nobel Peace Prize with former U.S. vice president Al Gore. He was also a lead author on the *Millennium Ecosystem Assessment,* the authors of which shared the 2005 Zayed Environment Prize with the UN Secretary-General Kofi Annan, and was a leading peer-reviewer of the *Stern Review of the Economics of Climate Change.*

Véronique Bugnion, PhD, manages Point Carbon's global trading analytics and research efforts, including the development of trading, data, and research products for the power, natural gas, and carbon markets. Véronique is a specialist in energy and environmental markets with experience in analyzing and modeling the U.S. energy markets, with particular focus on the deregulated power markets and the natural gas and emissions markets. Véronique also has a strong publication record in the areas of climate modeling and U.S. climate change policy. She holds a PhD in climate physics and an MSc in technology and policy, both from MIT.

Massimiliano Castelli, PhD, joined UBS in 2002 as senior economist in the international team of the Group Governmental Affairs (GGA). Massimiliano is in charge of the Middle East and supports investment bank, wealth management, and global asset management with economic and financial strategic analysis on the region. He is also responsible for analyzing global trends in areas such as capital flows, macroeconomic developments, and their business implications for the financial service industry. Massimiliano holds a PhD in economic policy and a master's in economics from the University of London. Until 1997, he was a lecturer in economics at the University of Rome La Sapienza, Italy. In 1997, he left academia and became a professional economist in the city of London. He subsequently relocated to Zurich, where he is currently based. When not working, Massimiliano spends his time skiing, reading, and visiting art exhibitions.

Tim Dixon, PhD, is professor of real estate and director of the Oxford Institute for Sustainable Development (OISD) in the School of the Built Environment at Oxford Brookes University, United Kingdom. The school, which includes OISD as its primary research vehicle, was recently ranked as one of the top five in the United Kingdom in terms of "research power" within the RAE 2008. With more than 25 years' experience in research, education, and professional practice in the built environment, he is a fellow of the RICS and of the Higher Education Academy, a member of SEEDA's South East Excellence Advisory Board, as well as the editorial boards of five leading international real estate journals. He has collaborated on research projects with U.K. and overseas academics and practitioners, and is working on a number of funded research programs relating to sustainable real estate.

Stefanie Engel, PhD, is professor of environmental policy and economics at ETH Zurich and a member of the Institute for Environmental Decisions (IED). She holds a master of science degree in agricultural and resource economics from the University of Arizona and a PhD in the same field from the University of Maryland, as well as a habilitation (postdoctoral degree) in resource economics and development economics from the Faculty of Agriculture, University of Bonn, Germany. She has held positions as a lecturer and research fellow at Universidad de Los Andes in Colombia. Before joining ETH in April 2006, she was senior researcher and group leader at the Center for Development Research of the University of Bonn, Germany. Stefanie has published widely in international refereed journals in the areas of environmental economics, new institutional economics, and development economics.

Christopher Fox is director of investor programs at Ceres. He coordinates the staff team that works on the Investor Network on Climate Risk (INCR) and has led Ceres's programs aimed at organizing investors to address climate change since 1998. Chris has played a lead role in organizing the three Investor Summits on Climate Risk at United Nations Headquarters that Ceres cohosted with the United Nations Foundation in 2003, 2005, and 2008. Chris is coauthor of *Questions and Answers on Climate Risk for Investors* (2004) and editor of several Ceres reports, including *Value at Risk: Climate Change and the Future of Governance* (2002) and *Corporate Governance and Climate Change: Making the Connection* (2003).

Prior to joining Ceres, Chris served as a program associate at the Heinz Family Foundation in Washington, D.C., and was cofounder and executive director of the Center for Environmental Citizenship. Chris received a master of divinity degree from Harvard University and a BA with honors from Yale University. Chris and his wife live in Brookline, Massachusetts, with their three children.

Mark Fulton, managing director and global head of climate change investment research, New York, joined DB Advisors in 2006 after 29 years of investment experience in senior roles in research and management at Citigroup in the United States, Salomon Smith Barney and NatWest in Sydney, Potter Partners in Melbourne, and James Capel in London. He received his BA in philosophy and economics from Oxford University.

David Gardiner is president of his own environmental consulting firm, David Gardiner & Associates. The firm helps organizations and decision makers marshal policy, technology, and finance to solve energy and climate challenges. It develops strategies, analyzes issues, and seizes communications opportunities for its clients, which include the American Public Transportation Association, Ceres, the Energy Future Coalition, Grounded Power, Pearson, and Recycled Energy Development.

Prior to founding his firm, he directed the White House Climate Change Task Force, the group established by President Clinton to coordinate the U.S. government's domestic and international policies on climate change. Before coming to the White House, Mr. Gardiner served for six years as assistant administrator for policy at the Environmental Protection Agency. At the EPA, he led the agency's climate change efforts, as well as programs to reinvent the EPA's approaches to key sectors, such as transportation, agriculture, metal finishing, and real estate development. Prior to joining the EPA, he was the Sierra Club's legislative director in Washington, D.C.

Mr. Gardiner has a bachelor of arts with honors from Harvard College.

Danyelle Guyatt is a principal in Mercer's Responsible Investment. Her focus is on intellectual capital development, consulting, and research for Mercer's global RI business. She has 10 years' experience, starting as an economist at the Commonwealth Treasury of Australia before moving to the financial sector as a strategist and later a global balanced fund manager with Deutsche Asset Management.

Danyelle graduated with an MSc in investment management with distinction at the Cass Business School and recently completed a PhD in economic psychology at the University of Bath, researching institutional investors and long-term responsible investing. Danyelle was part of a group of experts invited to contribute to the development of the UN Principles of Responsible Investment and was commissioned to carry out a study by the Marathon Club on investment beliefs related to the promotion of good corporate governance and responsibility. She has been involved in delivering workshops at the Rotman International Centre of Pensions Management on topics related to improving the pension fund management process.

Matthew Hale is head of environmental sustainability for the EMEA region of Bank of America Merrill Lynch. His oversees the development and implementation of the Bank's environmental strategy for the region. In particular, he works towards the integration of the Bank's carbon, debt and equity trading, banking and wealth management groups, while developing activities to reduce the firm's own carbon footprint and engage its employees in the challenge.

Matt is also Chair of UNEP FI's Biodiversity and Ecosystems Workstream. Prior to this position, he was head of the Treasury Group, overseeing capital, balance sheet, and liquidity for the EMEA region. Matt joined Merrill Lynch in 2004, and previously he worked in private equity and at Bankers Trust, where he was a partner and manager of their Global Securities Finance Business and, prior to that, their European treasurer.

Lisa Hayles is a senior client relationship manager at EIRIS based in London. Founded in 1983, EIRIS (Experts in Responsible Investment Solutions) provides research on corporate environmental, social, and governance (ESG) and other ethical performance indicators to more than 100 institutional investors around the world. EIRIS's clients range from those who use their research for stock selection or exclusion, to pension funds and other institutional investors applying an engagement or sustainability overlay to their investment strategy.

In her current role, Lisa supports institutional fund managers and pension funds in North America and Europe seeking to implement a variety of RI strategies in their investment processes. She also serves as a resource person on RI issues to several independent investment committees. She joined EIRIS in November 2003 and previously worked at the Social Investment Organization in Toronto, Ontario, where she was assistant director.

Lisa holds degrees from the University of Toronto and the University of Guelph in Canada and Université de Toulouse Le Mirail in France.

Tony Hoskins is chief executive of the Virtuous Circle, a management consultancy specializing in reputation and risk, corporate social responsibility (CSR), corporate reporting of nonfinancial information, risk management, and stakeholder communications. Working with major multinational companies, he has an in-depth understanding of the issues faced integrating them into business processes. He has published three CSR research reports and a research report used by the Accounting Standards Board on the quality of business reviews produced by listed companies in their annual reports.

In addition to his consultancy work, he speaks at conferences on CSR and the impact of nonfinancial reporting. He delivers training on CSR for the Institute of Chartered Secretaries and Administrators (ICSA) and Environmental Management for Imperial College. He has written five books, including The ICSA CSR Handbook, *CSR and Nonfinancial Reporting for Business* (2008), and contributed chapters for three others.

He has an MBA from London Business School, and graduated in economics and statistics from Bristol University.

Bruce M. Kahn, PhD, director and senior investment analyst for Climate Change Strategies: New York, joined the company in 2008 with 20 years of experience in environmental research, most recently as it relates to investments. Prior to joining Deutsche Bank, he managed assets for high-net-worth and institutional investors at CitiSmith Barney's Private Wealth Management Group. Previous experience includes investment and market research for IC Value, Inc. (previously Center for Sustainable Systems Studies, Miami University of Ohio); management consulting and corporate sustainability strategist for Cameron-Cole, LLC; and environmental research positions for the University of Wisconsin, Madison, the Ecological Society of America and Auburn University, and service in the U.S. Peace Corps as an agricultural agent and provincial representative.

He received his BA in ecology and evolutionary biology from the University of Connecticut; MS in fisheries and allied aquacultures from Auburn University; PhD in environmental science from the University of Wisconsin, Madison; and recipient of both a J. William Fulbright Scholarship and a National Science Foundation Fellowship in ecological economics.

Abyd Karmali is managing director and global head of carbon markets at Merrill Lynch, a wholly owned subsidiary of Bank of America. He is the bank's point person for carbon business opportunities and serves on Bank of America's Environmental Council. Mr. Karmali has worked for 18 years on climate change and the carbon markets and serves between 2008 and 2010 as elected president of the Carbon Markets and Investors Association, as well as a member of Her Majesty's Treasury Carbon Market Expert Group and of the World Economic Forum's (Davos) Steering Committee for Advancing Low-Carbon Finance. In 2008, Mr. Karmali's team won Carbon Finance Transaction of the Year and the Banker Award for Most Innovative in Sustainability. Mr. Karmali holds an MS in technology and policy from the Massachusetts Institute of Technology and was previously employed with ICF International in Washington, D.C., Toronto, and London, where he served most recently as managing director, Europe. In 1996–1997 he was climate change officer at the United Nations Environment

Programme's industry office in Paris and participated in the Kyoto Protocol negotiations.

Matthew J. Kiernan, PhD, was founder and chief executive of Innovest Strategic Value Advisors, the number one–rated sustainability investment research firm in the world. Innovest's work in the climate space has included doing the global research for the first five annual iterations of the Carbon Disclosure Project, as well as codesigning the world's first "climate risk–adjusted" bond index with JP Morgan. In 2009, Innovest was acquired by the RiskMetrics Group, where Dr. K became cohead of the Sustainability Solutions group. Prior to founding Innovest, Dr. Kiernan had served as director of the World Business Council for Sustainable Development in Geneva, and as a senior partner in the strategy consultancy at KPMG.

Dr. Kiernan holds a PhD in strategic environmental management from the University of London. His most recent book, *Investing in a Sustainable World,* is distributed by McGraw-Hill. In 2007, Dr. Kiernan received the UN Environment Program's Finance Initiative's executive award for "innovation and contribution in carbon finance." He is a frequent speaker at international investment conferences, and has addressed the World Economic Forum in Davos, Switzerland, on a number of occasions.

Hua Liu is a senior analyst at the Perella Weinberg Partners Capital Management LP, a position he has held since June 2007. He has primary responsibility for analyzing a global portfolio of public and private investments in the water sector. Prior to joining Perella Weinberg Partners, Mr. Liu spent six years in investment banking at Société Générale, Robertson Stephens, and Dain Rauscher Wessels, covering a diverse set of industries, including general industrials, telecommunications, and Internet software. In his capacity as an investment banker, Mr. Liu has had extensive exposure modeling and executing a variety of equity, equity-linked, fixed-income, and structured product transactions. He is a graduate of Columbia Business School and the University of California, Berkeley, and is fluent in Mandarin Chinese.

Mindy S. Lubber is the president of Ceres, the leading U.S. coalition of investors and environmental leaders working to improve corporate environmental, social, and governance practices. She also directs the Investor Network on Climate Risk (INCR), a network of more than 80 institutional investors representing nearly $7 trillion in assets that coordinates U.S. investor responses to the financial risks and opportunities posed by climate change.

Ms. Lubber is the recipient of the Skoll Social Entrepreneur Award, and under her leadership Ceres was awarded the Fast Company Social Capitalist

Award for 2007 and 2008. Before coming to Ceres, Ms. Lubber was the regional administrator of the U.S. Environmental Protection Agency and CEO of Green Century Capital Management, an investment firm managing environmentally screened mutual funds. She was recently voted one of "The 100 Most Influential People in Corporate Governance" by *Directorship-Magazine*, which noted Ceres' increasing influence in its field.

Andrew Mitchell is a leading international authority on forests and climate change. In 2001, he founded the Global Canopy Programme (GCP), an international network linking 38 leading scientific institutions in 19 countries and focused on the research and conservation of tropical forests. Andrew has been an impassioned advocate of forest conservation for over 30 years, with a career spanning scientific research, broadcast journalism, and environmental policy. Before founding the Global Canopy Programme, he was vice president of the Earthwatch Institute in Boston, overseeing 130 field research and conservation projects in 40 countries. He is a research associate at the zoology department, University of Oxford.

Andrew acts as an advisor to governments and international institutions, and is senior advisor to the Prince's Rainforests Project, which he established in 2007 with HRH The Prince of Wales. Andrew cofounded the U.K. Corporate Environment Responsibility Group, and has advised on environmental policy for companies including McDonald's, Barclays, and British Airways. He is the author of seven books, and a well-known and recognizable media figure. In 2007, he gave interviews to BBC World, Sky News, and Channel 4 News, and was quoted in publications from the *Independent* and the *Daily Telegraph* in the United Kingdom to the *Washington Post* and *Time* magazine in the United States.

Charles Palmer, PhD, is currently lecturer in environment and development at the London School of Economics in the United Kingdom. Previously, he was a senior researcher at the Chair of Environmental Policy and Economics, the Institute for Environmental Decisions (IED) at the Swiss Federal Institute of Technology in Zurich (ETH Zurich), Switzerland. He holds a BA in biological sciences from Oxford University, an MSc in environmental and resource economics from University College London and a PhD in agricultural economics from the Faculty of Agriculture, University of Bonn, Germany. He works in the fields of environmental and development economics, has published widely, and has, in the past, worked for a number of international organizations including the United Nations Development Programme (UNDP) and the Center for International Forestry Research (CIFOR).

Rod A. Parsley is a partner and portfolio manager at Perella Weinberg Partners Capital Management LP, where he manages a global portfolio of private and public investments in water, alternative energy, and clean technology. From 2004 to 2006, Mr. Parsley was the portfolio manager of the Water Fund, LP, a New York–based hedge fund that invested globally in water-related companies, including utilities, infrastructure, and technology companies. Mr. Parsley has extensive experience in securities valuation, risk management, private equity, and capital structure analysis. Previously, Mr. Parsley was a strategy consultant with McKinsey & Company.

Mr. Parsley serves on the board of directors of Underground Solutions Inc., a leader in trenchless water pipe technologies, and Blue Water HLD Corporation, an innovator in clean waste management services. From 2000 to 2004, Mr. Parsley served on the board of advisors of eSecLending, a securities lending firm that has auctioned over $600 billion in assets to many of the nation's largest financial institutions. He is a graduate of Harvard Business School and Pomona College.

Hylton Murray-Philipson first explored the Amazon basin at the age of 17 and subsequently established Morgan Grenfell in Brazil at the age of 23. He was later chief executive of the investment bank Henry Ansbacher in New York. As trustee of Rainforest Concern, Hylton has worked alongside the Commissao Pro-Indio do Acre in improving the quality of life of 650 Yawanawa people in the state of Acre, Brazil, thereby helping to conserve their traditional territory of 200,000 hectares.

Hylton is a senior advisor to the Prince's Rainforests Project, established in 2007 by HRH, the prince of Wales. Hylton established Wingate Ventures in 1990 providing corporate finance services to businesses making a positive contribution to the environment, including Agrivert, a leading U.K. recycler of organic waste; Geothermal International, a leading U.K. supplier of geothermal energy; Goingreen, suppliers of the G-Wiz electric car; and Carbon Conservation Pty Ltd and Canopy Capital Ltd, both at the forefront of initiatives to develop investment products to attract capital into the standing forests. In October 2008, Hylton addressed Finance Ministers assembled at the World Bank on the value of forest ecosystem services. He is a trustee of the Global Canopy Programme.

John T. Preston is a founder of C Change Investments and manages investments in clean energy and renewable resources. Mr. Preston has spent most of his career at MIT and was previously the director of technology development (and licensing) overseeing the commercialization of MIT inventions.

Mr. Preston is a director of Clean Harbors Corporation, Alseres Pharmaceuticals, and several private companies. He has also served in advisory positions for the governor of Massachusetts, the U.S. Department of Defense, and NASA, among others. Mr. Preston has advised the White House on energy and environmental issues in preparation for the Kyoto summit, as well as testifying seven times before the U.S. Congress on issues related to technology.

Mr. Preston received a BS in physics from the University of Wisconsin and an MBA from Northwestern University. Mr. Preston was awarded the rank of "Knight of the Order of National Merit of France" by President Mitterrand. He was awarded the "Hammer Award for Reinventing Government" by Vice President Gore.

Russell Read, PhD, served as chief investment officer for America's largest pension fund, the California Public Employees' Retirement System (CalPERS) prior to founding C Change Investments in 2008. During his tenure, he made CalPERS a leader in clean technology and environmental investments. Prior to CalPERS, Dr. Read served as deputy chief investment officer for Deutsche (Bank) Asset Management (Americas) and Scudder Investments.

Dr. Read is also a founding member of the P8, the world's largest pension funds, organized to meet the investment challenges associated with providing the world's emerging clean infrastructure. In 2007, he was recognized by *SmartMoney* in its Power 30 list of the most influential people in business and finance, and by *Institutional Investor* in 2008 as number 35 on its list of the 75 most effective chief executives.

He received his undergraduate degree in statistics and his MBA in finance and international business, both from the University of Chicago. He received his master's degree in economics and his doctorate in political economy from Stanford University. He is also a Chartered Financial Analyst (CFA), a Chartered Life Underwriter (CLU), and a Chartered Financial Consultant (ChFC).

Erica Scharn joined Ceres as an investor programs associate in October 2008. She helps to coordinate shareholder engagement with companies, which includes assisting Investor Network on Climate Risk (INCR) members with filing and cofiling shareholder resolutions on climate change and sustainability issues. She also works on INCR's disclosure program.

Erica has a master of public policy from Brown University. While at Brown, she spent almost a year as a Policy Fellow at the Office of the Rhode Island General Treasurer. Originally from western Massachusetts, Erica graduated from the University of Massachusetts–Amherst with a BA in sociology (departmental honors) and a minor in psychology. She was a teaching assistant and member of an intramural dance team at UMass.

Paul Q. Watchman has a well-established international reputation. Paul was named with Hank Paulson by Ethical Corporation as one of the six most influential international figures in respect of the development of sustainable finance. He is given credit for providing the United Nations with the foundations for changing the beliefs of the pensions fund investment industry. Paul's key areas of expertise include social and environmental assessment, structuring and allocation of environment risks and liabilities, climate change and carbon trading, land use and land development, project finance, sustainable finance, the Equator Principles, responsible investment, business and human rights, corporate and social responsibility, and corporate governance, generally.

Prior to establishing Quayle Watchman Consulting, he was a partner and established leading environmental and climate change practices at Brodies, Freshfields Bruckhaus Deringer and Dewey & LeBoeuf. He is a regular contributor to textbooks, monographs, reports, collections of essays, encyclopedias, and professional and academic journals. Recent published works include *Climate Change: A Guide to Carbon Law & Practice* (Globe Publishing, 2008) and "Carbon Capture and Storage: Burying the Problem of Climate Change" (*Utilities Law Review,* 2007).

Simon Webber joined Schroders in 1999 as an analyst on the Global Technology and U.S. teams before becoming a global sector specialist for consumer discretionary and telecom sectors. He currently focuses on the utility, auto, and telecom sectors, and manages investments for global portfolios in these areas. In 2006, Simon, along with Matt Franklin, began to develop the philosophy and investment universe behind the Schroders Global Climate Change strategy, and has managed the fund since its launch. With his experience in technology, utilities, and the consumer, Simon is ideally positioned to assess the relative strengths of the various new climate change technologies and how the consumer is likely to respond to rapidly increasing awareness of climate change.

Simon received a BSc (hons) in physics from the University of Manchester, and is a CFA charterholder.

Jurgen Weiss, PhD, is a cofounder and managing director of Watermark Economics. He has over 15 years of experience as a management consultant and economic expert in energy and environmental markets across the world. He has advised clients on the economics and the economic impact of climate change regulation and valued power plant and electric utility assets in multiple countries, has consulted on the design of incentives for renewable power and energy efficiency projects, and has published on a number of areas, including issues related to the design of carbon markets, the value of electricity storage, the role of demand in competitive electricity markets, and the design of incentive systems for renewable power. Prior to founding Watermark Economics, Mr. Weiss was the managing director of Point Carbon's advisory services as well as a director for LECG, a globally operating economic advisory services firm. He holds a PhD in economics from Harvard University and an MBA from Columbia University.

Dimitri Zenghelis recently joined Cisco's long-term innovation group as chief economist of the Climate Change practice in the Global Public Sector organization. He moved from heading the *Stern Review* team at the Office of Climate Change, London. Previously, he was a senior economist who spent a year working with Lord Stern on the *Stern Review on Economics of Climate Change,* commissioned by the then chancellor Gordon Brown. He continues to act as an external advisor to the U.K. government and works closely with Lord Stern at the LSE, where he is a Senior Visiting Fellow at the Grantham Institute on climate change. He is also an Associate Fellow at the Royal Institute of International Affairs (Chatham House). Dimitri joined HM Treasury in 1999, providing economic advice for the U.K. government as head of economic forecasting and head of the European Monetary Union Analysis Branch.

Index